REMEDIATION HYDRAULICS

REMEDIATION HYDRAULICS

Fred C. Payne

Joseph A. Quinnan

Scott T. Potter

CRC Press
Taylor & Francis Group
Boca Raton   London   New York

CRC Press is an imprint of the
Taylor & Francis Group, an **informa** business

CRC Press
Taylor & Francis Group
6000 Broken Sound Parkway NW, Suite 300
Boca Raton, FL 33487-2742

© 2008 by Taylor & Francis Group, LLC
CRC Press is an imprint of Taylor & Francis Group, an Informa business

---

Library of Congress Cataloging-in-Publication Data

Payne, Fred C., Ph.D.
    Remediation hydraulics / Fred C. Payne, Joseph A. Quinnan, Scott T. Potter.
        p. cm.
    Includes bibliographical references and index.
    ISBN 978-0-8493-7249-0 (hardback : alk. paper)
    1. Groundwater--Pollution. I. Quinnan, Joseph A. II. Potter, Scott T. III. Title.

TD426.P39 2008
628.1'68--dc22

2008001130

Visit the Taylor & Francis Web site at
http://www.taylorandfrancis.com

and the CRC Press Web site at
http://www.crcpress.com

Dedications

This book was made possible by the enduring love and support of Therese, Aaron, Nathan, Kate, and Meredith, and my parents, Harold and Dorothy Payne. Thank you all for your continuing encouragement and guidance.

– Fred C. Payne

To Christine, Niamh, and Andy for your patience and unending support during the writing of this book. This book was only possible because of your encouragement, commitment, and love.

– Joseph A. Quinnan

To Robin, Christina, Andy, and Rachel, thank you for your patience, support, and love.

– Scott T. Potter

Contents

Foreword

I am extremely privileged to have been invited to write the foreword for the book *Remediation Hydraulics* written by three of my world class colleagues. I have known the team of authors, led by Dr. Fred Payne and including Dr. Scott Potter and Joe Quinnan, for years. They have devoted the better part of the last decade to rediscovering the full scope of remediation hydrogeology. The authors' practical orientation, penetrating intelligence, curiosity, and independent spirit led them to question much of the accepted hydrogeological theory and practices of our time. It is my sincere belief that this book will have a significant impact in the remediation industry due to its originality, versatility, and universality of vision.

Guiding change may be the ultimate test of any leader — be it a business leader or a thought leader. Human nature being what it is, fundamental change is often resisted mightily by the people it most affects — those in the trenches of our industry. *Remediation Hydraulics* establishes a sense of extraordinary urgency in challenging accepted, long-term, industry-wide paradigms and in demonstrating the necessity for changing the methods by which we design remediation systems, particularly *in situ* systems. It is my belief that readers will recognize the basic scientific and fundamental truths as if they'd always known them. Dr. Payne and his team bolster the scientific fundamentals by developing an interdisciplinary know-how that is not taught today in university curricula. Readers will rediscover remediation hydrogeology and thus enable the authors' vocabulary and methodologies to enter the lexicon.

Groundwater is a key component of the ecosystem we live in and hydrogeologists always play an important role in the remediation and recovery of contaminated groundwater systems. Hydrogeologists are called on to play an important role in the ongoing debates of how to design successful *in situ* remediation systems. The debate in the *in situ* remediation arena has shifted and morphed from one scientific aspect to another during the last decade and a half. The topics have ranged from the type of reagents to geochemistry to microbiology to delivery of reagents during the different phases of evolution. The authors coined the phrase *Remediation Hydraulics* in their attempt to rely upon robust science and basic fundamentals to open the black box of reagent delivery systems.

You have to have passion to do something that will have an impact on a whole industry. Anyone who is passionate about the continuous evolution of successful implementation of *in situ* remediation systems will see the authors' perspective, vision, and conviction through their effort in putting together this comprehensive book. I have been very fortunate to watch the hard work put in by the authors and the countless hours of heated debates in trying to explain many field observations that contradicted today's accepted theories. It was an established wisdom that one could not question the validity and utility of bulk hydrodynamic dispersivity as a significant factor in analyzing and predicting contaminant and reagent transport. The entire remediation industry relied upon macro-scale hydrogeologic and hydrodynamic principles during the last quarter century. This was a legacy of the initial dependence on large-scale groundwater pump and treat systems for mass removal and containment of contaminant plumes.

The greatest tribute I can pay to Dr. Payne and his team is having the courage and determination to look deeper into finding scientific answers for pore-scale hydrogeologic questions. The universally disliked long-term tail of contaminant concentrations, the inability to reach stringent cleanup standards at most remediation sites, is a result of poor understanding of the hydro-bio-geo-chemical interactions that take place within the pore-scale groundwater environment. Despite this discouraging picture, hopeful opportunities do exist especially if hydrogeologists follow the call to action to strive for progress in both the fundamental scientific aspects as well as practical design concerns. I take some personal pride along with the authors for having been hyperaware of this reality and embarking on a journey to develop solutions at a very basic level. This necessity developed as a result of the business focus of our employer ARCADIS and the encouragement provided to us to develop a business platform that depended heavily on technical excellence.

It has been a valuable and enjoyable experience for my colleagues and me at ARCADIS to have had the opportunity to rely upon the expertise of Drs. Payne, Potter, and Quinnan to lead our industry in implementing cutting edge *in situ* remediation solutions. How, then, can you tell when you are dealing with genuine experts? First, their efforts must lead to performance that is consistently superior to that of the experts' peers. Second, real expertise produces concrete results. Finally, true expertise can be replicated and measured.

Skills in some fields are easy to measure. Standardization permits comparisons among individuals and organizations over time. I firmly believe that this book *Remediation Hydraulics* will lead the practitioners of *in situ* remediation in a journey towards superior performance which will result in technically and economically more efficient closures of contaminated sites. The technical excellence upon which this book is formulated is the result of practicing their beliefs and intuition again and again in the form of simple tasks such as tracer tests and data discipline. You need a particular kind of practice to achieve perfection. Deliberate practice is the only type of practice that leads towards developing true expertise. When most people practice their expertise, they focus on the things they already know how to do. I am truly proud of having been associated with the authors in their path of deliberate practice that has led to developing foundational scientific principles and breakthrough theories that will result in consistently successful *in situ* remediation projects. Transforming the practice of remediation hydrogeology from a macro-scale focus to micro-scale analysis and calling it *Remediation Hydraulics* will be the legacy of Dr. Payne and his team for years to come.

Suthan S. Suthersan, Ph.D., P.E.
Chief Technical Officer
ARCADIS Environmental Division

Preface

Every possible approach to groundwater restoration requires that we anticipate the movement of contaminants or treatment reagents through the subsurface. Containment methods such as groundwater extraction can be built on relatively low-resolution depictions of contaminant movement. *In situ* treatment methods, however, require a much more refined understanding of solute migration, as we try to arrange contact between prospective reactants in complex porous media. From the collective experience of the past 20 years, it is clear that our ability to predict or (more optimistically) control fluid movement in the subsurface needs significant improvement. For those of us who treat groundwater contamination using *in situ* remedial technologies, whether physical, chemical or biological, improving our understanding of contaminant and reagent mass transport in aquifers has become an urgent requirement — a much higher priority than further understanding of the treatment mechanisms themselves. The need to reliably predict and control fluid movement in the subsurface, to build realistic assessments of contaminant plume structure, and to achieve contact between injected reagents and target contaminants has been the main driving force for the development of *Remediation Hydraulics*.

The classic researches on fluid flow in the subsurface by Darcy, Dupuit, Meinzer, Muskat, Hubbert and others were undertaken to support development of underground mineral resources, primarily groundwater, crude oil, and natural gas [(Meinzer 1923), for example)]. Growth of the science of hydrogeology enabled development of groundwater resources as drinking water supplies in many regions of the world, reducing dependence on surface water which is generally less reliable in quantity and quality. By the 1950s, the science of groundwater resource development was well established and engineering firms such as Geraghty & Miller, predecessor to ARCADIS, flourished in the pursuit of groundwater supply development.

When the focus of drinking water protection shifted from surface water to groundwater in the 1980s, engineering practices that had evolved for water supply development were readily translated into groundwater extraction and containment strategies. Large-scale groundwater extraction, to remove contaminant mass and to prevent expansion of the contaminant plume, was the most common response to the discovery of contamination well into the 1990s. However, extraction alone is usually not effective for contaminant mass removal and the process has been generally abandoned as a treatment technology; only the containment function is considered effective today.

The calculus of contaminant migration developed at a time when groundwater extraction was the presumptive response to contamination. Because the groundwater extraction approach is applied at relatively large scales, there was no need for fine-scale site delineation or prediction of contaminant movement in the subsurface. The designs of remedial systems were built on pumping tests and capture zone analyses and the proof of a system's effectiveness was the successful re-direction of groundwater flows to the extraction points. Large-scale averaging of key system variables has been a highly successful strategy supporting water supply development and containment pumping systems, but the subsurface is strikingly heterogeneous and large-scale averaging and steady-state observations obscure details of aquifer structure that are essential for design and operation of *in situ* treatment systems.

In *Remediation Hydraulics*, we have abandoned the notion that representation of the subsurface through large-scale averages and steady-state observations can adequately support groundwater remedy designs.

- Even sites that are considered archetypically homogeneous have 1,000- to 10,000-fold variation in hydraulic conductivity at scales of 1 to 10 cm. When we inject tracers and map their movement at high resolution in space and time, we see that solute transport usually occurs in less than 10 percent of the aquifer volume, well below the 20-percent "effective porosity" typically assumed for porous media. This leads to a conceptual split of the aquifer matrix, between a small fraction that is transport-active and a much larger fraction that is effectively stagnant and serves as a reservoir of persistent contaminant storage.
- Investigative tools are generally scaled too large to observe and account for critical heterogeneities. Field studies are generally conducted at very low resolution, relative to the variability of natural aquifer systems. For example, pumping tests and the models developed from them correctly represent groundwater flow at scales of 100s of meters, but cannot match details observed at scales of 10s of meters.
- Investigative tools are susceptible to steady-state bias. Steady-state aquifer behaviors, such as pumping test drawdown and contaminant concentrations in mature plumes, mask aquifer heterogeneities that are critically important to effective remedy design and operation.
- Heterogeneities are unmasked under transient conditions. We have learned that aquifer heterogeneities can be observed when systems are displaced from steady-state and these transient conditions can be induced by tracer or reagent pulsing and pressure transient propagation in groundwater. Transients are also observed during pre-steady-state drawdown in pumping testing (and systematically discarded) and in contaminant concentration patterns in early-stage contaminant plume development.

In the current marketplace, where a significant premium is placed on achieving regulatory compliance in short time frames, it is essential to realistically map contaminant source mass and to account for heterogeneous groundwater, solute, and injected reagent movements.

Dispersivity and diffusivity are two other topics that receive close examination in *Remediation Hydraulics*. Groundwater practitioners commonly assumed that bulk hydrodynamic dispersivity was a significant factor in contaminant and reagent distributions and that diffusivity was relatively unimportant in flowing groundwater. Detailed studies of contaminant plume structure and reagent injections suggest the opposite — transverse dispersivity, in particular, is quite limited and diffusivity is a major contributor to the longitudinal dispersion normally observed in solute mass transport. Revising expected transverse dispersivities to minimal levels has had a significant impact on reagent-based remedy designs, causing us to significantly increase injected volumes to achieve adequate coverage. Over-estimates of transverse dispersivity also allowed practitioners to rationalize contaminant concentration discontinuities in groundwater plumes, when more closely spaced investigative sampling would have uncovered high-concentration flow conduits that require intervention for groundwater resource protection.

We believe that the prospects for success in the application of any *in situ* remedy are largely controlled by the variability of the subsurface and its impact on our ability to find contaminant sources and to deliver reagents in the correct quantities to the correct locations. There is a very strong incentive from stakeholders to make the *in situ* reactive zone technologies work, despite the complexities of the subsurface. One of the most important functions of this book is to guide students and practitioners through the development of an understanding of the complex fine structure of the subsurface and to understand how those structures and associated transport and storage processes control the distribution and migration of contaminants and the reagents we inject to destroy them. To do this, we map out a view of hydrogeology and reagent-dependent remedial processes at scales smaller than the representative elementary volume[1] [(REV) conceived by M. King Hubbert (Hubbert 1937; Hubbert 1956) and continued by Jacob Bear (Bear 1972)]. Scales

---

[1] The REV defines the minimum scale at which averaging groundwater flow hydraulic analysis (Darcian) becomes most effective.

larger than the REV comprise what we term the classical Darcian hydraulics domain and at smaller scales we enter the remediation hydraulics domain, where aquifer heterogeneities cause flow and contaminant mass distributions to vary greatly over very short distances, often confounding our efforts at characterization and treatment.

The heterogeneities and anisotropies that we're observing in contaminated aquifers can easily overwhelm remedial system designers responsible for delivery of reagents to the subsurface. But a designer choosing to ignore aquifer architecture is more likely to suffer the humiliation of a failed remedy than one who tackles formation complexities head-on. We categorize the potential responses to aquifer heterogeneity into three broad types:

1) *Overwhelm the heterogeneity* — A simple approach is to factor aquifer heterogeneities into project planning and employ treatment strategies that allow us to overwhelm the heterogeneous distribution of contaminants and hydraulic conductivities. This approach is sometimes the only means of achieving remedial objectives in a timely manner and the costs can be high, as a significant portion of the remedial effort can be unavoidably directed at aquifer volumes that don't require further work.

2) *Take advantage of the heterogeneity* — In many cases, we can focus our remedial efforts on the more conductive portions of the aquifer formation, understanding that some contaminant mass will remain in the system, stranded in the less conductive zones. It is often possible to achieve remedial objectives without elimination of the entire contaminant mass and formation heterogeneities are the key to achieving that outcome.

3) *Create useful heterogeneities* — When we encounter mature contaminant plumes in lower-conductivity strata, it often isn't possible to perfuse the entire contaminated cross-section with reagents. In these cases, it may be possible to insert a high-conductivity path into the formation using hydraulic fracturing techniques. The fracture strata provide a quasi-permanent by-pass of the contaminant mass, concentrating the groundwater flow in a small fraction of the aquifer volume and limiting contact between the flowing groundwater and the stagnated contaminant mass.

Aquifer heterogeneities and the challenges they entail provide an opportunity for practitioners to distinguish themselves.

Remediation Hydraulics is organized along four main themes: 1) Flow in Porous Media, 2) Mass Transport Processes, 3) Investigation Tools, and 4) Applications. Our presentation of flow in porous media begins with the fundamental characteristics of fluids and surfaces (Chapter 1) and the porous matrix (Chapter 2), then combines the two, describing fluid flow in porous media (Chapter 3). Chapter 4 introduces the behaviors of non-aqueous liquids in porous media and handles two- and three-fluid systems. Then, we shift to the second major theme, the analysis of the movement of dissolved substances in porous media, or mass transport. First, we consider processes that don't involve chemical reactions and are generally viewed as arising from random movements of dissolved substances within a water mass (whether moving or stagnant) or by random motions of a flowing water mass in a porous matrix (Chapter 5). We closely examine the role of diffusion in solute mass transport through realistic porous media and discover that the role of diffusion has been understated by many observers and that a phenomenon known as "non-Fickian" or "anomalous" transport is actually a predictable consequence of Fickian diffusion. The discussion then shifts in Chapter 6 to the movement of compounds that interact chemically with aquifer matrix materials (through sorption processes or reduction-oxidation reactions), or with aquifer bacteria populations. In Chapter 7, we introduce several conceptual approaches to contaminant mass transport at larger scales and lay the foundation for tracer study design and interpretation, the single most important tool in application of remediation hydraulics.

Later chapters turn to practical topics, including investigation tools and remedy applications. Chapter 8 stresses the importance of conceptual site models and exhorts the reader to view conceptual site models as dynamic documents that are modified to accommodate new observations, not to be used as a filter that allows in only observations that agree with the model. Chapter 9 provides a synopsis of hydrostratigraphy, a valuable tool for the characterization of depositional environments from lithologic logs. Chapter 10 covers principles of well design and encourages the use of high-quality well materials, construction and development practices, noting that quality standards seem to have been downgraded during the 1990s and need to be re-established to meet the challenges of *in situ* remediation technologies. Chapter 11 summarizes aquifer and vadose zone characterization testing, with sections on slug and pump tests, pressure transient analysis, and pneumatic testing protocols for aquifer sparge and soil vapor extraction technologies. The discussion of investigation tools concludes with Chapter 12, Tracer Study Design and Interpretation, the most data-rich chapter of this volume. We wrap up the major themes by examining two classes of *in situ* remedial action from the perspective of remediation hydraulics: injection-based reactive zone designs (Chapter 13) and flow-concentrating remedies (Chapter 14). We conclude *Remediation Hydraulics* with an overview of the major issues and their implications (Chapter 15).

Throughout the book, we have minimized the use of acronyms and abbreviations to maintain clarity. We have also used the "cgs" or centimeter-gram-second units convention wherever possible, to maintain a uniform frame of reference for our readers.

Fred C. Payne, Ph.D.
Joseph A. Quinnan, P.E., P.G.
Scott T. Potter, Ph.D., P.E.

REFERENCES

Bear, J. (1972). *Dynamics of Fluids in Porous Media*. New York, Dover Publications, Inc.

Hubbert, M. K. (1937). "Theory of scale models as applied to the study of geological structures." *Geol. Soc. Amer. Bull.* 48: 1459–1520.

Hubbert, M. K. (1956). "Darcy's Law and the field equations of the flow of underground fluids." *Trans. Amer. Inst. Min. Met. Eng.* 207: 222–239.

Meinzer, O. E. (1923). *The Occurrence of Groundwater in the United States, With a Discussion of Principles*.

Acknowledgments

This book was written as a direct result of support and encouragement from the Site Evaluation and Remediation management group at ARCADIS. Curt Cramer, Dr. Suthan Suthersan and Mike Maierle have created a team environment in which critical thinking and innovative problem-solving are both expected and rewarded. Suthan Suthersan, in particular, has provided critical guidance and encouragement. Dr. Suthersan developed the *in situ* reactive zone concept that was introduced in his book *Natural and Enhanced Remediation Systems* and he fostered the introspective analyses of reactive zone performance that provided the basis for development of *Remediation Hydraulics*. He has been a mentor and a friend to all of us and we are eternally grateful.

Many of these ideas grew from concepts initially introduced in the ARCADIS Advanced Remedial Technology Group (ARTG) workshops and we thank Pete Palmer for providing us with a venue at the ARTG to get these ideas their early tests. The ARCADIS advanced hydrogeology training camps allowed us to engage in a series of spirited exchanges with the many very bright hydrogeologists and remediation engineers at ARCADIS and these thoughtful discussions led to improvements in our understanding of the subsurface. We have come to appreciate Dave Willis' assertion from those sessions that it would be necessary to tackle the problem of aquifer heterogeneity to correctly understand and (hopefully) control reagent injection.

We could not have made the observations that underlie this book without access to the hundreds of sites serviced by ARCADIS and we are grateful to our clients and all the project teams who have supported our data extraction efforts. Special thanks go to the many individuals who helped us in our data-gathering, including Craig Divine, Abigail Faulkner, John Horst, Kelly Houston, Alison Jones, Marc Killingstad, Mark Klemmer, Jack Kratzmeyer, Mark Lupo, Elena Moreno-Barbero, Denice Nelson, Eric Panhorst, Sandra Russold, Kimberly Schrupp, Song Wang, Nick Welty, and Kevin Wilson.

Stephan Foy of BioGenie in Quebec engaged us in excellent discussions of hydrogeology and provided data from his exquisite field tracer studies.

We wish to thank Lee Ann Doner and Dr. Tom Sale of Colorado State University for sharing their ideas and excellent graphics showing diffusive interaction between mobile and immobile porosities. Dr. Ralph Ewers of Eastern Kentucky University provided photos of epikarst formations.

Several individuals read chapter drafts and their input has been invaluable. Khandaker Ashfaque, Craig Divine, John Horst, Mark Klemmer, Kelly Houston, Denice Nelson, Nick Welty, Kevin Wilson, and Mike Wolfert all took time from their busy schedules to read chapters and provide helpful guidance. Therese Payne proofread the manuscript and helped us construct more coherent explanations. Amy Dant graciously agreed to typeset the book for us, which simplified all our lives tremendously.

About the Authors

Fred C. Payne, Ph.D., is a vice president and director of *In Situ* Remediation Services at ARCADIS. He has been with ARCADIS for nine years, and he provides technology development and design support for physical, chemical, and biological *in situ* remedial systems throughout North and South America and Europe.

Dr. Payne received his B.S. in biology and botany and his M.S. and Ph.D. in limnology from Michigan State University. In 1982, Dr. Payne founded the company Midwest Water Resource Management (later MWR, Inc.), which became an international leader in physical *in situ* remedial technologies, including soil vapor extraction and aquifer sparging. His enhanced-mode soil vapor extraction process was responsible for many challenging site closures since its inception in 1985, including the first reported detection-limit closure of vadose zone soil for perchloroethene (Purdue Industrial Waste Conference, May 1986). During his 17 years at MWR, he was the lead inventor of six and the co-inventor of a seventh U.S. patent for physical *in situ* remedial technologies and subsurface fluid delivery systems.

Since joining ARCADIS in 1999, Dr. Payne has supported the company's development of cost-effective chemical and biological *in situ* remedial technologies, including biostimulation for enhanced reductive dechlorination and chemical oxidation for destruction of miscible as well as hydrophobic organic compounds.

In addition to his remedial technology responsibilities, Dr. Payne led the ARCADIS design team for sustainable stormwater treatment systems, which recently designed the 67-acre Phase I Sustainable Stormwater Management System at the Ford Rouge Complex in Dearborn, Michigan.

Joseph A. Quinnan, P.E., P.G., is a principal hydrogeolgist at ARCADIS, a leading, global, knowledge-driven service provider in the areas of environment, infrastructure and facilities. He has been with ARCADIS seven years and has over 17 years professional experience in environmental consulting. He is a discipline leader in remediation hydrogeology and subsurface evaluation and characterization techniques and serves as chief hydrogeologist in ARCADIS' North Region.

Mr. Quinnan received his B.S. and M.S. in geological engineering from Michigan Technological University. During summer internships with petroleum companies during his college years, he applied reservoir characterization techniques that now influence his practice and approach to hydrogeology at ARCADIS.

Prior to joining ARCADIS, he was involved in *in situ* technology development and demonstrations for the Department of Defense and several CERCLA sites. He has been involved in innovative *in situ* technology applications since the 1990s, having successfully applied diverse techniques including nano-scale iron to treat chlorinated solvents, propane biostimulation to treat MTBE and *in situ* remediation of perchlorate via biostimulation.

During his tenure at ARCADIS, Mr. Quinnan has focused on developing site characterization and reagent distribution techniques that complement the ARCADIS' innovative *in situ* remediation capabilities. He has supported the design of *in situ* remediation solutions in North and South America and Europe for a wide range of contaminants including nitroaromatics, chlorinated solvents, metals, and PCBs. He is actively engaged in the development of new tracer testing methods, DNAPL characterization and mobility assessments and hydrostratigraphy methods.

Scott T. Potter, Ph.D., P.E., is a vice president and director of hydrogeology and modeling services at ARCADIS. He has been with ARCADIS for 18 years, and has more than 25 years of experience in groundwater hydrology. In his current role, he is chief hydrogeologist of ARCADIS providing technical leadership in hydrogeologic assessments and groundwater remediation projects throughout North and South America, and Europe.

Dr. Potter received his B.S. and M.S. in environmental engineering, and his Ph.D. in civil engineering from The Pennsylvania State University. Prior to working for ARCADIS, he spent nine years with the U.S. Department of Agriculture at the Northeast Watershed Research Center as a hydrologist and a research assistant investigating groundwater flow and solute transport problems in agriculture.

Since joining ARCADIS in 1989 as a hydrogeologist for Geraghty & Miller, Dr. Potter has supported the development of practical methods to more effectively characterize groundwater conditions and assess remedial technologies. Dr. Potter led the ARCADIS design team for the evaluation and implementation of alternative landfill remedial solutions, which resulted in construction of some of the earliest vegetative covers approved by the USEPA. He is currently involved in developing more effective strategies to deliver reagents *in situ*.

1 Fluids and Surfaces

Fluids are large-scale aggregations of molecules whose interactions at the molecular scale lead to emergent behaviors of the fluid at the macroscopic scale. Potter and Wiggert (1991) define fluids as *"those liquids and gases that move under the influence of shear stress, no matter how small that shear stress may be"* (italics in the original). In simpler terms, when we apply a force that moves molecules in one part of a fluid, intermolecular cohesion transmits that motion to other molecules in the fluid. The fluids of interest in remediation hydraulics may be liquid, gaseous or mixtures of liquids, gases and solid particles. We also encounter polymers and other specialty fluids in remediation hydraulics. It is the fundamental physical-chemical properties that determine how any fluid will interact with the solid phase of an aquifer or with other, dissimilar fluids.

For both liquids and gases, the molecular densities and frequencies of molecule-to-molecule interactions (collisions or near-misses) are immense. In a liquid, water for example, the molecules are packed at densities in the range of 10^{21} per cm^3 (5.6 x 10^{21} molecules per cm^3 for water at 4°C). For gases, the molecular densities are more than 1,000-fold lower (4.4 x 10^{18} molecules per cm^3 for an ideal gas, such as air, at 4°C). The behaviors of these large-scale aggregations of molecules and particles are determined by their interactions at the molecular scale.

Liquid behaviors at solid surfaces (wetting), at the interface between liquid and gas (surface tension), and at the contact between dissimilar fluids (interfacial tension), all play an important role in the distribution and mobility of water and other fluids in aquifers. For remediation hydraulics, we will place emphasis on water and its interactions with hydrophobic fluids such as air, fuels and organic solvents. We are especially interested in the interaction between groundwater and porous or fractured aquifer media and we will contrast its behavior to other aquifer fluids of interest, such as fuel hydrocarbons that form lighter-than-water non-aqueous phase liquids (LNAPLs)[1] and chlorinated solvents that form denser-than-water non-aqueous-phase liquids (DNAPLs).

1.1 COHESION AND SURFACE TENSION

Molecules condense to form liquids when temperatures are low and attractive forces between molecules prevent the (approximately) inelastic collisions that occur at high temperatures in gaseous aggregations of the same molecules. There are two sources of attractive force at work in the formation of liquids: 1) dipole moments, especially in organic molecules, generate van der Waals forces that cause very weak attractions between molecules and, 2) hydrogen bonding, which generates much stronger intermolecular attraction. Each of these mechanisms contributes to a liquid's cohesive (internal) and adhesive (liquid-to-solid-surface) attractions, as well as its solvent behaviors.

---

[1] Throughout *Remediation Hydraulics*, we severely restricted the use of acronyms. NAPL, LNAPL, and DNAPL are among the rare exceptions.

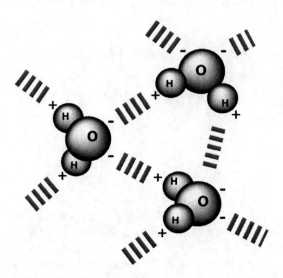

FIGURE 1.1 The structure of water molecules, showing positive charge concentrations at the hydrogen atoms (H) and negative charge concentrations at the oxygen atoms (O). The dashed lines indicate hydrogen bonding between molecules.

Water is the fluid of greatest interest in remediation hydraulics. It has a very strong dipole moment and extensive hydrogen bonding capacity, relatively unique among the fluids we will examine. **Figure 1.1** shows the concentration of negative charge at the oxygen atoms in water molecules and the corresponding positive charges that develop at the hydrogen atoms. Attraction between a hydrogen atom of one molecule and the oxygen atom of a neighboring molecule is called hydrogen bonding and it contributes to the cohesive strength of water and its adhesive attraction to negatively charged surfaces. Negative charge concentration at water's oxygen atom provides adhesive attraction to positively charged surfaces. The strong dipole moment makes water an excellent solvent for ionic solids (salts) and also renders it immiscible with non-polar fluids such as oils and degreasing solvents. It has a high surface tension, yet has good surface wetting capabilities with many solid surfaces, especially those of aquifer matrix materials.

Attraction between molecules at a liquid surface causes the liquid to resist increases in its surface area. The resistance to deformation that would increase the liquid's surface area is termed **surface tension** and variations in molecular-level physical and chemical characteristics generate a large range of surface tensions for liquids of interest in remediation hydraulics. For water, hydrogen bonding links molecules throughout the liquid mass, as depicted earlier in **Figure 1.1**. This loose lattice permits movement of water and solute molecules throughout the liquid. At a water surface, the hydrogen bonding lattice is truncated, forcing an alternative molecular packing. The surface layer packing pattern is jointly determined by the chemical nature of molecules on each side of the interface. At an air-water interface, for example, the topmost layer of water molecules has a net hydrogen-outward orientation (Goh, et al., 1988).

To illustrate surface tension, consider the following experiment: If we suspend a small volume of water in a zero-gravity habitat, it will form a spherical droplet. The sphere has the smallest surface area per unit volume and, therefore, the smallest amount of work expended in forming a surface on the droplet. If we wish to flatten the spherical droplet, its surface area must be increased, requiring that we apply force to overcome the cohesive attraction between molecules at the surface (and to move molecules from the body of the liquid to the surface). **Figure 1.2** diagrams the application

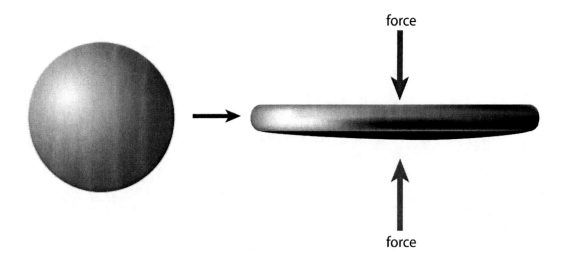

FIGURE 1.2 Surface tension resists increases in the surface area of a liquid. For example, force must be applied to flatten a spherical water droplet, because the flatter shape requires a larger surface area.

of force to increase a droplet's surface area. The fluid's surface tension resists the surface area expansion and force in excess of the surface tension must be applied to increase the surface area.

$$\sigma = \frac{Force \times Distance}{Area} \quad \frac{dynes \times cm}{cm^2} = \frac{dynes}{cm} \tag{1.1}$$

The work performed (force x distance) to expand a liquid surface area by one unit is used to define surface tension, σ, as shown in Equation (1.1).

For the example in **Figure 1.2**, if the droplet was water and the surface area increased from 1 to 4 cm^2 (a 3 cm^2 expansion), the work expended was approximately 225 dyne-cm, indicating that the surface tension of the water in the experiment was 75 dynes/cm. **Table 1.1** provides values for surface tension and other properties of water, as a function of temperature and **Table 1.2** gives physical properties for many other fluids of interest in remediation hydraulics.

Up to this point, we have considered the surface of a liquid in contact with air and we introduced the concept of surface tension to describe cohesion of molecules on the water side of the interface. Surface tension is a special case of the broader class of interactions that occur at fluid boundaries of all types. The general interaction is termed interfacial tension, which describes the work performed to increase the area of an **interfacial surface** that forms between dissimilar fluids[2]. We can observe the effects of interfacial tension at liquid-liquid (e.g., water-oil), gas-liquid (e.g., surface tension), gas-solid and liquid-solid interfacial contacts. The most interesting of interfacial tension effects in remediation hydraulics occur between fluids and aquifer matrix solids as various fluids wet an aquifer matrix and between water and organic liquids as they compete to occupy pore space in the aquifer matrix. **Table 1.3** introduces fluid dynamic properties of several non-aqueous liquids (NAPLs) and liquid mixtures that can be encountered at groundwater contamination sites. Before

---

[2] Surface tension is equivalent to the air-liquid interfacial tension for the liquid of interest.

TABLE 1.1
Properties of pure water. Data from Weast[1] unless otherwise noted.

Temperature (°C)	Density (g·cm^{-3})	Viscosity – cP (10^{-2}·g·cm^{-1}·s^{-1})	Surface Tension Water-air interface (dynes·cm^{-1})
0	0.99987	1.787	75.6[2]
4	1.00000	1.567	
10	0.99973	1.307	74.4[2]
20	0.99823	1.002	72.7[2]
30	0.99568	0.7975	71.2[2]
40	0.99225	0.6529	69.6[2]
50	0.98807	0.5468	67.9
60	0.98323	0.4665	66.2
70	0.97779	0.4042	64.4
80	0.97182	0.3547	62.6
90	0.96534	0.3147	
100	0.95838	0.2818	58.9

[1] Weast, R. C., *Handbook of Chemistry and Physics*, 50th ed. The Chemical Rubber Company, Cleveland, Ohio, 1969.
[2] Wetzel, R. G., *Limnology*, 2nd ed. Saunders College Publishing, Orlando, Florida, 1983.

TABLE 1.2
Properties of selected aqueous solutions used in remedial technology applications at 20°C. Data from Lide (2004)[1], unless otherwise noted.

Solution	Solution Strength (Mass %)	Density (g·cm^{-3})	Viscosity – cP (10^{-2}·g·cm^{-1}·s^{-1})	Interfacial Tension, Air Interface (dynes·cm^{-1})
Water	100.0	0.99823	1.002	72.7[2]
Ethanol	1.0	0.9973	1.023	
	5.0	0.9893	1.228	
	10.0	0.9819	1.501	47.53[1]
	100.0	0.7893	1.203	21.82[3]
D-Fructose	1.0	1.0021	1.028	
	5.0	1.0181	1.134	
	10.0	1.0385	1.309	
	48.0	1.2187	9.06	
Lactate	1.0	0.9992	1.027	
	5.0	1.0086	1.138	
	10.0	1.0199	1.296	
	60.0	1.1392	6.679	
Potassium permanganate	1.0	1.0051	1.000	
	2.0	1.0118	0.998	
	4.0	1.0254	0.992	
	6.0	1.0390	0.985	
Sucrose	1.0	1.0021	1.028	
	5.0	1.0178	1.146	
	10.0	1.0381	1.336	
	60.0	1.2864	58.487	

[1] Lide, D. R., *Handbook of Chemistry and Physics*, 85th ed. CRC Press, Boca Raton, Florida, 2004.
[2] Wetzel, R. G., *Limnology*, 2nd ed. Saunders College Publishing, Orlando, Florida, 1983.
[3] Value for 25 °C.

TABLE 1.3

Properties of selected non-aqueous-phase liquids encountered during remedial technology applications. Data sources as noted.

Liquid	Solution Strength (Mass %)	Density (g·cm⁻³)	Viscosity – cP (10^{-2}·g·cm⁻¹·s⁻¹)	Interfacial Tension, Air Interface (dynes·cm⁻¹)
Trichloroethene (neat)	100	1.47	0.444	34.5 [1]
Perchloroethene (neat)	100	1.63	0.844	44.4 [1]
Gasoline	100	0.75	0.4	25
Weathered JP-4 LNAPL	0.04 % CVOC	0.78	0.96	23.5
Mixed DNAPL[2]	17% CVOCs	1.05	10.4	11.6
Weathered Mixed DNAPL[2]	<0.1 % CVOCs	1.04	51.7	14.3
NAPL collected at a manufactured gas plant[3]	MGP tar mix	1.03	1,389	83.75

[1] Lide, D. R., *Handbook of Chemistry and Physics*, 85th ed. CRC Press, Boca Raton, Florida, 2004.

[2] Samples collected from a former waste disposal site at the Lake City Army Ammunition Plant, Independence, Missouri.

[3] Samples collected from a former manufactured gas plant in the Midwestern United States.

we can explore the effects of interfacial tension in multi-fluid systems, the concept of surface energy and wetting must be introduced.

1.2 SURFACE ENERGY AND WETTING

When a fluid comes into contact with a solid or dissimilar liquid, some of the liquid molecules can adhere to the opposing surface in the process we know as **wetting**. The **wetting fluid** spreads on the opposing surface, which may be a solid or another fluid. Many of the large-scale fluid behaviors we observe in aquifers originate at the contact between solid matter of the aquifer matrix and the perfusing fluid. The wetting process is driven by differences in surface energy between the wetting and wetted phases. **Surface energy** is synonymous with and has the same force-per-length units (e.g., dyne/cm) as surface tension. **Table 1.4** shows solid surface energy values for materials that range from low (Teflon at 18 dynes/cm) to high energies (soda lime glass at 47 dynes/cm).

Wetting entails an increase in the liquid surface area of the wetting fluid and that increase requires the exertion of force, as we saw in **Section 1.1**. When adhesive bonds that form between the wetting fluid and the wetted surface are strong enough to overcome cohesive binding between

TABLE 1.4.
Surface energy values for various materials. Data from Diversified Enterprises, Claremont, New Hampshire.

Material	Surface Energy (dynes/cm)
Polytetraflouroethylene (PTFE or Teflon)	18-20
Paraffin	23-25
Polyethylene	30-31
Polyvinyl chloride (rigid)	39
Aluminum	45
Iron	46
Glass, soda lime	47
Styrene butadiene rubber	48

molecules in the wetting fluid surface, the fluid is drawn to cover (wet) the opposing surface. There is a continuum of wetting interactions:

Perfect wetting — In "perfect" wetting, the wetting fluid is drawn so strongly to the surface that it is drawn into a thin film covering the wetted surface. One example of perfect wetting occurs when a drop of gasoline is placed on a water surface. The surface tension of the gasoline is overcome so completely that it spreads into a film only a few molecules in thickness.

Partial wetting — In partial wetting, the fluid body may be distorted, but the interaction is not strong enough to significantly increase the interfacial surface area. An example of partial wetting occurs when a drop of vegetable oil is placed on a water surface. In contrast to the gasoline case, the vegetable oil droplet remains largely intact, indicating partial wetting of the water surface. Another example of partial wetting is the behavior of pure water droplets on glass.

Perfect non-wetting — In a "perfect" non-wetting contact, adhesion does not occur between the liquid and the solid surface, so the liquid surface area is unchanged. A droplet of liquid mercury on a glass surface provides a good example of a (typically) non-wetting contact.

The energy that drives the wetting process comes from the adhesive bonds that form between molecules or atoms in the wetting fluid and the wetted surface. The surface area of the wetting fluid increases, so it is reasonable to infer that its surface tension must be lower than that of the wetted surface. **Plate 1** shows droplets of three fluids (water, 70-percent isopropanol and vegetable oil) on a low-energy "non-stick" surface. Although none of the droplets shows perfect wetting or non-wetting behaviors, there are clear differences among the liquid-solid interactions. The water is essentially non-wetting on the low-energy non-stick surface, while it would be partially wetting

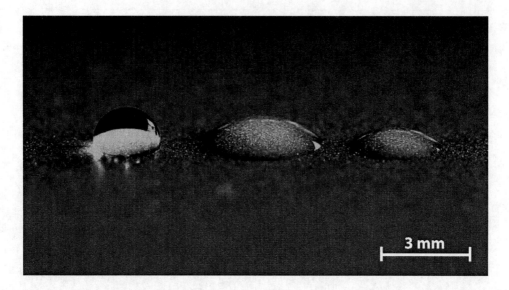

PLATE 1 Droplets of three fluids on a low-energy surface (l to r): water, 70-percent isopropanol and vegetable oil. The contact angle for the water droplet exceeds 90 degrees, so it is a non-wetting fluid on this surface. The isopropanol and vegetable oil contact angles appear to be less than 90 degrees, although the surface hasn't drawn either fluid fully out of its droplet, indicating they are partially wetting fluids on this surface. *(See Plate section for color version of Plate 1.)*

on glass, a higher energy surface. The alcohol and vegetable oil are partially wetting on the non-stick surface are more completely wetting than water on glass. Surface tension data for water, isopropanol and vegetable oil in **Table 1.2** explains the differences in droplet behavior — the non-stick surface cannot overcome the high surface tension of the water, but is able to partially draw the lower surface tension fluids out of the droplet shape. A clean glass surface would be much more effective at drawing the water out of its droplet form because the adhesive attraction between glass and the tested fluids is greater than for the non-stick surface.

Analysis of wetting interactions between fluids and surfaces was formalized by Thomas Young in 1805. **Figure 1.3** shows the contact angle formed between the gas phase (normally air), a liquid and a solid surface. Young's Equation (1.2) describes the contact angle, θ_γ, formed at a three-phase (liquid-vapor-solid) junction, as a function of the tensions, γ, at each of the three interfaces [solid-vapor (SV), solid-liquid (SL) and liquid-vapor (LV)] of a liquid droplet at equilibrium on a solid surface.

$$\cos\theta_\gamma = \frac{\gamma_{SV} - \gamma_{SL}}{\gamma_{LV}} \qquad (1.2)$$

In the upper panel (A) of **Figure 1.3**, the adhesive attraction between the fluid and the solid surface completely overcomes the liquid surface tension and the liquid spreads, wetting the surface and generating a near-zero contact angle. In the middle panel (B), the surface tensions (also known as surface energies) of the solid and liquid phases are approximately equal, so wetting does not occur and the resulting contact angle is 90 degrees. The lower panel (C) shows a droplet on a low-

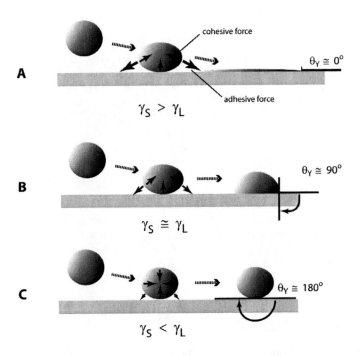

FIGURE 1.3 Wetting is determined by relative surface energies of the liquid and solid phases (γ_L and γ_S, respectively). Wetting can be viewed as the spreading of a liquid onto a solid surface that occurs when adhesive attraction between the solid and liquid surfaces overcomes the cohesive forces within the liquid. A. When the solid-phase surface energy exceeds that of the liquid phase, perfect wetting can occur and the contact angle approaches 0°. B. When the liquid and solid surface energies are approximately equal, partial wetting occurs and the three-phase contact angle is approximately 90°. C. When the liquid surface energy exceeds that of the solid phase, no wetting occurs and the three-phase (gas-liquid-solid) contact angle (θ_γ), tangent to the liquid surface, may approach 180° (perfect non-wetting condition).

energy surface is unable to exert a noticeable adhesive attraction on the fluid droplet. The result is near-perfect non-wetting and the contact angle approaches 180 degrees.

1.3 INTERFACIAL TENSION, CAPILLARITY, AND ENTRY PRESSURE

When two dissimilar liquids come in contact, physical and chemical characteristics of the fluids jointly determine the surface tension that develops at the fluid boundary. This is termed the interfacial tension. We are most often interested in the impact of interfacial tension when a non-aqueous fluid such as a fuel or solvent intrudes into a porous medium already occupied by water as the wetting fluid. **Figure 1.4** provides an example of a two-fluid contact in a small-aperture cylindrical conduit. We can determine which is the wetting fluid by examining the side wall contact — the wetting fluid has a contact angle less than 90°. If the interfacial tension is high, the non-wetting fluid pushes strongly against the wetting phase and the radius of the contact interface is small (strongly convex). If the interfacial tension is low, the radius of the contact interface is large (weakly convex). The forces acting on the interface are the wetting and non-wetting fluid pressures, each acting over the cross-section of the cylindrical container, and the interfacial tension, acting over the circumference of the cylinder, where the wetting fluid attaches to the solid surface.

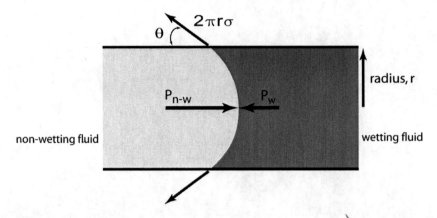

FIGURE 1.4 Force balance for the interface between dissimilar fluids in a cylindrical container. Adhesion of the wetting fluid to the containing surface, along with the surface tension of the wetting fluid, act to balance the pressure exerted by the non-wetting fluid against the two-fluid interface (the pressure of non-wetting fluid, $P_{n\text{-}w}$, exceeds that of the wetting fluid, P_w).

If the interface is not moving (i.e., the system is at steady-state), we can infer that the force per area pushing the intruding fluid (non-wetting fluid pressure) is in balance with the forces acting to hold the water in place (wetting fluid pressure per area and surface tension per unit length of contact). The force balance shown in **Figure 1.4** leads to Equation (1.3),

$$\Delta P = P_{non-wetting} - P_{wetting} = \frac{2 \cdot \sigma}{r} \cdot \cos(\theta) \tag{1.3}$$

where ΔP is the pressure difference between the wetting and non-wetting fluids (force/area), σ is the interfacial tension at the interface (force per length) and r is the radius of curvature of the interface (length). ΔP is also known as the capillary pressure, P_c.

The adhesion of a wetting fluid on aquifer matrix surfaces causes the wetting fluid to resist displacement by a dissimilar fluid. The pressure required for a non-wetting fluid (air, oil or solvent, for example) to displace the wetting fluid (typically water) is called the entry pressure and it can be calculated from surface tension data in the same manner as shown in Equation (1.3).

Table 1.3 provides examples of organic-water interfacial tensions that are used in entry pressure calculations. As can be seen in the table, these values are dramatically affected by the composition of the NAPL and groundwater. To support the analysis of fluid dynamics in NAPL source zones, it is critical to obtain laboratory values for NAPL and groundwater collected in the source zone. In fact, our observations indicate that there are significant variations in NAPL fluid behaviors at the fringe of source zones, where weathering effects are greatest, relative to the core of NAPL source zones, where weathering impacts are minimal. In Chapter 4, we will utilize the concept of entry pressure to calculate the mobility of non-aqueous liquids in aquifers.

When water or other wetting liquid enters a small-dimensioned air-filled space, adhesion binding at the solid surface draws liquid into a wetting film and surface tension transmits that attractive pull to the rest of the liquid. This capillary pressure differential elevates the liquid surface in a phenomenon known as **capillary rise**. In **Figure 1.5**, two capillary tubes are inserted into a liquid. The left tube is Teflon, a low-surface-energy material and the liquid cannot wet the tube surface, inside or out. Consequently, fluid doesn't enter the tube and the contact angle approaches

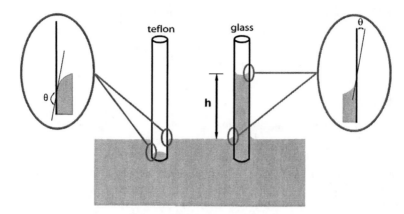

FIGURE 1.5. Behavior of water in capillary tubes of contrasting surface energies. On the left, water does not wet the low-energy (teflon) surface and cannot enter the tube, even when it is pushed below the water surface. On the right, water is drawn above the free liquid surface by energy released by the adhesive attraction between water and the high-energy surface (glass). The inset detail shows the contact angles, θ, that form between the fluid and wetted surfaces.

180 degrees. The right-side tube is composed of silica glass, a high-surface-energy material and the fluid is drawn into the capillary tube as its surface is wetted.

Energy is expended in the capillary rise, to increase the surface area of the fluid and to overcome the force of gravity, indicating that the adhesion process entails an exothermic reaction. The height of rise is a function of the strength of the adhesion binding and surface tension, balanced by the gravitational force acting on the lifted water mass. The force balance gives Equation (1.4):

$$2\pi \cdot r \cdot \sigma \cdot \cos\theta = \pi \cdot r^2 \cdot h \cdot \rho \cdot g \quad dynes \qquad (1.4)$$

Solving for the height of rise:

$$h = \frac{2 \cdot \sigma \cdot \cos\theta}{r \cdot \rho \cdot g} \quad cm \qquad (1.5)$$

From Equation (1.5), we can see that the capillary rise, h, is inversely proportional to the radius of the conduit. In aquifers, coarse-grained soils (relatively large values of conduit radius) generate limited capillary rise. Fine-grained aquifer soils (small values of conduit radius) generate very large values for capillary rise.

1.4 SURFACTANTS

The interaction between fluids and surfaces can be interrupted by surface-active agents, which are broadly referred to as surfactants. These agents may be dissolved in a liquid phase, modifying its surface tension, coat a solid surface and intervene between the solid and liquid surfaces to enhance or suppress adhesion, or modify the interfacial tension between dissimilar liquids changing entry pressures and other behaviors of two-fluid systems. Surfactants may be used in remedial systems to reduce interfacial tension between groundwater and non-aqueous liquids, enhancing the recoverability of the non-aqueous liquids. Surfactants are also produced by aquifer bacteria populations and groundwater surface tension values are often reduced significantly in the presence of contaminant-degrading bacteria. Surfactant flooding strategies and the impact of bacterial surfactants on DNAPL dynamics are discussed more fully in later chapters.

1.5 VISCOSITY

In every fluid, whether liquid or gas, attraction between molecules in the fluid generates cohesion which is observed in the fluid's resistance to parting or shearing. The resistance to shear is termed *viscosity*, and the strength of that resistance is one of the very obvious macroscopic behaviors that distinguish one fluid from another. When pouring various fluids, for example, we notice differences in their behavior that we refer to as their "thickness" or "stiffness," a consequence of their respective viscosities.

When a fluid comes into molecular-scale contact with a solid surface, a thin layer of the fluid adheres to the solid in the process we know as wetting. The wetting layer is effectively immobile, even when the nearby fluid molecules are moving at relatively high velocities. Potter and Wiggert (1991) refer to this as the no-slip condition. When a fluid flows through a simple conduit, a velocity profile is formed, from the zero-velocity fluid at the wetted surface to a zone of maximum velocity at the center of the conduit. For flow to occur in the conduit, cohesive forces with the fluid must be overcome. The moving portion of the fluid must be continuously torn, or sheared, from the wetting layer. The combined adhesion of the fluid to the solid surface through wetting and the fluid's internal resistance to shearing (viscosity) leads to frictional drag as the fluid moves through the conduit and the fluid's viscosity transmits the drag of side wall friction throughout the entire fluid.

Figure 1.6 shows a generalized velocity profile for a fluid moving through a simple conduit. Viscous resistance to shearing transmits frictional drag from the side wall to the remainder of the fluid mass. As the resistance to shearing (viscosity) increases, the energy required to push a fluid along a conduit also increases. Viscosity reflects how a fluid transmits sidewall drag through the entire fluid body: high-viscosity fluids strongly resist tearing away from wetting layers at solid surfaces and sidewall drag is transmitted very strongly; low-viscosity fluids offer little resistance to tearing away from wetting layers and sidewall drag is transmitted weakly into the moving fluid. The flow profile shown in **Figure 1.6** indicates parallel movement of fluid at all points in the cross-section. This condition is termed laminar flow. Because high-viscosity fluids resist shear, they are less susceptible to perturbations that cause chaotic movement (turbulence), which occurs more easily in lower-viscosity fluids.

Viscosity (sometimes called *dynamic* viscosity or *absolute* viscosity) is defined as the ratio of the shear stress to the velocity gradient. The shear stress is the force applied parallel to the flow axis, per unit area, and the velocity gradient is the change in velocity per unit length perpendicular to the flow axis, as shown in **Figure 1.6**. Equation (1.6) shows the calculation of viscosity, μ, from the shear stress and velocity gradient.

$$\mu = \frac{F/A}{dv_x/dz} \quad \frac{dynes/cm^2}{cm \cdot sec^{-1}/cm} = dyne \cdot sec \cdot cm^{-2} \tag{1.6}$$

The two curves in **Figure 1.6** show how viscosity determines the velocity profile that develops under a specified shear stress (F/A). The average velocity is greater for the lower-viscosity fluid (curve 1), because the lower-viscosity fluid "tears" more easily and drag of the side wall is not transmitted as effectively across the moving fluid body. As a consequence, the velocity gradient (dv_x/dz) is greater for the low-viscosity fluid (curve 1) than for the higher-viscosity fluid (curve 2).

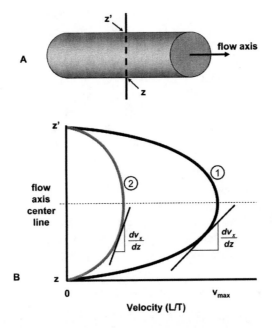

FIGURE 1.6 Fluid velocity profiles of two fluids in a simple conduit. Curve 1 represents the lower viscosity, relative to the fluid in curve 2, with each fluid subject to identical shear stress (force per unit area, parallel to the flow axis). A. Orientation of a cross-sectional transect, perpendicular to the flow axis. B. Velocity profiles along the transect, y - y'. Wetting of the conduit wall leads to a zero-velocity layer at the side wall for each fluid. Movement of the fluids through the conduit requires continuous shearing, which exerts a drag on their movement. For the lower-viscosity fluid (curve 1), the slope of the velocity gradient is greater than for the higher-viscosity fluid (curve 2).

The cgs viscosity unit shown in Equation (1.6) is called a poise (P) and it reduces to $g \cdot cm^{-1} \cdot sec^{-1}$ (a dyne is the cgs unit of force, $g \cdot cm \cdot sec^{-2}$). One one-hundredth of a poise, or one centipoise (cP), equals one one-thousandth of the SI unit for viscosity, the pascal-second (Pa·s).

$$10^{-2}\ P = 1\ cP = 10^{-3}\ Pa \cdot sec = 1\ mPa \cdot sec \tag{1.7}$$

Equation (1.7) relates various units for viscosity. The centipoise is the most commonly-used unit for viscosity ($10^{-2} g \cdot cm^{-1} \cdot sec^{-1}$).

An alternative measurement of viscosity is the kinematic viscosity, calculated as the ratio of the viscosity to density (η/ρ). The standard unit of kinematic viscosity is the Stoke (S) with the cgs units cm^2/sec and values are typically reported as **centistokes** (cS), ($10^{-2} \cdot cm^2 \cdot sec^{-1}$).

Table 1.1 shows viscosity values of pure water for temperatures ranging from 0 to 100 °C. The range of interest for most remediation applications is 4 to 20 °C and viscosity decreases from 1.787 to 1.002 cP over that range. **Table 1.2** provides viscosity values for several solutions that might be used in association with remedial technologies. Most of the solutions have a higher viscosity than pure water and their viscosities increase with increasing concentration. Most of these fluids require greater force to establish flow in conduits or porous media than for comparable flows of pure water. Permanganate solution is an exception, as its viscosity is less than water and decreases with increasing concentration.

For many fluids of interest in remediation hydraulics, the viscosity is a constant for all levels of shear stress. Such fluids are termed Newtonian. Water, for example, is a Newtonian fluid. There are also many fluids for which viscosity is not a constant, and non-Newtonian fluids may also be encountered in aquifer restoration activities. Mixtures of solid particles or fine bubbles and water may behave as shear-thickening fluids (viscosity increases under increasing shear stress), while others may behave as shear-thinning fluids (viscosity decreases under increasing shear stress). Thixotropic fluids are a special case of shear-thinning fluids: they are solid when little or no shear stress is applied then liquefy when shear stress exceeds a critical value. Toothpaste is an example of a thixotropic fluid. **Figure 1.7** shows the conceptual relationship between viscosity and shear stress for Newtonian and non-Newtonian fluids.

For remediation hydraulics, shear-thickening behavior can pose a significant difficulty. Bubbles can form in reactive solutions such as oxidants or in biodegradable solutions such as lactate or carbohydrates and the entrained bubbles can cause the fluids to thicken under pumping stress. Particle suspensions such as those encountered in zero-valent iron injections can also thicken under pumping stress. Remedial system designers need to be aware of potential non-Newtonian fluid behaviors that arise under pumping stresses and account for them in the design process.

1.6 LAMINAR AND TURBULENT FLOWS

The cross-sectional velocity profiles for fluid flows confined to conduits can be very stable along the flow path. These well-organized flows are termed laminar, and a molecule that enters the conduit on the centerline [or near the fringe] is likely to remain on the centerline [or near the fringe] as it moves along the conduit. Laminar flows can be maintained in a fluid if the conduit is relatively straight and smooth and the shear stress is limited (i.e., low velocity). Byrd et al. (2002) provide equations that describe laminar viscous flow in tube segments of known radius, r: The maximum (centerline) velocity, vmax, is:

$$v_{max} = \frac{(P_0 - P_L) \cdot r^2}{4 \cdot \mu \cdot L} \quad cm/\sec \tag{1.8}$$

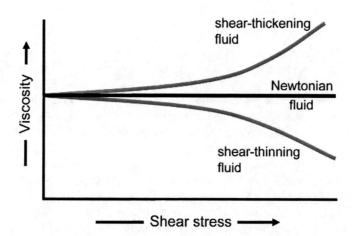

FIGURE 1.7 Relationship between shear stress and viscosity for Newtonian and non-Newtonian fluids. The viscosities of non-Newtonian fluids either increase (shear-thickening fluids) or decrease (shear-thinning fluids) when shear stresses are applied.

where $(P_0\text{-}P_L)$ is the pressure drop over a conduit segment of length, L, and μ is the viscosity of the fluid. Average velocity is defined by the Hagen-Poiseuille Equation (1.9):

$$v_{avg} = \frac{(P_0 - P_L) \cdot r^2}{8 \cdot \mu \cdot L} = \frac{1}{2} \cdot v_{max} \quad cm/\sec \tag{1.9}$$

and the mass flow rate is:

$$w = \frac{\pi \cdot (P_0 - P_L) \cdot r^4 \cdot \rho}{8 \cdot \mu \cdot L} \quad g/\sec \tag{1.10}$$

where ρ is the fluid density (g/cm³).

In laminar flows, cohesive viscous forces in the moving fluid exceed inertial forces and the flow remains well organized. At high velocities, inertial forces exceed viscous forces and flows become chaotic and disorganized, a turbulent flow condition. In turbulent flow, the position of a molecule at one point in the conduit cannot be used to predict that molecule's position farther along the flow path. The cross-sectional velocity profiles break down in turbulent flow conditions and the side wall drag of viscous flow is not transmitted uniformly through the moving fluid mass as it was in laminar flows.

The ratio of inertial to viscous forces provides a method to assess whether flows are likely to be laminar or turbulent. The **Reynolds number** calculates this ratio, as shown in Equation (1.11)

$$\mathrm{Re} = \frac{v_{avg} \cdot 2 \cdot r}{\upsilon} = \frac{v_{avg} \cdot 2 \cdot r \cdot \rho}{\mu} \quad dimensionless \tag{1.11}$$

where υ is the kinematic viscosity, μ is the dynamic viscosity and ρ is the fluid density; v_{avg} and r are the average velocity and conduit radius, respectively. It is important to use dimensionally consistent units for this calculation. The common cgs units for υ (cS) expands to ($10^{-2}\cdot cm^2 \cdot sec^{-1}$) and for μ (cP) expands to ($10^{-2}\cdot g \cdot cm^{-1} \cdot sec^{-1}$). It is important to divide the cP or cS value by 100 to scale the units correctly when entering it into calculation of Re with other cgs denominated variables.

A large value for Re indicates that the inertial forces are large, relative to viscous forces and turbulent flows are likely to develop. The Re value at the laminar-to-turbulent transition is termed the critical Reynolds number, Re_{crit}. Values for Re below Re_{crit} indicate laminar flow and values above Re_{crit} indicate transition to turbulent flow regimes. The value for Re_{crit} varies according to conduit roughness and other factors. Potter and Wiggert (1991) indicate that Re_{crit} is unlikely to fall below 2,000, which can be used as a conservative estimate for the onset of turbulent flow behavior.

1.7 FRICTION FACTORS AND PIPE PRESSURE LOSS

Viscous drag generated at a conduit wall causes pressure loss in pipes and other conduits. The Darcy-Weisbach equation for head loss is:

$$h_L = f \cdot \frac{L}{D} \cdot \frac{V^2}{2 \cdot g} \quad cm \tag{1.12}$$

where f is the friction factor, L is the pipe segment length (cm), D is the pipe diameter (cm), V is the fluid velocity (cm/sec) and g is the acceleration due to gravity (981 cm/sec^2). The head loss is expressed in relative elevation of the fluid in the conduit. If the fluid is water, the head loss can be converted to pressure as follows:

$$\Delta P = h_L \cdot \rho_{water} \cdot g \quad dynes \, / \, cm^2 \qquad (1.13)$$

The friction factor varies according to the roughness of the conduit wall and the type of flow (laminar or turbulent). For laminar flows (Re < 2,000), the friction factor is estimated by Equation (1.14):

$$f = \frac{64}{Re} \quad dimensionless \qquad (1.14)$$

For turbulent flows, there are numerous methods for estimating friction factors, each of which is applicable to a range of flow and conduit conditions. The Swamee and Jain friction factor approximation can be applied to smooth-walled conduits and values of Re up to 10^8:

$$f = 1.325 \left\{ \ln\left[0.27\left(\frac{e}{D}\right) + 5.74\left(\frac{1}{Re}\right)^{0.9} \right] \right\}^{-2} \quad dimensionless \qquad (1.15)$$

where e is the conduit wall roughness (1.5 x 10^{-4} cm for smooth-walled pipe) and D is the pipe diameter in cm.

1.8 DENSITY AND COMPRESSIBILITY

The mass density of a fluid is expressed in weight per unit volume (g/cm^3) and is denoted by the variable ρ. Fluid densities reflect how tightly the molecules (combined with suspended particles or colloids, in some cases) pack into space and vary in direct response to changes in pressure, according to Equation (1.16):

$$\frac{\Delta\rho}{\rho} = \frac{\Delta P}{B} \qquad (1.16)$$

where B is the bulk modulus of elasticity (Potter and Wiggert, 1991).

The densities of liquids are much greater than those of gases and they are much less susceptible to density changes that can be induced by changes in pressure (compression or decompression). Gases, in contrast, are highly compressible and knowledge of system pressure is essential for understanding the densities and other basic fluid properties of gases.

Equation (1.17) is most often applied to determine the change in fluid pressure (ΔP) required to increase the density of a fluid by a specified amount ($\Delta\rho/\rho$):

$$\Delta P = B \cdot \frac{\Delta\rho}{\rho} \quad dynes \, / \, cm^2 \qquad (1.17)$$

When the fluid of interest is an ideal gas, the modulus of elasticity (B) is equal to its pressure (Potter and Wiggert, 1991) — to double the density, double the pressure; to decrease the density two-fold, decrease the pressure to one-half of the starting value. In contrast, water and most other fluids are effectively incompressible. The modulus of elasticity, B, for water is approximately 2.4 x 10^{10} dynes/cm^2 (21,000 atmospheres). In Chapter 2, we will examine the compressibility of aquifer formations and it will be clear that water compressibility is not a significant contributor to aquifer compressibility.

1.9 IMPACTS OF TEMPERATURE AND DISSOLVED SOLIDS ON FLUID PROPERTIES

Changing the composition or temperature of aquifer fluids, such as water or non-aqueous organic liquids such as fuels and solvents, can dramatically influence their fluid properties and, consequently, their behavior in aquifers. There are many circumstances in remediation hydraulics where we need to understand the movement of distinct fluids in aquifers. The movements of non-aqueous liquids or treatment reagents are two obvious examples and the fluid properties of these materials are generally quite different from those of the groundwater.

Temperature has an influence on each of the fluid dynamic properties we have examined: density, surface tension and viscosity and its impacts are summarized in **Figure 1.8**, for temperatures ranging from 0 to 100 °C. Density differences are limited to less than 0.5-percent decrease over the range of temperatures normally observed in groundwater, 5 to 30 °C. The surface tension of water decreases by 7 percent over that same temperature range. Much greater temperature-induced changes occur in viscosity: between 0 and 30 °C, viscosity decreases almost two-fold. As we saw in earlier sections, fluid viscosity is a critical variable in transmitting drag from solid surfaces to flowing fluids. Decreasing viscosity significantly decreases the resistance to flow in conduits and porous media, such as aquifers.

The temperature effect on viscosity of non-aqueous liquids can be much greater than for groundwater. **Figure 1.9** compares temperature-viscosity relationships for groundwater and non-aqueous liquid collected from an aquifer underlying former waste oil/solvent pit at a groundwater contamination site in the Midwestern United States. Based on the standard laboratory procedures,

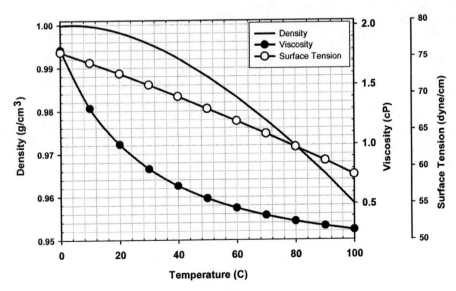

FIGURE 1.8 Effect of temperature on the fluid properties of water: temperature, viscosity and surface tension. Data from Lide (2004).

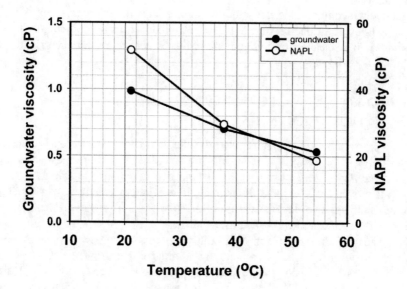

FIGURE 1.9 Comparison of temperature effects on the viscosity of groundwater and near-neutral-density non-aqueous liquid (NAPL) from a groundwater contamination source zone.

viscosities were measured at three temperatures from 20 to 55 °C. Over that range, groundwater viscosity decreased by one-third and the non-aqueous liquid viscosity decreased 2.5-fold. There are at least two important implications of this data for remediation hydraulics: 1) remedial actions that increase aquifer temperatures increase the mobility of groundwater and greatly increase the mobility of (at least some) non-aqueous liquids, and, 2) because of the high temperature sensitivity of viscosity in these non-aqueous liquids, efforts should be made to obtain laboratory viscosity results at the native aquifer temperature. Otherwise, the mobility of non-aqueous liquids in the subject aquifer may be greatly over-estimated.

Addition of dissolved solids to water or other fluids increases the fluid density, one of the common modifications to fluid properties encountered in remediation hydraulics. **Figure 1.10** uses the mix of salts in seawater to show the impact of salinity, an expression of total dissolved solids, on the density of water. At 10 °C and 40 parts per thousand salinity (4 percent dissolved solids by weight), the density of water is approximately 3 percent greater than for water less than 1 part per thousand total dissolved solids at 10 °C (typical of temperate climate groundwater). The increased density affects pumping dynamics and may cause reagents injected into aquifers to sink due to density contrast with the native groundwater. We will examine the effects of temperature and dissolved solids content of injected fluids in Chapter 13, where we provide detailed discussion of fluid injections.

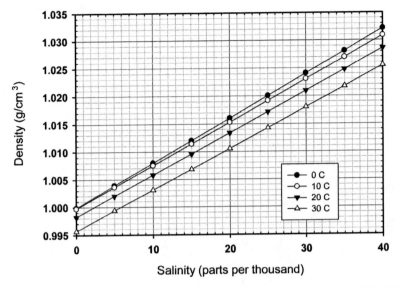

FIGURE 1.10 Effects of salinity and temperature on seawater density. Data from Lide (2004).

2 Properties of Porous Media

Each porous medium is composed of a skeletal matrix, consisting of solid minerals or mineral grain particles, and fluid-filled (gas or liquid) pores that occupy the interstices between the solids. To enable fluid flow, at least a fraction of the pore spaces must be interconnected and the pore sizes must be large enough to transmit water molecules (Bear, 1988; Corey, 1994). An individual **pore** is formed by the void space between three or more mineral grains, as shown on **Figure 2.1**. When adjacent pores are interconnected, the term **pore channel** or **pore throat** is often used to describe the path that connects the individual pores to form a three-dimensional network, which enables fluid flow through the porous media. As the pore channels anastomose around and between adjacent grains of different shapes and sizes, the size and shape of the pore channel varies along its path. **Figure 2.2** illustrates schematically how multiple pore channels combine to form a complex, three-dimensional network around a cluster of individual grains.

An **aquifer** is a water-saturated porous medium composed of geologic unit(s) that can store and transmit significant quantities of water under ordinary conditions. The term **reservoir** has the same general meaning, but it is often used when the fluid of interest is petroleum. In this text, the primary focus is the assessment of soils and unconsolidated sediments comprising aquifer systems. However, bedrock aquifer systems consisting of consolidated sediments, igneous, and metamorphic rocks also contain interconnected pore spaces due to depositional and genetic processes, as well as physical and chemical weathering. In unconsolidated sediments and soils the characteristics of the matrix skeleton are a direct result of the depositional environment, which dictates the type and volume of the sediment, the shapes of the mineral grains, and distribution of particle grain sizes. The more dynamic the nature of the depositional process that created the deposit, the higher the degree of variability to be expected in the grain sizes composing the skeletal matrix. The depth of emplacement and loading history subsequent to deposition also affect the density and packing of the skeletal matrix. To this end, we recommend a multidisciplinary approach to aquifer characterization that incorporates quantitative geology and hydrogeology, as well as engineering methods.

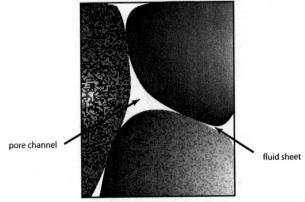

FIGURE 2.1 Pore channels form at the intersection of three or more matrix particles. Thin sheets of fluid link to the pore channels along their length. Bulk water movement in thin sheets is limited by viscous forces.

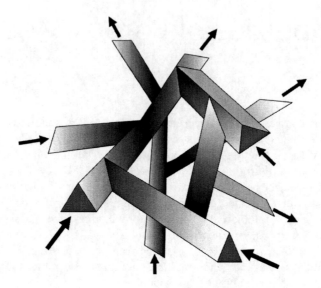

FIGURE 2.2 Flow network developed by the intersection of channel segments that occur in the contact interstices between three or more aquifer matrix particles. Arrows suggest possible groundwater flow paths.

2.1 SKELETAL MATRIX

Granular porous media consist of an assemblage of mineral grains that is dictated by the depositional environment. At the megascopic scale, the depositional environment determines the cycle of physical and chemical weathering that makes the mineral grains available for transport through gravity-driven processes that redistribute the sediment. At the microscopic scale, the mineral grains form the basic building blocks for the skeletal matrix of porous media that comprise aquifer systems.

The fundamental basis for describing the skeletal matrix is grain size. Sediment grains range in size from gigantic boulders to minute specs of wind-blown dust (Leeder, 1982). The standard unit of measure for grains is typically the millimeter (mm), although some engineers continue to use thousandths-of-an-inch. **Table 2.1** shows the Udden-Wentworth grain-scale system, which forms the basis for geologic and engineering classification of soils. In sedimentology, practitioners often use phi-units (φ), which is a geometric series based on the negative log-base 2 ($-\log_2$) of the grain diameter (d) in millimeters, to characterize sediments:

$$\phi = -\log_2 d \tag{2.1}$$

There are practical reasons for evaluating grain size from the mm-diameter or phi-unit perspective. The first is that the conventions for grain-size classification are based on a geometric series using "fractional diameters," as shown on **Table 2.1**. The corresponding entry in phi-units for the grain-size class is a whole number — 2 phi-units for a diameter of 1/4 mm is the threshold for separating medium-grained sand from fine-grained sand, 3 phi-units (1/8 mm diameter) demarcates the threshold between fine-grained sand and very-fine-grained sand, and so on. From a practical perspective, it is more intuitive to evaluate grain-size classifications in this logical progression than looking-up classifications based on thousandths-of-an-inch.

TABLE 2.1
Summary of the Udden-Wentworth grain-size classification for sediments (after Leeder, 1982). The millimeter grain-size scale is broadly used by sedimentologists and geologists in the field. The phi-unit scale, which is based on the geometric progression of the grain diameter, d, provides a natural means of classifying sediment size fractions as transport is a function of mass (d^3) and settling is a function drag (d^2). Phi-unit is $-\log_2$ (d), where d is mean grain diameter in millimeters.

	US Standard sieve mesh	Millimeter scale decimal	Millimeter scale fraction	Phi Units	Udden-Wentworth size classification
GRAVEL		4096		-12	
		1024		-10	boulder
		256	256	-8	
		64	64	-6	cobble
		16		-4	pebble
	5	4	4	-2	
	6	3.36		-1.75	
	7	2.83		-1.5	granule
	8	2.38		-1.25	
	10	2	2	-1	
SAND	12	1.68		-0.75	
	14	1.41		-0.5	very coarse sand
	16	1.19		-0.25	
	18	1	1	0	
	20	0.84		0.25	
	25	0.71		0.5	coarse sand
	30	0.59		0.75	
	35	0.5	1/2	1	
	40	0.42		1.25	
	45	0.35		1.5	medium sand
	50	0.3		1.75	
	60	0.25	1/4	2	
	70	0.21		2.25	
	80	0.177		2.5	fine sand
	100	0.149		2.75	
	120	0.125	1/8	3	
	140	0.105		3.25	
	170	0.088		3.5	very fine sand
	200	0.074		3.75	
	230	0.0625	1/16	4	
SILT	270	0.053		4.25	
	325	0.044		4.5	Coarse silt
		0.037		4.75	
		0.031	1/32	5	medium silt
		0.0156	1/64	6	
		0.0078	1/128	7	fine silt
		0.0039	1/256	8	very fine silt
CLAY	use pipette or hydrometer	0.002		9	
		0.00098		10	
		0.00049		11	clay
		0.00024		12	
		0.00012		13	
		0.00006		14	

Apart from the convenience that the mm-scale or phi-unit scales provide in classifying sediments, research in sediment dynamics has shown that transport and deposition processes can be quantified according to geometric progressions of grain diameter.

Sediment transport is initiated when the inertial force of a particle, F_I (due to its mass), is overcome by the drag force, F_D, and the lift force, F_L (due to the fluid, air or water, flowing over it) (Knighton, 1987; Raudkivi, 1990). For a spherical grain of diameter, d, incipient motion occurs when the forces balance as in Equation (2.2):

$$F_I = F_D + F_L \tag{2.2}$$

The downward inertial force is due to the particle's submerged weight, which stems from the relative density of the particle, ρ_p, and the density of the fluid, ρ_f, as shown in Equation (2.3):

$$F_I = \pi g(\rho_p - \rho_f)\left(\frac{d}{2}\right)^3 \quad \left(\frac{cm}{s^2} \cdot \frac{g}{cm^3} cm^3\right) or\ (dynes) \tag{2.3}$$

The drag force, F_D, and lift force, F_L, exerted on the grain depend on several factors including: the shape of the grain; the roughness and slope of the bed, where the sediment transport is occurring; the density and viscosity of the fluid, including potential suspended sediments already entrained in the fluid; the depth of water, which imparts shear stress on the grain; and the velocity and flow regime of the fluid.

Sedimentation or settling of particles is described by Stokes' Law, after George Gabriel Stokes, an Irish mathematician from the 19th Century. Stokes' Law shows that the terminal settling velocity, V_t, of small spheres[1] is also related to the square of the sphere (grain) diameter, d:

$$V_t = \frac{gd^2(\rho_g - \rho_f)}{18\mu} \quad \frac{cm}{s} \tag{2.4}$$

where g is the gravitational constant and μ is the fluid viscosity. Stokes' Law was derived based on a force balance between the downward gravitational force, due to the particle's mass, and the resistance to settling due to the fluid's frictional resistance caused by drag on the particle in a stationary fluid. Because the settling velocity in Equation (2.4) is proportional to the square of the grain-size diameter, the settling velocity for fine-grained sand (diameter of 1/4 mm) is 4-times greater than very-fine-grained sand (diameter of 1/8 mm). The implication of Stokes' Law is that the sedimentary depositional processes are also governed by a geometric progression of the grain-size diameter.

GRAIN-SIZE DISTRIBUTION

As a portion of a sediment sample may consist of thousands of grains that span several orders of magnitude in diameter, it is common to characterize the assemblage of grains using statistical methods. By convention, grain-size distributions are measured using standardized sieve analysis, where the weight fraction retained on successively finer mesh-size openings is used to determine the primary- and lesser-order grain-size fractions. Histogram plots and cumulative-weight-retained

---

1 Stokes' Law applies strictly only to grains with diameters less than 0.7 mm because turbulent eddies form in the lee of larger particles, reducing drag and increasing the terminal settling velocity.

plots are used to characterize the grain-size distribution. **Table 2.1** illustrates the US standard sieve opening-size (thousandths-of-an-inch) versus the Udden-Wentworth classification system (mm).

In sedimentology and engineering, statistical methods and moment analysis are used to classify and describe the sediment grain-size distribution that characterizes a particular deposit. The mean, μ, or average is defined as the sum of the measurements, x_i, divided by the number of measurements, n:

$$\mu = \sum_{i=1}^{n} \frac{x_i}{n} \qquad (2.5)$$

For sieve analysis, the measurement, x_i, corresponds to the individual weight-fraction retained on each sieve, or grain-class based on sieve opening-size, and the number of measurements is based on the number of sieves used. The mean grain size is a useful comparator of sediments, as the weight of the particle must be overcome by applied shear-stress before transport by wind or water is possible. Therefore, the mean grain size of a particular deposit reflects the flow strength, or energy associated with the depositional process (Leeder, 1982). **Sorting** is a geological term that describes the uniformity of the grain sizes within a particular deposit. A well-sorted deposit consists of a narrow range of grain sizes, whereas a poorly sorted deposit consists of a wide range of grain size.

FIGURE 2.3 Sorting and grading concepts in aquifer matrix particle assemblages. Particle assemblage A has uniform particle sizes and is well sorted or poorly graded. Assemblage B comprises a range of particle sizes and is poorly sorted or well graded.

PLATE 2 Well sorted sand, collected at Makaha Beach, Oahu, Hawaii (scale approximate). Particle sizes ranged from 0.5 to 1.5 mm. *(See Plate section for color version of Plate 2.)*

PLATE 3 Well sorted sand, collected at Nordhouse Dunes on the eastern shore of Lake Michigan (scale approximate). Grain sizes ranged from 0.2 to 0.5 mm. *(See Plate section for color version of Plate 3.)*

PLATE 4 Well sorted sand, collected from the Pacific Ocean intertidal zone at Carmel, California (scale approximate). Grain sizes ranged from 0.2 to 0.5 mm. *(See Plate section for color version of Plate 4.)*

PLATE 5 Poorly sorted sand and fine gravel, collected 20 meters below ground surface from an aquifer formation near Valcartier, Quebec, Canada (scale approximate). Particle sizes ranged from approximately 0.01 to 5 mm. The field-determined hydraulic conductivity for this material was 1.5×10^{-1} cm/sec. Grain sizes ranged from 0.2 to 0.5 mm. *(See Plate section for color version of Plate 5.)*

Figure 2.3 illustrates the concept of sorting schematically. **Plates 2 through 5** show photographs of several well-sorted sands and a poorly-sorted sand and gravel, respectively.

A fluvial formation consisting of interbedded laminae (alternating thin layers) of sands and silts records the cycles of waxing and waning energy in the depositional environment during the accumulation of sediment. The characteristic distribution and mean grain size within each of these thin laminae reflects the **hydraulic-** or **transport-sorting** as a result of the cyclic energy associated with the depositional process. A composite sample collected across multiple laminae reflects the **bulk sorting** of the sample, which is simply a measure of the uniformity of grain sizes comprising the distribution, but not a good indicator of the depositional process. Therefore, careful consideration should be given to ensure collecting representative samples within a depositional unit, if the objective is to evaluate hydraulic sorting.

In sedimentology the standard deviation (σ), or square-root of the variance (σ^2) of the grain-size distribution, is the statistical measure of sorting:

$$\sigma^2 = \sum_{i=1}^{n} \frac{(x_i - \mu)^2}{n} \qquad (mm^2) \qquad\qquad (2.6)$$

Practitioners often utilize graphical methods in combination with cumulative grain-size plots to estimate the mean and standard deviation in a more convenient way. Selected values of grain-size diameter are read from the cumulative-weight-retained plot to estimate the mean, μ, and standard deviation, or sorting, σ:

$$\mu = \frac{(d_{16} + d_{50} + d_{84})}{3} \qquad (mm) \qquad\qquad (2.7)$$

$$\sigma = \left(\frac{d_{84} - d_{16}}{4} \right) + \left(\frac{d_{95} - d_5}{6.6} \right) \qquad (mm) \qquad\qquad (2.8)$$

In Equations (2.7) and (2.8), the subscripts denote the cumulative-percentile grain-size diameters for the graphical equations. Sedimentologists typically use grain diameters in phi-units for the statistics and the grain-size plots to further simplify the analysis.

Figure 2.4 compares a phi-unit grain-size plot (**Plot A**) with a conventional grain-size plot (**Plot B**), commonly used for geotechnical purposes. When sorting values are computed using phi-units, the result can be plotted directly on phi-unit grain-size plot to evaluate the physical spread of the grain-size distribution, or used to compute confidence intervals about the mean. Conventional statistical methods tell us that one standard deviation about the mean includes 68.3 percent of the normal distribution, two standard deviations include 95.5 percent and three include 99.7 percent.

Table 2.2 presents the sedimentological ranking of sorting, σ, using phi-units. Based on Table 2.2, the Aeolian sand shown on Figure 2.4 would be characterized as well-sorted. Intuitively, smaller values imply higher-degrees of sorting because each phi-unit represents a halving or doubling of the grain size about the mean. The sorting statistic loses its meaning when computed in millimeters for the conventional plot without transformation, because the conventional diagram plots the grain diameter using a log-scale.

Sorting is the opposite of grading[2], which is widely used in the geotechnical and engineering disciplines — a well-sorted deposit consists of near uniform grain-size distribution, whereas a

---

[2] We prefer to use grading to describe mean grain-size trends in the stratigraphic (vertical) profile of sedimentary deposits. In this vernacular, a fining-upward sequence associated with fluvial depositional systems is a "graded formation," as the mean grain size decreases from the stratigraphic bottom of the bed to its top. The grading can be expressed in a continuous fashion, where the reduction in grain size is gradational, or in stratified layers that show progressive decreases in grain size.

well-graded deposit exhibits all grades of grain-size classes. **Figure 2.5** illustrates the concept of sorting in aquifer matrices by comparing conventional cumulative-weight-retained grain-size plots for glacial till, dune sand and hydraulically-placed silt. Very-well-sorted deposits, such as the dune sand, are characterized by cumulative-weight-retained (percent finer versus log diameter) distribution curves with near vertical slopes; poorly-sorted deposits such as the till and silt are characterized by broader, flatter cumulative distribution curves.

Geotechnical engineers prefer to use the uniformity coefficient, C_u, to evaluate the spread of grain-size distributions.

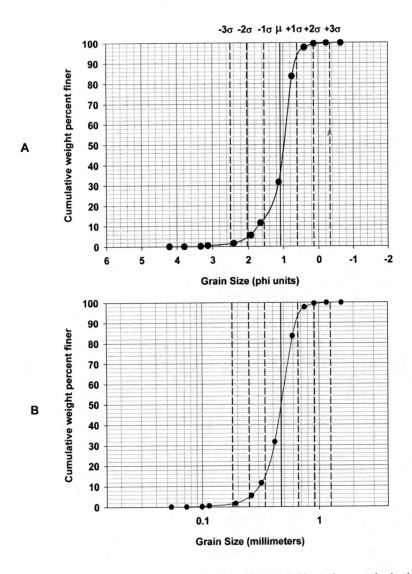

FIGURE 2.4 Comparison of cumulative grain size plots for aeolian sand. Plot A shows grain size in phi-units. Plot B shows grain size in millimeters using log scale. The mean grain size, μ, was computed at 1.08 phi-units or 0.47 mm using Equation (2.7). The sorting or standard deviation, σ, was computed at 0.47 phi-units or 1.39 mm using Equation (2.8). Confidence intervals (dashed lines) about the mean were obtained using the mean (solid line) +/- 1, 2 and 3σ in phi-units. For Plot B, phi-unit results were converted to millimeters and plotted.

TABLE 2.2
Descriptive measures of sorting based on graphical statistics
methods from Equation (2.8) (after Folk, 1974).

Sorting (standard deviation)	Verbal description
0 to 0.35 φ	very well sorted
0.35 to 0.50 φ	well sorted
0.50 to 0.71 φ	moderately well sorted
0.71 to 1.00 φ	moderately sorted
1.00 to 2.00 φ	poorly sorted
2.00 to 4.00 φ	very poorly sorted
4.00 φ +	extremely poorly sorted

$$c_u = \frac{d_{60}}{d_{10}} \tag{2.9}$$

Because the uniformity coefficient reflects the ratio of d_{60} to d_{10} diameters, increasing values of C_u imply a greater spread in the grain-size distribution. By convention, deposits are considered well-sorted for values C_u between 1 and 3 and poorly-sorted for values above 4 to 6 (Perloff, 1976).

The descriptive statistics including mean, sorting and uniformity coefficient are also shown on Figure 2.5. Inspection of the distribution plots indicates that the glacial till shows more than three orders of magnitude variability in grain size, the silt two orders of magnitude — without accounting for the fact that approximately 10-percent by weight is finer than 0.001 mm (10 phi-units), and the dune approximately one-half order of magnitude. The sorting parameters calculated for the till (2.8 phi-units), silt (2.0 phi-units) and dune sand (0.3) imply that the till is very-poorly-sorted, the silt is poorly-sorted and the dune sand is very-well-sorted. The calculated uniformity coefficients for the till (17.4), silt (10.9) and dune sand (1.3) result in similar classifications of sorting.

There are two major drawbacks associated with the conventional interpretation of the cumulative grain-size distribution curves. First, the measured grain-size distribution obscures the bedding morphology of the deposit, as in the example of a composite sample taken across multiple thin laminae. Evaluating grain-size distributions and statistics without consideration of the depositional features provides a superficial characterization of the aquifer and can result in poor interpretations regarding flow and transport potential of the system. Second, the statistical descriptors based on weight-percent-finer do not intuitively reflect the number of grains contained within each grain-size fraction. Because the mass/weight of the individual grains is proportional to the diameter cubed, the perspective is skewed dramatically toward the larger grain-size fraction, even though the finer grains outnumber the coarser grains by several orders of magnitude. See **Plot B** on **Figure 2.5** for a graphical illustration of this point.

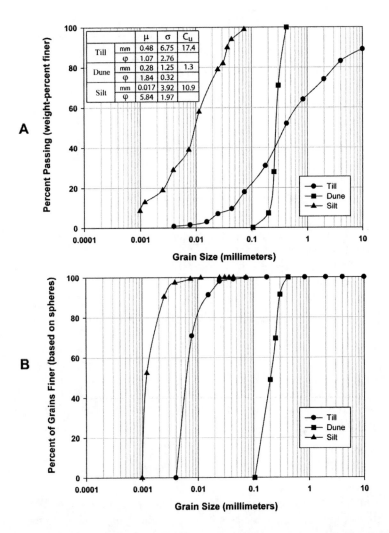

FIGURE 2.5 A - Comparison of grain-size distributions for glacial till, dune sand and hydraulically dredged silt (adapted from Perloff and Baron, 1976). Mean grain size, μ, sorting, σ and uniformity coefficient, C_u, were calculated using Equations (2.5) through (2.7). B - Plot showing percentage of grains, assuming spheres for till, sand and silt.

2.2 COMPOSITION

The composition of the system can be quantitatively described using a phase diagram based on an elemental volume as shown in **Figure 2.6**. In this three-phase system commonly used in soil mechanics, the total volume, V, of the porous media is comprised by the volume of the solids, V_S, and the volume of the voids, V_V:

$$V = V_S + V_V \qquad (cm^3) \qquad (2.10)$$

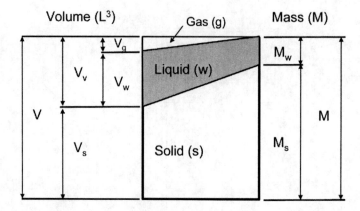

FIGURE 2.6 Porous media phase diagram (after Perloff and Baron, 1976). Left side of diagram is based on volume, V; the right side is based on mass, M.

The voids may consist of air-filled voids, V_A, and/or liquid (water) filled voids, V_w, such that:

$$V_V = V_A + V_W \qquad (cm^3) \tag{2.11}$$

From these basic volume relations, we can define porosity, θ, as the ratio of the volume of the voids to total volume:

$$\theta = \frac{V_V}{V} \quad \left(\frac{cm^3}{cm^3} \right) \tag{2.12}$$

In unconsolidated sediments and soils, the porosity of the system is a direct result of the depositional environment, which dictates the source and type of the sediment, the shapes of the mineral grains and distribution of particle grain sizes. The depth of emplacement and loading history subsequent to deposition also affect the density and packing of the skeletal matrix.

Void ratio, e, is the ratio of the volume of voids to the volume of solids:

$$e = \frac{V_V}{V_S} \quad \left(\frac{cm^3}{cm^3} \right) \tag{2.13}$$

The *degree of saturation* or simply *saturation*, S_r, is defined as the ratio of water-filled voids to void volume:

$$S_r = \frac{V_W}{V_V} \left(\frac{cm^3}{cm^3} \right) \tag{2.14}$$

Note that saturation is a volumetric term, which is not to be confused with *water content*, **w**, that is defined in terms of the ratio of the mass of water, M_w, to the mass of the solids, M_s, as shown below:

$$w = \frac{M_W}{M_S} \qquad \left(\frac{g}{g}\right) \qquad\qquad (2.15)$$

Several other descriptive terms are defined using the mass side of the phase diagram. Often, it is convenient to extend phase density (mass per unit volume) to "unit" weight terms when evaluating soils that make-up porous media. This soil mechanics convention facilitates calculation of overburden pressures and stresses. At four degrees Celsius, the density of water is 1.000 g/cm³; for a unit volume of 1 cubic centimeter, the mass is 1.000 gram. Multiplying the mass by the acceleration the gravitational constant of 9.80 cm/s² yields the weight of the water — 9.80 dynes (or g/cm²s²), which for the unit area of one square centimeter leads to pressure in barye (ba or dynes/cm²).

DENSITY

The density of a porous medium, ρ, is derived from the mass of solids, M_s, comprising the skeletal matrix, and the mass of water, M_w, contained in a unit volume:

$$\rho = \frac{M_s + M_w}{V} = \frac{\rho_s V_s + \rho_w V_w}{V} \qquad \left(\frac{g}{cm^3}\right) \qquad\qquad (2.16)$$

In aquifer sediments, the solids typically consist of the mineral quartz, with a density of 2.6 to 2.7 g/cm³. By convention, geotechnical engineers assume a **solid density**, ρ_s, of 2.65 g/cm³. **Table 2.3** provides density values of various porous matrix materials.

For quartz sand with porosity, θ, of 35 percent and saturation, S_r, of 10 percent, the density of the soil can be obtained by combining Equations (2.12) and (2.14) with Equation (2.16) above as follows:

$$\rho = \rho_s(1-n) + \rho_w \theta S_r$$

$$\rho = 2.65(0.65) + 1.00(0.35)(0.10) \qquad\qquad (2.17)$$

$$\rho = 1.76 \left(\frac{g}{cm^3}\right)$$

Because the density of the soil depends on saturation or water content, it is common to refer to **dry density**, ρ_d, when the saturation or water content of the soil is zero. Similarly, the **saturated density**, ρ_{sat}, is defined when the soil is saturated (e.g., S_r=1). For the example cited in Equation (2.17), the calculated dry density and saturated density are 1.72 and 2.07 g/cm³, respectively. Note that the water content, Equation (2.15), for the saturated case is calculated at 0.20, as it is based on the ratio of the mass of the water (0.35 g) to mass of solids (1.72 g).

The **unit weight of the solids** or skeletal matrix, γ_s, is defined by the solid density, ρ_s, and the acceleration due to gravity, g, as shown in Equation (2.18). The resulting units are force per unit volume (mass x acceleration).

$$\gamma_s = \rho_s \cdot g \qquad \left(\frac{g}{cm^3} \cdot \frac{cm}{s^2}\right) \qquad or \qquad \left(\frac{dynes}{cm^3}\right) \qquad\qquad (2.18)$$

TABLE 2.3
Densities, porosities and compressibilities of aquifer matrix materials. Soil density, ρ, refers to the common ranges of density for unsaturated conditions and various degrees of compaction. Porosities cited reflect the typical range for each material considering compaction and sorting. Actual soil properties should be determined based on site conditions.

Material	Soil Density, ρ (kg/m³)	Porosity, θ (m³/m³)	Compressibility, α (cm²/dyne)
Gravel	2,000 – 2,350 [c]	24 – 38 [d]	$10^{-11} - 10^{-9}$ [a]
Coarse sand	1,400 – 1,900 [c]	31 – 46 [d]	
Fine sand	1,400 – 1,900 [b]	26 – 53 [d]	$10^{-10} - 10^{-8}$ [a]
Silt	1,300 – 1,920 [c]	34 – 61 [d]	
Clay	600 – 1,800 [c]	34 – 60 [d]	$10^{-9} - 10^{-7}$ [a]
Glacial tills	1,700 – 2,300 [c]	20 [b]	
Silts and clays (inorganic)	600 – 1,800 [c]	29 – 52 [c]	
Silts and clays (organic)	500 – 1,500 [c]	66 – 75 [b]	
Peat	100 – 300 [c]	60 – 80 [e]	
Water (β)	1,000 [a]	–	4.2×10^{-11} [a]

[a] Data reported by Freeze and Cherry (1979). *Groundwater.* Prentice Hall, Englewood Cliffs, New Jersey.
[b] Data reported by Perloff, W.H. and W. Baron. 1976. *Soil Mechanics: Principles and Applications.* John Wiley & Sons, New York.
[c] Data Reported by Holtz, R.D. and W.D. Kovacs. 1981. *An Introduction to Geotechnical Engineering.* Prentice Hall, Englewood Cliffs, New Jersey.
[d] Data reported by Zheng, C. and G.D. Bennet. 1995. *Applied Contaminant Transport Modeling.* Van Nostrand Reinhold, New York.
[e] Data reported by Bear. 1988. *Dynamics of Fluids in Porous Media.* Dover Publications, New York, New York.

The cumulative force due to soil mass at any depth can then be expressed as a linear function of depth, **h**, as shown in Equation (2.19). The units reduce to force per unit area, or stress, when one considers a unit area:

$$\sigma_s = \gamma_s \cdot h \quad \left(\frac{g \cdot cm}{cm^3 \cdot s^2} \cdot cm \right) \quad or \quad \left(\frac{dynes}{cm^2} \right) \tag{2.19}$$

The unit weight of water is defined by Equation (2.20).

$$\gamma_w = \rho_w \cdot g \quad \left(\frac{dynes}{cm^3} \right) \tag{2.20}$$

The cumulative pressure due to the weight of overlying water, or **pore pressure**, σ_w, can be quantified by Equation (2.21).

$$\sigma_w = \gamma_w \cdot h \quad \left(\frac{dynes}{cm^2} \right) \tag{2.21}$$

2.3 EFFECTIVE STRESS

The **total stress**, σ_t, at depth in an aquifer is the sum of the weight of the unsaturated soils above the water table and the weight of the saturated soils below the water table as shown in Equation (2.22), where h_d and h_s refer to unsaturated (dry) and saturated soil depths, respectively.

$$\sigma_t = \rho_d g h_d + \rho_s g h_s = \gamma_d h_d + \gamma_s h_s \tag{2.22}$$

Consider the example cited in Equation (2.17) for the development of these concepts. Refer to **Figure 2.7** for a graphical illustration of stresses in the system with the water table at 500 cm, h_d, and the point of interest 1,500 cm below the water table, h_s. For dry soils, the dry weight, γ_d, is 1.72 g/cm^3 x 980 cm/s^2 or 1,686 dynes/cm^2 and the saturated weight, γ_s, is 2.07 g/cm^3 x 980 cm/s^2 or 2,029 dynes/cm^2. Substituting these values into Equation (2.22) yields a total stress of 3.89 x10^6 dynes/cm^2 1,500 cm below the water table.

In an aquifer, buoyancy forces associated with pore pressure counteract the stress induced by the weight of the overlying soil mass in saturated, permeable soils. The intergranular, or **effective stress**, σ_e, acting between soil grains in an aquifer is the difference between the total stress and the pore pressure as defined by Equation (2.23) after Terzaghi (Terzaghi, 1943; Terzaghi, 1960).

$$\sigma_e = \sigma_t - \sigma_w \tag{2.23}$$

The relationship between total, effective and pore pressure stresses is shown in **Figure 2.7**. The buoyant force increases linearly with depth as a function of pore pressure ($\gamma_w h_s$) because pore pressure is transmitted through the saturated soils. In the example, the pore pressure 1,500 cm below the water table is 1.47 x10^6 dynes/cm^2 and the effective stress is 2.42 x10^6 dynes/cm^2.

Effective stress is important in considering strength of soils in geotechnical problems and in remediation hydraulics when injection or withdrawal of fluids is used to distribute reagents or establish hydraulic control of contaminant plumes. Consider the case where reagents are injected

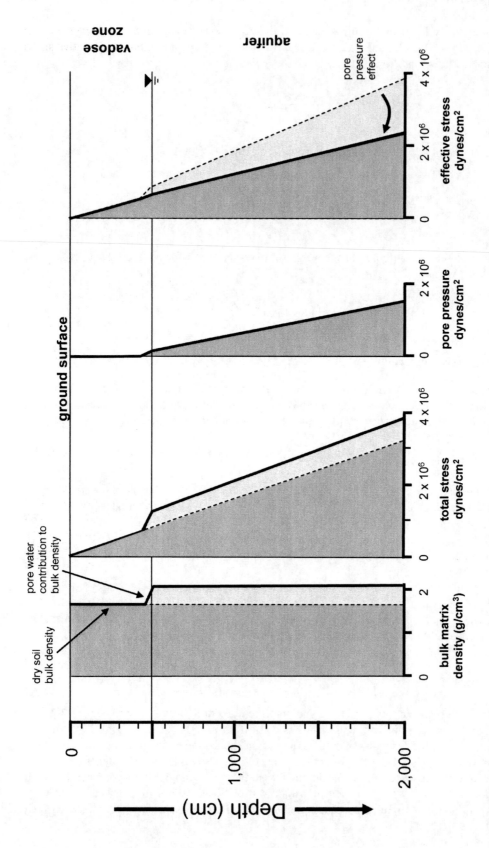

FIGURE 2.7 Relationship between bulk matrix density, total stress, groundwater pore pressure and effective stress as a function of depth in a soil and aquifer profile. For these calculations, the dry bulk soil density was 1.7 g/cm³ and the dry bulk density was 1.7 g/cm³ and the total porosity was 35 percent. The density of water was assumed to be 1.0 g/cm³ and the acceleration due to gravity was 980 cm/sec².

into the aquifer in the example, increasing the pore pressure incrementally and raising the water table. Under this circumstance, the total stress in the system is constant but the effective stress decreases with the increasing pore pressure as shown in Equation (2.24):

$$d\sigma_e = -d\sigma_w = -\gamma_w dh \qquad\qquad (2.24)$$

Dropping the water table 500 cm decreases the pore pressure 4.90×10^5 dynes/cm^2, and proportionately increases effective stress to 2.91×10^6 dynes/cm^2. The issue of effective stress and aquifer matrix failure is a critical design factor in remediation hydraulics, as it is used in evaluating strength and failure in soil mechanics. For example in clay soils, where pore pressure dissipates slowly due to low permeability, adding too high a surcharge load can increase the pore pressure and dramatically reduce the effective stress in the soils causing catastrophic failure. This topic will be revisited in Chapter 10 – Principles of Well Design.

2.4 POROSITY

The constitutive definition of porosity in Equation (2.12) provides little insight regarding the nature of the porous medium and how its composition controls its porosity. In addition, two types of porosity are generally used to describe the nature of the void-volume fractions in porous media — primary porosity and secondary porosity, where the modifiers "primary" and "secondary" denote the genesis of the voids. **Primary porosity** stems from several factors associated with the depositional processes and nature of the deposit including grain packing, sorting, grain shape and sedimentation rates. **Secondary porosity** voids form through physical and chemical processes that follow deposition (or emplacement of igneous rocks) including compaction, fracturing, faulting, dissolution and mineralization. Descriptive terms of secondary porosity include **fracture porosity** (in fractured media) and **vuggy porosity** (in carbonates).

Grain packing bounds the upper- and lower-ends of the porosity spectrum for equi-dimensional spheres: maximum porosity of 48 percent occurs when spheres are stacked one on top of another in a cubic geometry; minimum porosity of 26 percent occurs when spheres are off-set in a rhombohedral geometry. Note that the same theoretic bounds apply whether considering bowling balls, golf balls, marbles or well-rounded fine-grained sand particles of ¼ mm diameter. Most sedimentary deposits exhibit porosities in the intermediate range but often are much lower due to poor-degrees of sorting, compaction and secondary mineralization. Introduction of multiple grain-size fractions, as in a poorly-sorted sand and gravel, reduces the porosity well below the equi-dimensional limits, as the progressively finer particles occupy the voids between the larger particles. The result is that well-sorted sands typically exhibit a 25 to 50 percent higher porosity than poorly-sorted sands of the same mean grain-size diameter (Bear, 1988; Leeder, 1982).

Grain shape and form is probably the most important factor in determining porosity in uncompacted, cohesionless sedimentary deposits. Very-high porosities are found in coquina deposits comprised of irregularly-shaped shells or engineered crushed rock fill materials. The most common measure of grain shape and form are sphericity and roundness. **Sphericity** measures how closely a grain approximates a sphere, while **roundness** measures the smoothness of the grain surfaces. Well-rounded, spherical sediments tend to exhibit lower porosity than assemblages of irregularly-shaped, very angular sediments because the latter resist packing due to interference and increased friction between grains.

2.5 COMPRESSIBILITY AND CONSOLIDATION

Aquifers consist of a skeletal matrix of grains and voids that can be saturated with water. When stresses are applied to aquifers, deformation occurs that results in distortion of the skeletal matrix and a corresponding change in total volume. The relationship between the change in the effective stress, $d\sigma_e$, and **volumetric strain**, dV_t/V_t, is given by the definition of **bulk modulus, B,** of the material in Equation (2.25):

$$B = -\frac{d\sigma_e}{dV_t/V_t} \qquad (2.25)$$

The change in total volume in porous media occurs from a rearrangement of grains in the matrix and resulting reduction in volume of void space (Freeze and Cherry, 1979; Holtz, 1981; Perloff, 1976). The **compressibility** of a porous medium, α, is defined as the reciprocal of the bulk modulus: $\alpha = B^{-1}$. Based on the assumption that compressibility of the grains comprising the porous medium is negligible, this implies that the change in the solid volume, dV_s, is zero such that change in the total volume, dV_t, is attributable to a change in the void volume, dV_v. If the porous medium has an **initial void ratio, e_0** (where $e = V_v/V_s$), Equation (2.25) can be expressed as follows:

$$\alpha = \frac{-de/(1+e_0)}{d\sigma_e} = \frac{-dV_t/V_t}{d\sigma_e} \qquad (2.26)$$

Equation (2.26) implies that the deformation of the porous medium is elastic or reversible. However, the stress-strain relationship exhibited by distorting porous media is non-linear and depends on loading history. **Figure 2.8** illustrates the stress versus void ratio curve for a saturated clay sample subjected to an increased effective stress through an applied load using an odeometer[3]. The non-linear relationship between stress and strain is reflected by the changing slope on the void ratio versus effective stress curve as effective stress is incrementally increased to some predetermined load. The change in slope reflects a "hardening" of the material and the grains are re-oriented and packed more closely together. Once the load (effective stress) is incrementally reduced, the specimen follows a different path on the void ratio versus effective stress curve. The noticeably reduced slope on the stress relaxation curve is most common in cohesive soils such as clays, where the ratio of compressibility under compression to expansion are typically 10:1. Deformation of clay soils under these circumstances is largely irreversible. Cohesionless sands typically exhibit compressibility ratios near 1:1, implying a near elastic behavior unless significant deformation results in crushing of the grains.

The concept of compressibility can be extended to aquifers by considering the one-dimensional case in the vertical direction. In an aquifer at depth with saturated thickness, **b**, the total stress, σ_t, at any point is due to the weight of the overlying soils and water as shown in **Figure 2.7**. If water is pumped out of the aquifer, reducing the water level by -dh, the resulting pore pressure is decreased as given in Equation (2.24). Rewriting Equation (2.26) in terms of **saturated thickness, b,** and the change in saturated thickness, **db,** yields Equation (2.27) after Freeze and Cherry (Freeze and Cherry, 1979):

---

[3] Odeometers are used in geotechnical engineering to conduct one-dimensional drained compression tests. Soil specimens are subject to incremental loads, where the pore water is drained and excess pore-pressure dissipated.

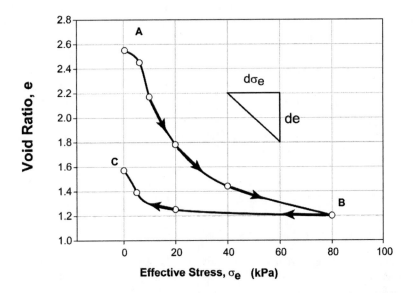

FIGURE 2.8 Stress-strain relationship for compressible soil. Line A-B illustrates a virgin compression curve, the initial response to increased effective stress. Line B-C illustrates the relaxation curve after applied effective stress is removed. Adapted from Holtz and Kovacs (1981).

$$\alpha = \frac{-db/b}{d\sigma_e} \tag{2.27}$$

One-dimensional consolidation/compressibility tests use Equation (2.27) to represent the change in void ratio versus effective stress as shown in Equation (2.26) because the actual change in voids is measured through the change in specimen thickness. The tests are normally conducted using an apparatus such as the odeometer, which constrains the sides of the sample, limiting the distortion in the lateral directions. In an aquifer undergoing changes in effective stress induced by changes in the water table elevation, the surrounding matrix is thought to minimize the lateral distortions in the matrix. Under the simplifying assumption that the distortions are primarily vertical, it is possible to approximate the compaction/expansion of an aquifer undergoing changes in effective stress using Equation (2.27). Substituting the definitions of effective stress and pore pressure in Equation (2.27) and rearranging to solve for the change in aquifer thickness, db, yields the one-dimensional equation for aquifer compaction as shown in Equation (2.28):

$$db = -\alpha b \, d\sigma_e = -\alpha b \rho \, g \, dh \tag{2.28}$$

Consider the example on **Figure 2.7**, where the water table was decreased 500 cm in a 1,500 cm thick aquifer yielding an incremental increase in effective stress, $d\sigma_e$, of 4.90×10^5 dynes/cm^2. Substituting the values into Equation(2.28) and assuming compressibility, α, for sand of 10^{-9} cm^2/dyne (10^{-8} Pa^{-1}) results in a compaction of the aquifer of about 0.7 cm. In permeable materials like sand, the changes in pore-pressure and effective stress are readily transmitted through the aquifer. As a result, deformation occurs contemporaneously with the change in effective stress. Low permeability materials such as silt and clay cannot transmit the water readily, so pore-pressure equilibration and compression in clays might require several months or years. The time-dependent compression is termed **consolidation** in geotechnical engineering.

As shown on **Table 2.3**, the compressibility of clay strata can be several orders of magnitude higher than the sand used in this example. If the example above were evaluated using compressibility of clay at 10^{-7} cm²/dyne (10^{-6} Pa^{-1}), the resulting compaction of a 1,500 cm saturated clay would ultimately be about 70 cm or nearly 5 percent of its initial thickness. The implication is that aquifer systems consisting of sand and clay can exhibit significant compaction through overpumping and resulting depressurization of the more compressible clay strata. On a regional-scale, compaction due to over-pumping in the subsurface is manifested as subsidence at the land surface. On the remediation hydrogeology scale, compressibility of the aquifer matrix (especially concerning silt and clay strata within the aquifer) contributes to aquifer storage for water use, but also plays a key role in accommodating injected fluids.

2.6 AQUIFER STORAGE CONCEPTS

Water is stored in the voids and interstices between the grains comprising porous media. As discussed above, the volume of pore space per unit volume of aquifer is defined as the porosity — this is where the water is stored. However, the volume of water that can be released from storage in an aquifer depends on the connectedness of the pores and whether the aquifer is confined or unconfined.

A **confined aquifer** is bounded above by a much less permeable stratum, which enables the aquifer to become pressurized – that is the hydraulic head in the confined aquifer is above its stratigraphic top. A **flowing artesian aquifer** is a special case of a confined aquifer where the system is pressurized sufficiently to enable flow at the ground surface. An **unconfined aquifer** or **water table aquifer** is bounded above by the free surface of the water table.

SPECIFIC STORAGE

The **specific storage, S_s,** of a saturated aquifer is defined as the volume of water that a unit volume of aquifer releases from storage under a unit decline in hydraulic head (Bear, 1988; Freeze and Cherry, 1979). From the discussion of effective stress and compressibility, it is clear that two mechanisms control the release of water under declining head conditions: 1) compaction of the aquifer matrix due to increasing effective stress generated by reduction in pore-pressure and 2) expansion of water caused by the decreased pore pressure. The magnitude of the volume released from storage depends on the compressibility of the porous medium, α, and compressibility of water, β following the derivation by Freeze and Cherry (Freeze and Cherry, 1979):

For volume of water released from compaction of the aquifer, we know that total volume decrease in the aquifer, $-dV_t$, will equal the volume of water released, dV_w. For a unit volume and unit head decline, dh= -1, Equation (2.26) can be combined with Equation (2.24) to yield the volume of water released by compaction of the aquifer:

$$dV_w = \alpha V_t\, d\sigma_e = \alpha \rho g \qquad (2.29)$$

The volume released by the expansion of the water is similarly defined from the compressibility of water from Section 1.9:

$$dV_w = -\beta V_w\, dp \qquad (2.30)$$

For a saturated aquifer, the volume of water, V_w, is defined by the product of the porosity, θ, and Vt. Recognizing that the **change in pressure, dp**, is given by Equation (2.31),

$$dp = \rho g \, dh \qquad (2.31)$$

For dh= -1, Equation (2.30) can be rewritten to show that the volume of water released from expansion is:

$$dV_w = \beta \theta \rho g \qquad (2.32)$$

Summing Equations (2.29) and (2.32) yields the specific storage, S_s, from both mechanisms:

$$S_s = \rho g(\alpha + \theta \beta) \quad (cm^{-1}) \qquad (2.33)$$

It is important to note that the compressibility of water, β, is a constant-valued parameter, 4.2 $\times 10^{-11}$ cm^2/dyne (4.2 $\times 10^{-10}$ Pa^{-1}). Compressibility values for sand and gravel aquifer matrices can be several orders of magnitude higher than water, ranging from 10^{-6} to 10^{-9} cm^2/dyne.

STORATIVITY

The **storativity** of a confined aquifer is the volume of water released from storage per unit area of aquifer per unit decline in the hydraulic head. For a confined aquifer of thickness, b, the **storativity** or **storage coefficient**, **S**, is defined as given in Equation (2.34).

$$S = S_s b = \rho g b(\alpha + \theta \beta) \quad (cm/cm) \qquad (2.34)$$

In confined aquifers, which are pressurized, the predominant mechanism for storage is deformation of the porous matrix. Typical storativity values of sand and gravel aquifers range from 1 $\times 10^{-3}$ to 1 $\times 10^{-5}$ (dimensionless) for a range of saturated thickness of 5 to 100 meters. The implication of these low numbers is that substantial water level drop is required to sustain high rates of pumping in confined aquifers. In fact, one of the distinguishing characteristics of a confined aquifer is that low storativity leads to rapid propagation of drawdown during pumping tests.

Practitioners have also realized that significant long-term aquifer storage is also derived compaction of compressible clays and organic materials within or adjacent to the system. Compressibility values for clay, organic rich silts and peats may be as high as 10^{-5} to 10^{-7} cm^2/dyne, which would yield storativity values on the order of 0.01 to nearly 10 for strata on the order of 100 meter thick. While it is true that the rate of water yield is time dependent for such low-permeability materials, the potential contribution from these low permeability units to aquifer storage capacity is significant. Strata that demonstrate low permeability but significant storage capacity are termed **aquitards**. Strata that have very low storage and very low permeability are termed **aquicludes**.

SPECIFIC YIELD

Under shallow conditions, the storage capacity of unconfined aquifers is much higher than confined aquifers because the volume of water released from storage in an unconfined aquifer represents gravity drainage of the pore spaces. The storage term used for unconfined aquifers is **specific yield**, S_y. Specific yield values for shallow unconfined aquifers range from 0.01 to 0.25. In deep unconfined aquifer systems of 50 meters or more, the mechanisms associated with confined aquifer storage dominate under conditions of minimal water table change, and storage values are much more in line with lower storativity values.

As water is drained from the pores during a water table decline, a certain amount is retained in the interstices between the grains by molecular forces and surface tension, or trapped in dead-end pores. The volume of water that is retained in the porosity, θ, under gravity drainage is termed the **specific retention**, S_r, while the volume that drains freely under gravity is the specific yield, S_y (Bear, 1988):

$$n = S_r + S_y \tag{2.35}$$

This definition implies that specific yield is the portion of porosity that is mobile under natural gradient flow conditions. By definition, the water contained in the specific retention portion of porosity does not readily drain under gravity flow, but it is generally not mobile under natural gradient flow conditions, either. From a hydraulics perspective, specific yield provides a good initial approximation to the term **effective porosity**, θ_e. Effective porosity is widely used by practitioners to denote the portion of porosity that contributes to groundwater flow and transport — topics that will be discussed in Chapter 3 and Section II of this book. This derivation of effective porosity is certainly better than assuming 20 percent, as is recommended by the EPA in many guidance documents, or applying the nebulous correction factor, ε, which is between 0 and 1, to total porosity, θ_t, as recommended in many text books as in Equation (2.36) from Freeze and Cherry (Freeze and Cherry, 1979):

$$\theta_e = \theta_t \varepsilon \tag{2.36}$$

Hydraulics-based estimates are well suited to providing parameters for water supply and other large-scale flow problems. We distinguish the hydraulics-based term, effective porosity, from the transport term, **mobile porosity**, θ_m, to denote the portion of total porosity that contributes to advective flow and transport in aquifers. Using this convention, we also define **immobile porosity**, θ_i, as the portion of the pore-space that does not contribute to advective flow in aquifers, but rather acts as a reservoir of immobile or much more slowly moving groundwater when compared to the mobile pores. Thus, the total porosity is the sum of the mobile and immobile porosity as shown in Equation (2.37) below:

$$\theta_t = \theta_m + \theta_i \tag{2.37}$$

Plate 6 shows a cross-section through a three-dimensional model of porous media that was constructed using a well-sorted gravel (grains 0.5 to 1.5 cm) infused with a silicone rubber matrix. After the silicone vulcanized, the grains were removed from the mold to enable close examination of the pore structure network. Note that the pores are shown in the photo as solids while the voids represent the volumes filled by the grains. The pores shown on the photo are annotated with arrows to illustrate potential flow paths along the continuous pores (mobile porosity). The pores that are constricted or blocked by adjacent grains transmit less flow because of higher frictional losses as discussed in Chapter 1. The highlighted path is one of two or three routes groundwater may travel freely around each grain (mobile pores) and there are numerous dead-end or highly constricted pores connecting to the main flow path, where water is essentially stored (immobile pores).

2.7 AQUIFER HETEROGENEITY

Aquifer systems are composed of coarse-grained, high-energy deposits such as sand and gravel, which readily transmit water, and fine-grained, low-energy deposits such as silt and clay, which do not. The three-dimensional shape and size of each depositional unit, as well as the grain-size distribution comprising the deposit, are characteristics that are imparted by the depositional processes that formed it. The assemblage of depositional units creates the *unique*, three-dimensional aquifer architecture with variability and complexity that stem from processes that operate on the continuum of temporal and spatial scales. The complexity of aquifer architecture is manifested

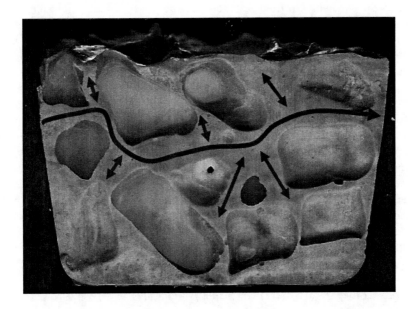

PLATE 6 Cross-section through a physical pore model, showing flow channels and connecting dead-end pore spaces that provide solute mass storage capacity at the pore-scale. (*See Plate section for color version of Plate 6.*)

from the **mega-scale** – the geologic provenance and basin; to the **macro-scale** — variability of the formation comprising the aquifer; to the **fine-scale** — variability within a depositional unit; and the **micro-scale** — variability of grain sizes and distributions within depositional forms and features.

Erosion and sedimentation processes span the continuum of time and spatial dimensions. Regional uplift creates the parent rocks and provides the gravitational potential necessary to move the weathered sediments from uplands to the lowlands, while subsidence at the locus of deposition makes room for more sediment. The geologic provenance dictates the type of parent rock material and sediment transport processes that move the materials through the local scale of the basin or drainage system. Energy levels decrease systematically away from the uplands as the slope of land flattens, limiting the size of the particles that can be moved and yielding a progressively finer matrix of mineral grains as the sediments are deposited.

In the time scale, energy levels wax and wane, both gradually — through global sea-level changes, and episodically — through catastrophic events such as debris flows at the continental margin or cyclic, daily variations in meltwater flow from glaciers. Dynamic fluctuations in the energy levels associated with depositional processes create the lithologic variability and structure that leads to sedimentary stratigraphy and sequences, which contribute to regional basin architecture at the mega-scale. **Sedimentary sequences** are the megascopic building blocks of aquifer architecture at the regional scale; they provide the record of deposition of many hundreds-of-meters to kilometers in thickness spanning tens-of-thousands to millions-of-years.

Spatial and temporal variability in the depositional processes across the basin give rise to **sedimentary lithofacies** — the characteristic grain-size distribution that is indicative of consistency in the depositional process through space and time at particular segments of the basin or depositional environment. The lateral dimension and vertical thickness of lithofacies depend on the nature depositional environment: In coastal marine settings where energy is driven by global sea-levels, high-energy sand petroleum reservoirs can be traced several kilometers and might be hundreds of meters thick; Lithofacies trends, or simply facies trends record the variability across the aquifer formation, while **depositional forms,** the organization of sediment grains during deposition, reflect the dynamic nature of the process and impart structure arising from the localized variations in energy and sediment type. Depositional forms, such as point bar gravels and cross-laminated sand beds, give structure to the sedimentary sequences and yield three-dimensional aquifer architecture. Stratigraphically, aquifer architecture is expressed at the macroscopic scale, which is typically 10's to 100's of meters in thickness. Still finer-scale processes, such as the pool and riffle associated with a single meander in fluvial systems, dictate the characteristics of bed-forms — sub-meter to centimeter-scale stratification (e.g., lenses) and micro-scale variability in bedding planes, on the scale of individual grains.

An aquifer is termed **homogeneous** when the degree of variability is low and material properties appear uniform in the formation. When the variability is high and the characteristic of the formation depends on location, the aquifer is termed **heterogeneous**. Isotropy and anisotropy are analogous terms that describe systems with structures or material properties that depend on direction (e.g. Cartesian axes) in the system, rather than simply location. Systems with structures or material properties that are independent of direction are termed **isotropic**. If the material properties depend on the direction within a system, the system is termed **anisotropic**.

The greater the degree of contrast in material properties in heterogeneous or anisotropic aquifers, the stronger the bias imparted by the structure of the system on groundwater flow and transport behavior. The majority of aquifer systems are neither homogeneous nor isotropic because the depositional processes that form them are dynamic and imprint structures at several scales. Consider a simple three-layered aquifer system consisting of gravel, sand and silt from bottom to top. One might characterize each layer as homogeneous if the grain-size distributions were uniform, but the system would be anisotropic in the vertical direction because grain size changes from gravel to sand to silt. Close examination of the grain-size distribution within the three "layers" comprising the formation might indicate that the deposit is actually a fluvial deposit that shows a fining-upward sequence with characteristic trends in lithofacies and bedforms in the vertical direction from bottom to top. Fining-upward sequences typically exhibit a transition from high-energy, coarse-grained gravel bars and planar sand beds at its bottom to moderate-energy foreset sand beds in the middle to interbedded sand and silt and ultimately silt at its top. The high-energy deposits are associated with the erosion and deposition at the active fluvial channel and the river bend, while the progressively lower-energy deposits toward its top are related to deposition that occurs away from the channel, both in the upstream-downstream direction and in the direction of the channel migration. As the energy of the system varies in space and time, the high-energy deposits are well-developed in certain areas and less well-developed in others, leading to gradational changes in lithofacies and thickness of each facies unit. There are also discontinuities in these systems, as the channel is developed by eroding prior formations away, or as the river abruptly jump its banks during storm events and abandons its former channel. The implication is that an aquifer created by a fluvial system would show dramatic variability in grain-size distribution that depends both on location and direction within the formation. At the stratigraphic scale, the system might appear homogeneous and isotropic; however, at the fine-scale, the facies trends and abrupt facies changes would impart a complex aquifer architecture that is both heterogeneous and anisotropic.

The scale at which these classifications are applied to aquifers is important, particularly in the evaluation of transport behavior. Aquifer heterogeneity at the fine- and micro-scales is often below the detection limit of conventional characterization techniques, or too difficult to map and interpret when the degree of variability is high. As a result, the tendency is to oversimplify our interpretations and neglect the heterogeneities below the macro-scale in aquifers. For regional water supply problems, the influence of pumping averages out the heterogeneities below the macro-scale, and the simplified approach can lead to appropriate interpretations. However, at the scale of site remediation, the remediation hydrogeology scale, heterogeneities at the fine- and micro-scale have a significant and controlling influence on reagent distribution and contaminant transport.

Practitioners have long been taught that the simplest interpretation is best — that is, we should strive to invoke the principle of parsimony. As a result, all too often we treat aquifers as homogenous and isotropic systems and forget that we have simplified the system to make our interpretations and calculations more tractable. In the practice of remediation hydrogeology, we have to adapt to the fact that aquifers are complex, due to the dynamic nature of the processes that created them. Most aquifers are strongly heterogeneous and typically exhibit anisotropic behavior during mass transport. Recognizing and adapting to aquifer heterogeneities and anisotropies is a continuing theme through the remainder of *Remediation Hydraulics*.

3 Groundwater Flow Concepts

The interactions between porous media and fluids flowing through them are very intense, because the wetted perimeter to fluid cross-section ratios are very large in porous media, relative to other examples of viscous flow in conduits. The principles of viscous fluid flowing in contact with confining surfaces (such as tubular conduits) were described in the early through mid-1800s, in a progression of theoretical developments by Navier, Hagen and Pouseuille and Stokes[1]. Collectively, they laid down a mathematical framework to calculate the resistance to viscous fluid flow through porous media. However, natural aquifer matrices are too irregular to allow explicit mapping of the contact between flowing groundwater and the porous medium. Without a matrix map, the mathematics of viscous flow cannot be applied directly to groundwater flow. Even for mappable matrices, such as packed beds of glass beads[2], the mathematics of a Navier-Stokes analysis is quite daunting, beyond the capacities of desktop computers.

Empirical formulations for fluid flow in porous media were developed by Darcy and others, beginning the mid-1800s and continued to progress in parallel with the viscous fluid theories. We now express the empirical relationship between flow and the porous medium as Darcy's Law. The empirical approach, which we will refer to broadly as "Darcian," provides excellent representation of groundwater flow at scales large enough to 1) allow us to define a hydraulic conductivity value for the volume of interest, and 2) smooth local variations of the porous matrix structure. **Figure 3.1** depicts the progression of development for the Navier-Stokes and Darcian theories which, together, stand as the basis for a comprehensive understanding of fluid flows in regular conduits and porous media.

Darcy's Law can be applied at scales ranging from a small-cross-section tubular conduit to an aquifer at cubic kilometer scales. It has typically been applied in the development of groundwater resources for reliable water supply systems and, more recently, in support of groundwater pumping systems designed to halt the migration of dissolved-phase contaminants. The main prerequisite for Darcian analysis is that we provide an empirical characterization of the fluid permeability for the medium. This is a *de facto* averaging approach, because the field measurement of permeability inevitably blends the contributions of dissimilar porous materials that span volumes smaller than the scale of measurement. For example, an aquifer slug test conducted in a well with a 1.5-meter long screened interval provides a hydraulic response that we interpret as representative of the entire screened interval, even though that response may be transmitted through only a small portion of the screened interval. This analysis correctly represents the average condition for the screened interval, but systematically understates the maximum groundwater conductance for the aquifer. For large-scale analyses, like those supporting water supply development, the average condition is what's needed. When we're trying to understand contaminant mass transport and the distribution of reagents from an injection well, the average condition can be (and usually is) seriously misleading.

---

[1] We refer to the theory of viscous fluid flow in conduits as the Navier-Stokes theory.
[2] Graton and Fraser (1935) provided a detailed mapping of the interstitial spaces formed by spheres packed in various configurations.

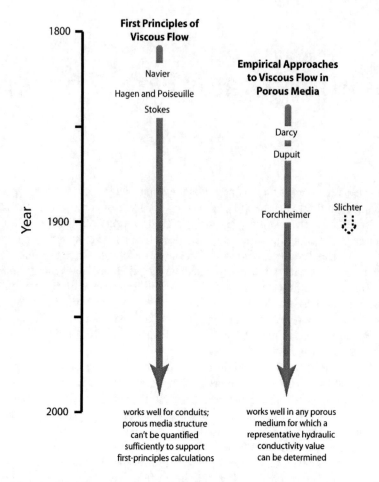

FIGURE 3.1 Historical development of fluid flow from a first-principles (Navier-Stokes) and empirical (Darcy) perspective.

In the following examination of groundwater flow through aquifers, we briefly trace the historical development of the Navier-Stokes analysis, then move to groundwater flow theory, arriving at the standard empirical description given by Darcy and others. We examine, in some detail, the loss of critical information that results from matrix averaging analyses and suggest a solution, through bifurcation of the aquifer matrix into two broad permeability classes – the mobile and immobile pore fractions. Even the most homogeneous systems show several orders of magnitude variability in permeability over very short distances, as the facies comprising the aquifer show both gradational and abrupt changes in grain-size distributions, at all scales. The resulting three-dimensional permeability structure of typical aquifers controls the distribution of flow to a much higher degree than hydraulic gradients do, because the rate of change of permeability with distance is typically 10 to 1,000-fold greater than the rate of change for groundwater elevation. Moreover, because permeability gradients are often not aligned with the groundwater elevation gradient, predictions of flow direction from elevation gradients are often incorrect. Recognizing that flow is concentrated in the mobile porosity allows us to focus our efforts on characterizing these segments of the aquifer to better interpret the hydraulic data that we collect. The bifurcation of porosities will be utilized frequently in the remainder of the book.

3.1 VISCOUS FLOW

The general equations of motion for a real fluid can be developed from analysis of the forces acting upon the fluid and application of the conservation principles for mass, momentum and energy. The resulting series of equations is called the Navier-Stokes equations. They were first proposed by L.M.H. Navier in 1827 (Navier, 1827), not as the general equations of fluid movement, but as a logical extension of the Euler equations to the flow of viscous fluids (the Euler equations relate acceleration to displacement and velocity). They are referenced as the Navier-Stokes equations because G.G. Stokes (1845) first successfully applied the equations to a problem, the movement of spheres through a fluid.

The Navier-Stokes equations evolved to provide a foundation for analysis of the movement of any fluid, liquid or gas. Flows of viscous fluids through porous media are described by a subset of the Navier-Stokes equations, in which changes in the momentum of the fluid are small and the fluid's response to external forces (elevation potential or pressure variations) is controlled by its resistance to shear stress (viscosity).

The viscous flow of Newtonian fluids (incompressible and constant viscosity) was first characterized in 1839, through experimentation conducted independently by Hagen and Poiseuille. Each of these researchers demonstrated that the flow of fluids through small diameter tubes was linearly related to the pressure difference across the length of the tube. This was an important observation that ultimately led to the development of equations describing groundwater flow.

The variables associated with viscous flow in conduits are illustrated in **Figure 3.2**. Fluid of viscosity, μ (cP), is flowing through a tube of length, **L** (cm), and inside radius, r_0 (cm), under a pressure difference, $p_1 - p_2$ (dynes per cm^2). If the fluid wets the tube surfaces, the fluid velocity will be zero at the wall, increasing to a maximum at the center (recall from Chapter 1 that the wetting process is essentially an exothermic binding of liquid to solid). The flow can be visualized as a series of concentric shells moving at different velocities exerting viscous forces on each other, expressed as follows in Equation (3.1):

$$F = \mu \cdot A \frac{dv}{dr} = \mu \left(2 \cdot \pi \cdot r \cdot L \right) \frac{dv}{dr} \tag{3.1}$$

where **A** is the area of the wall of the tube, *dv/dr* is the velocity gradient across the radius and **F** is the force in dynes. The driving force moving the water in this tube is a pressure differential *(p₁ − p₂)* acting on the cross-sectional area perpendicular to flow *[(p₁ − p₂) πr² dynes]*. If the fluid is not accelerating, the sum of the driving and viscous forces must be zero, as shown in Equation (3.2):

$$\mu \left(2 \cdot \pi \cdot r \cdot L \right) \frac{dv}{dr} + \left(p_1 - p_2 \right) \pi \cdot r^2 = 0 \tag{3.2}$$

Multiplying by *dr* and dividing by $\mu(2 \cdot \pi \cdot r \cdot L)$ yields Equation (3.3):

$$dv = -\frac{\left(p_1 - p_2 \right)}{2 \cdot \mu \cdot L} r \cdot dr \tag{3.3}$$

Integration between r_0 and **r** yields the velocity at a radius r and is expressed as shown in Equation (3.4):

$$v = -\frac{\left(p_1 - p_2 \right)}{4 \cdot \mu \cdot L} r^2 + \beta_0 \tag{3.4}$$

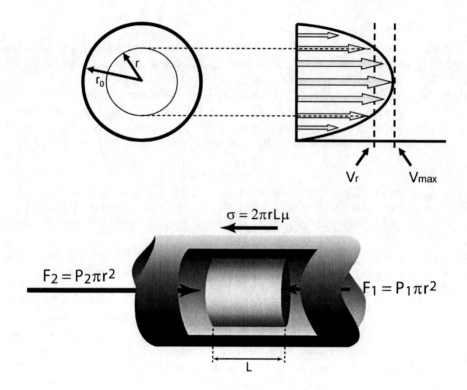

FIGURE 3.2 Force balance for a cylindrical mass of fluid under the influence of a pressure gradient (P_2-P_1/L) and side wall drag, σ.

where β_0 is an integration constant. The constant in Equation (3.4) may be evaluated by applying the boundary conditions, $v = 0$ at $r = r_0$, leads to:

$$\beta_0 = \frac{(p_1 - p_2)}{4 \cdot \mu \cdot L} r_0^2$$

which, through substitution, leads to a general relationship for velocity in Equation (3.5):

$$v = \frac{\left(r_0^2 - r^2\right)\left(p_1 - p_2\right)}{4 \cdot \mu \cdot L} \tag{3.5}$$

Equation (3.5) describes velocity as varying parabolically from a maximum at the center, to zero at the wall as shown in **Figure 3.2**. The flow through the tube equals the integral of Equation (3.5) across the cross-sectional area of the tube.

$$q = \int_0^{r_0} v \cdot dA = \int_0^{r_0} \frac{(p_1 - p_2)\left(r_0^2 - r^2\right)}{4 \cdot \mu \cdot L} d\left(\pi r^2\right)$$

$$q = \frac{\pi \cdot r_0^4 \left(p_1 - p_2 \right)}{8 \cdot \mu \cdot L} = \frac{\pi \cdot r_0^4}{8 \cdot \mu} \frac{dp}{dx} = C \frac{dp}{dx} \tag{3.6}$$

Equation (3.6) confirms the experimental results of Hagen-Poiseuille that flow of a Newtonian fluid through a small diameter tube is proportional to the pressure gradient.

The boundary conditions applied to Equation (3.4) are important to note. They arise from the observation that at the contact between a moving fluid and a solid, or the contact between two dissimilar fluids, the three components of stress across the surface must be continuous. In this case the shear stress exerted by the pipe on the fluid and the fluid on the pipe must be equal. Therefore if the pipe is not in motion, the fluid in contact with the pipe is also not in motion. This conclusion was first postulated by Stokes (1845), which, when combined with the exactness of the data collected by Poiseuille to the above equations, was considered to be sufficient proof that there is no slippage between these points of contact[3].

Equation (3.6) can be extended to soils by conceptualizing soil as a bundle of Poiseuille tubes, parallel to the direction of flow, and the number of pores in a unit cross-sectional area perpendicular to flow is **m**. The total flow may be expressed as the sum of the flow through each tube in Equation (3.7):

$$Q = m \cdot q \tag{3.7}$$

Consider a real soil where each pore has a different size. If m_i represents the cumulative area of pores with a radius, r_i, then q_i is the flow through each pore of radius, r_i, and pressure loss is expressed as a differential. Equation (3.7) can be expressed as follows in Equation (3.8) :

$$Q = \sum_{i=1}^{n} m_i \cdot q_i = \frac{1}{8 \cdot \mu} \frac{dp}{dx} \sum_{i=1}^{n} \pi \cdot m_i \cdot r_i^4 \tag{3.8}$$

Equation (3.8) assumes all of the pores are parallel to the direction of flow. This is a significant deviation from flow through soils, with the true path length 1.25 to 1.8 times longer than the direction of flow (Carman 1937; Bear 1972). This increase in the length of the flow path introduces the concept of tortuosity. A tortuosity factor was first proposed by Carmen (1937), to account for the longer pathways of interstitial water moving around soil grains. The net effect of tortuosity is to reduce the potential gradient, but mostly it allows for the introduction of a correction factor to account for deviations between Hagen-Poiseuille equation and flow through porous media. For this discussion we will introduce the correction factor as $1/\tau$ with tortuosity, τ, varying between 1.25 and 1.8. Incorporating tortuosity, Equation (3.8) is rewritten as:

$$Q = -\frac{1}{8 \cdot \mu} \cdot \frac{1}{\tau} \cdot \frac{dp}{dx} \sum_{i=1}^{n} \pi \cdot m_i \cdot r_i^4 \tag{3.9}$$

Since the quantity $\pi \cdot m_i r_i^2$ is the incremental cross-sectional area for flow in tubes, we can equate it to the incremental contribution of the mobile porosity in porous media, $\Delta\theta_i$. The flow in an equivalent porous media can be expressed as shown in Equation (3.10):

---

3 The validity of the no-slip assumption was also critical to the calculation of mass transfer between fluid phases, i.e., between a NAPL and water (Lamb, 1932).

$$Q = -\frac{1}{8 \cdot \mu \cdot \tau} \sum_{i=1}^{n} (\Delta \theta)_i \, r_i^2 \, \frac{dp}{dx} \tag{3.10}$$

Dividing Equation (3.10) by the cross-sectional area, $\Delta\theta_i$, results in the average velocity as shown in Equation (3.11):

$$v = -\frac{1}{8 \cdot \mu \cdot \tau} r_i^2 \frac{dp}{dx} = C \frac{dp}{dx} \quad where \quad C = \frac{1}{8 \cdot \mu \cdot \tau} r_i^2 \tag{3.11}$$

Equations (3.10) and (3.11) show that flow through a porous medium is related to the size distribution of water filled pores as well as the viscosity of the fluid and provides a simple explanation as to why soils with small pores transmit less water than soils with large pores.

Equation (3.6) is the Hagen-Poiseuille equation for viscous flow of liquids through capillary tubes. This equation was developed for horizontal flow — the discharge is driven only by pressure differentials. A more general form of the pressure differential is expressed in term of fluid potential or total head, where flow is driven by the change in mechanical energy per unit mass of the fluid. As flow through soils can be interpreted as an extension of the Navier-Stokes equations, it is natural to define fluid potential, Φ, consistent with three components of Bernoulli's equation: elevation potential, pressure potential and velocity potential. Potential is commonly expressed in terms of energy per unit mass for assessing fluid movement with a free surface, as shown in Equation (3.12):

$$z + \frac{p}{\rho \cdot g} + \frac{v^2}{2 \cdot g} = z + h + \frac{v^2}{2 \cdot g} = \Phi \tag{3.12}$$

For groundwater flow it is more useful to consider Bernoulli's equation as energy per unit volume, as shown in Equations (3.13) and (3.14):

$$\rho \cdot g \cdot z + \frac{\rho \cdot v^2}{2} + p = \Phi \tag{3.13}$$

$$\rho \cdot g \left(z + h + \frac{v^2}{2 \cdot g} \right) = \Phi \tag{3.14}$$

The velocity head (kinetic energy) in groundwater is very small and can be neglected in Equation (3.13). To demonstrate this, consider a fast groundwater flow system with an average velocity of 10 m/d (1.16×10^{-4} m/s). The velocity head for this system is:

$$\frac{v^2}{2 \cdot g} = \frac{\left(1.16 \times 10^{-4} \, \frac{m}{s} \right)^2}{2 \cdot \left(9.81 \, \frac{m}{s^2} \right)} = 6.83 \times 10^{-10} \, m \tag{3.15}$$

which is a *de minimis* quantity, despite the extremely high groundwater velocity and we can safely eliminate it from consideration for groundwater flows. Eliminating the velocity head and substituting $H = z + h$, Equation (3.13) can be substituted into Equation (3.11) yielding Equation (3.16) below:

$$v = -\frac{\rho \cdot g}{8 \cdot \mu \cdot \tau} r_i^2 \frac{dH}{dx} = K \frac{dH}{dx} \tag{3.16}$$

$$K = \frac{\rho \cdot g}{8 \cdot \mu \cdot \tau} r_i^2 \tag{3.17}$$

The constant, K, in Equation (3.17), is analogous to the Darcian hydraulic conductivity of a porous medium, relating the concepts of pore size, total porosity, fluid properties and path length to ability of a soil to convey a fluid. Equation (3.16) represents a theoretical basis for Darcy's Law, providing a link with the Navier-Stokes equation and can be conceptualized as a vector, with **tortuosity** or **mobile porosity** varying with direction.

The concept of total potential was explored in depth by Muskat (1937) and Hubbert (1940). Muskat (1937) presented the concept in terms of driving force expressed as a velocity potential — "a function whose negative gradient gives the velocity vector." Muskat expressed the velocity potential as Equation (3.18):

$$\Phi = \frac{k}{\mu}(p - G) \tag{3.18}$$

where **k** is a soil property (intrinsic permeability with units of cm^2), and G is the gravitational force, so that $G = \rho \cdot g \cdot z$. The units for Equation (3.18) are length per time, hence the term velocity potential. Equation (3.18) is an awkward presentation of the concept of potential gradient because it combines both fluid and soil properties.

The modern concept of fluid potential was developed by Hubbert (1940). Hubbert's work was undertaken because widely used hydrology texts of the time often stated Darcy's Law as a proportionality between flow rate and pressure gradient alone, neglecting the effects of potential energy or elevation head (Hofmann and Hofmann, 1992). Hubbert conceptualized fluid potential as a measurable physical quantity at any location within a porous medium where groundwater flow always occurs from regions of high potential towards regions of lower potential. Hubbert proved that if the force due to gravity was constant, a water level elevation was the measurable physical quantity representing total potential: $\Phi = z + h$. While this seems intuitive (water flows downhill), it was an important observation, establishing the validity of using water levels to understand groundwater gradients. Analytically, Hubbert expressed the total potential consistent with Bernoulli's equation.

3.2 DARCY'S LAW

The experiments of Hagen and Poiseuille were repeated by Darcy (1856), who noted similar results with his low velocity pipe experiments. These relationships may have also provided Darcy the inspiration for understanding the results of his other experimental studies on the flow of water through sand beds. These classic experiments gave rise to *Darcy's Law;* that the flow **Q** (cm^3/s) of water through a sand bed is proportional to the area, **A** (cm^2), of the sand bed and to the difference in water levels, ΔH (cm), at either face of the sand bed, and inversely proportional to the thickness, L (cm), of the bed,

$$Q = K \cdot A \cdot \frac{\Delta H}{L} \qquad (3.19)$$

where K is the hydraulic conductivity (cm/sec), a constant of the porous media and the fluid. Equation (3.19) is often presented in terms of the specific discharge, q (cm/sec), as shown in Equation (3.20):

$$q = \frac{Q}{A} = K \cdot \frac{\Delta H}{L} \quad or \quad q = K \cdot \frac{dH}{dx} \qquad (3.20)$$

The specific discharge is often referred to as the Darcy velocity because of its units; however, Equation (3.20) is a unit discharge or flow per unit area (cm³/s cm²). Darcy's law is empirical because, although it is rooted in the viscous flow theories of Navier, Stokes, Hagen and Poiseuille, it relates the average behavior of an observable volume of porous medium, without attempting to additively characterize each element of the matrix with a Navier-Stokes analysis.

All porous media are heterogeneous at the smallest scales (grains versus voids) and natural aquifers are heterogeneous over all scales. As a result, the volume over which we make observations for Darcian analysis controls the result we obtain. Starting at the lowermost volume limit (measured as a point within the soil matrix), porosity is either 0.0 or 1.0 depending upon whether the measurement point falls on a grain or in the pore between grains. As the volume we examine expands from that initial point, spanning a greater number of grains and pores, average porosity across the volume follows a dampened oscillation, eventually settling at the average for the soil being characterized. The volume at which sampling of porosity (or other porous matrix characteristic) becomes uniform is the **representative elementary volume (REV)** (Hubbert, 1937, 1956; Bear, 1972).

Figure 3.3 shows the oscillation of measurements, decreasing from the wide swings at smaller scales to the dampened observation of the average value when the REV is reached. We denote the region of volumes greater than the REV as the "classical Darcian" domain, because the resulting observations represent aquifer average conditions, with the heterogeneities (and important information content) dampened out of the characterization. This is a desirable result for water supply development and other groundwater management practices; however the range of groundwater behavior at scales below the REV are critically important in remediation engineering. We identified the domain at scales smaller than the REV as the remediation hydraulics domain, to reflect the need for analysis at the smaller scales. Darcy's Law remains valid and necessary at the smaller scales, but we will need to set aside the average aquifer parameters in many cases.

Darcy's Law applies across the wide range of scales and groundwater velocities for which laminar, viscous flow occurs. There are two conditions that could arise for which laminar viscous flow does not occur and Darcy's Law is invalid:

Lower limit — In extremely fine-grained soils, the pore apertures are small enough that the electrostatic interaction between ions dissolved in the water cannot be ignored. **Figure 3.4** is a conceptual diagram of large and small-aperture spaces. The matrix particle surfaces are net negatively charged, attracting cations dissolved in the groundwater (or the positive hydrogens in the dipole structure of water, itself)[4]. On both sides of the figure, ionic

---

[4] An interesting consequence of these interactions is that electrokinetic potentials, sometimes known as streaming potentials, are generated when groundwater containing dissolved ions flows across mineral surfaces and soil grains. Electrokinetics is an emerging geophysical tool (Kim, et al., 2004).

charge in the water is balanced over all locales. On the right hand side of the figure, the charge imbalance caused by collection of cations on the mineral surfaces averages across the pore aperture and generates only a small impedance for anion (and groundwater) flow. On the left side, the aperture is so small that anions are effectively locked in place by the charge balance requirement and the force required to push water through the gap does not follow Darcy's Law. Von Engelhardt and Tunn (1955) described this electrokinetic effect

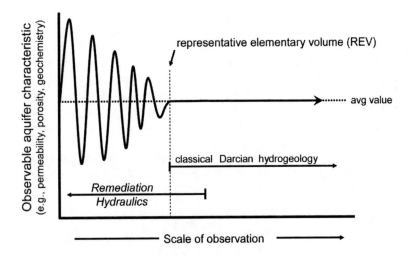

FIGURE 3.3 Graphical depiction of the domains of classical (averaging) Darcian hydrogeology and remediation hydraulics, where the information lost from averaging becomes critical to understanding details of groundwater flow and contaminant mass transport. Note that Darcy's Law remains valid and useful at scales below the REV.

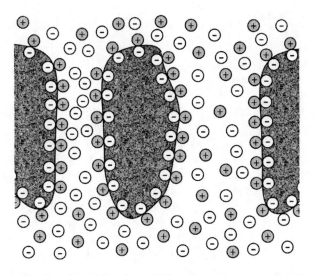

FIGURE 3.4 Electrokinetic effect on viscous flow in small-aperture spaces. Soil particles with high negative surface charge densities attract positive ions from the groundwater solution. Small-aperture spaces, as in the left channel, can cause charge "gridlock", preventing groundwater flow.

as increasing the apparent viscosity of water. Consequently, Darcy's Law is valid for soils with pore apertures large enough to overcome the ion adsorption. All cohesionless soils probably meet this criterion.

Upper limit — The upper limit of Darcy's Law is marked by the transition from laminar to transitional flow. At low velocities the viscous forces are much larger than the inertial forces and flows are laminar. At high velocities, inertial forces become a significant factor causing transition and turbulent flows.

LAMINAR AND TURBULENT FLOWS

In laminar flows the macroscopic motion of any water molecule is strictly along the bulk fluid flow axis. In other words, there is no transverse motion (other than molecular-level Brownian motion) in the moving fluid. These are well-ordered flows and, in theory, if we know the position of a water molecule at one point on its path, we can predict its location at a later time. In turbulent flows there is chaotic mixing of the water mass and the position of a water molecule is completely unpredictable through time. Transition flow is a discernable zone of flow with mixed laminar and turbulent characteristics. All flows we encounter in groundwater are laminar and the following analysis provides the basis for that conclusion.

The transition from laminar flow to turbulence is characterized using the Reynolds Number, a dimensionless number characterizing the ratio of the inertial forces to the viscous forces. The Reynolds Number, $\mathbf{R_e}$, is defined as follows:

$$R_e \propto \frac{inertial\ forces}{viscous\ forces} \tag{3.21}$$

$$R_e = v \frac{d \cdot \rho}{\mu} = \frac{q}{\theta_m} \cdot \frac{d \cdot \rho}{\mu} = \frac{1}{\theta_m} \cdot \frac{k \cdot \rho \cdot g}{\mu} \cdot \frac{dh}{dx} \frac{d \cdot \rho}{\mu} \tag{3.22}$$

$$R_e = i \cdot \frac{k \cdot g}{\theta_m} \cdot \frac{d \cdot \rho^2}{\mu^2} \quad or \quad R_e = i \cdot \frac{K}{\theta_m} \frac{d \cdot \rho}{\mu} \tag{3.23}$$

where $\mathbf{v}$ is the velocity through the bed, $\mathbf{d}$ is the mean grain size, $\mathbf{q}$ is the Darcy flux, θ_m is the mobile porosity, k is the intrinsic permeability, μ is the dynamic viscosity, i is the hydraulic gradient, $\mathbf{g}$ is the gravitational constant and $\mathbf{K}$ is the hydraulic conductivity.

The Reynolds Number associated with transition to turbulence is a function of the flow conduit size and shape. For porous media, the onset of turbulence occurs at Reynolds Numbers between about 1 and 10 (Bear 1972), marking the upper limit of validity for Darcy's Law. Muskat (1937) reported a series of experiments from which we can deduce the transition to turbulence. **Figure 3.5** shows several plots of friction factor versus Reynolds Number. The friction factor effectively represents fluid conductivity and the Reynolds Number accounts for fluid velocities. Darcy's Law has predictive capacity over the linear range of each of the curves in **Figure 3.5** and the transition to turbulence is noted at the highest values of Re, where the onset of non-linearity occurs. For the porous media reported by Muskat (1937), the transition to turbulence began in the range of Re = 1 to 10. The Re value at which laminar flow begins to break down in porous media is quite low, compared to values in excess of 10^3 typical of macroscopic conduit flow. This tends to give the sense that an Re value of 1 to 10 could easily be exceeded. Let us explore Equation (3.21) to assess the appropriateness of the laminar flow assumption and under what conditions it may not be valid.

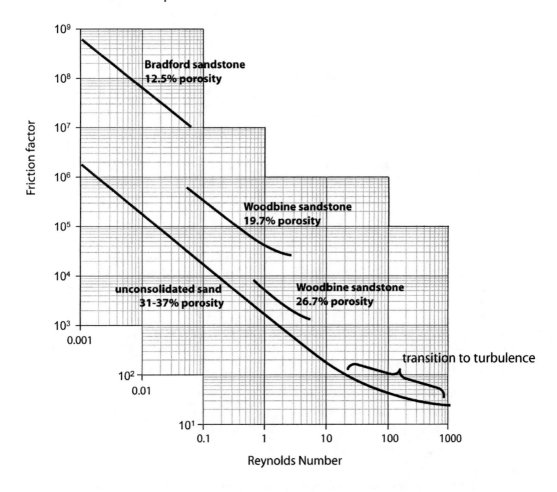

FIGURE 3.5 Friction factors developed by Muskat (1937) for fluids moving through sand, both consolidated and unconsolidated.

Figure 3.6 presents the Reynolds Number as a function of groundwater velocity for different porous media. This figure shows that ambient groundwater flow is laminar in unconsolidated sediments. The onset of turbulence will only occur in very coarse soil, consisting primarily of large pebbles or cobbles, at high groundwater velocities. To illustrate the concept, **Figure 3.7** presents groundwater velocities resulting from fluid being injected or extracted at 10 gpm (0.64 L/s) near a 50 mm (2-inch) well with 1.5 meters (5 feet) of screen. The Reynolds Number for this case is less than 10 even in the gravel pack of the well, where velocities are highest due to the limited flow area in annulus around the well. These data suggest that the pressures associated with delivering sufficient flow to generate turbulent conditions near a remedial injection well would likely result in structural failure of the aquifer, before the flow became turbulent.

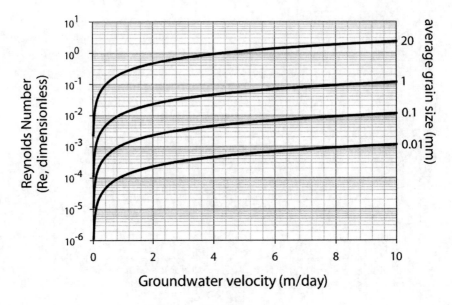

FIGURE 3.6 Reynolds Number as a function of groundwater velocity, calculated for aquifer pore-scale conduit dimensions, according to Equation (3.23). The grain sizes represent coarse gravel (20 mm), coarse sand (1 mm), medium sand (0.1 mm) and fine sand (0.01 mm).

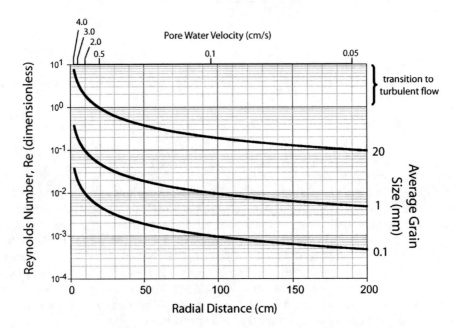

FIGURE 3.7 Reynolds Numbers calculated for 10 gpm (0.63 L/s) radial flow near a 2-inch diameter, 150-cm length well screen. Average grain size dimensions correspond to coarse gravel (20 mm), coarse sand (1 mm) and fine sand (0.1 mm).

3.3 HYDRAULIC CONDUCTIVITY AND PERMEABILITY

Hydraulic conductivity refers to the ability of a saturated soil to conduct water under an induced hydraulic or pressure gradient. In Darcy's Law, Equation (3.19), hydraulic conductivity is the constant of proportionality relating water flow with pressure drop. An expression for K was also presented in Equation (3.17) which considered flow through porous media as analogous to flow through bundles of tubes of various sizes. That derivation was useful, because it showed that hydraulic conductivity depends on the properties of both the soil and the fluid. From Equation (3.17), we develop the following:

$$K = \frac{\rho g}{8\mu\tau} \sum_{i=1}^{n} (\Delta\theta)_i \, r_i^2 \qquad (3.24)$$

$$K = \frac{\rho g k}{\mu} \qquad (3.25)$$

The units of Equation (3.25) are length per time, sometimes leading to confusion, but it is a conveyance term not a velocity term. Hydraulic conductivity represents the volume of water that can be conveyed through a unit cross-section of a porous media and has unit of $cm^3/cm^2 s$. The lumped portion of Equation (3.25), k, is the intrinsic permeability, or simply permeability. The intrinsic permeability is a porous medium property, whereas the viscosity and density are fluid properties.

The intrinsic permeability of a soil depends on the size and connectedness of the pores comprising the porous media. A conceptual view of an aquifer pore channel is given in **Figure 3.8** showing a hypothetical cross-section of an aquifer pore channel, with wetting of the grain surfaces building a zero-velocity layer on the matrix and viscous drag slowing groundwater flow in all areas near the side wall. This view provides a conceptual connection between the Navier-Stokes quantification of viscous flow in simple conduits and the complex patterns of viscous flow that develop in porous media. Factors that affect permeability include properties of the solid matrix particles (i.e., grain-size distribution, sorting, grain shape, packing, and stratification in sedimentary deposits) and the pores that form between particles (i.e., connectedness, cementation and compaction).

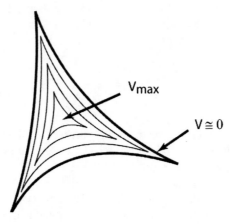

FIGURE 3.8 Conceptual velocity profile for a pore channel that forms at the intersection of three or more aquifer matrix particles, as shown on Figure 2.1. Viscous forces and side wall drag generate significant zones of near-zero flow velocities.

Practitioners have devoted considerable effort to establish a quantitative link between the grain size, grain shape, and porosity of the porous matrix and permeability. Two of the most widely used methods are the Hazen formula and the Kozeny-Carman formula. In the late 19th Century, Hazen (Hazen 1892; Hazen 1911) developed an empirical relationship for estimating permeability when designing sand beds. The formula, known as the Hazen Equation, is shown below in Equation (3.26):

$$k = d_{10}^2 \, C_H \qquad (cm^2) \tag{3.26}$$

where d_{10} is the 10-percent finer grain size (cm) and C_H is the Hazen coefficient. According to Carrier (Carrier 2003), Hazen coefficient values published in the literature range between 1 and 1,000, making it difficult to select the appropriate coefficient value. The practical limitations of the approach are that the method was developed for well-sorted granular media (C_u less than 2) and is suitable for a narrow range of d_{10} grain sizes between 0.10 mm and 0.70 mm (very-fine to coarse-grained sand).

The Kozeny-Carman formula extended the work of Hazen, incorporating the influence of specific surface area of the grains, S_o, Equation (3.27) and void ratio, e, or porosity, θ (Carrier 2003):

$$k = \frac{e^3}{C_{K-C} s_o^2 (1+e)} = \frac{\theta^3}{C_{K-C} s_o^2 (1-\theta)^2} \tag{3.27}$$

The Kozeny-Carman coefficient, C_{K-C}, is an empirical value that ranges between 4.5 and 5.1. The **specific surface area**, which is area of the grains divided by the volume or mass, can be measured using soil science methods involving nitrogen gas adsorption; however, the utility in the method is that grain-size analysis can be used to estimate the specific surface area, facilitating permeability estimates without additional indirect testing methods. The following relationships establish the link with grain-size distribution after Carrier (2003) in Equation (3.28):

$$s_o = \frac{area}{volume} = \frac{\pi d_{eff}^2}{\pi d_{eff}^3 / 6} = \frac{6}{d_{eff}} \tag{3.28}$$

where d_{eff} is the effective grain size. To account for the fact that grain shapes are not spherical, investigators have proposed an alternative form of Equation (3.28), which incorporates the shape factor, **SF**, as shown in Equation (3.29) below:

$$s_0 = \frac{SF}{d_{eff}} \tag{3.29}$$

Recommended shape factors range from 6 to 8.4, to account for increasing angularity and irregularity in shapes. Carrier summarized the ranges based on grain shapes: rounded – 6.1 to 6.6, medium angular – 7.4 to 7.5, and very angular – 7.7 to 8.4.

The effective grain size is estimated by summing the pair-wise harmonic means computed sequentially from the largest to the smallest sieves. The sum of the average grain size between adjacent sieves yields the effective grain size as shown in Equation (3.30):

$$d_{eff} = \left\{ 100\% \sum_{1}^{n-1} \left(\frac{f_i}{d_{li}^{0.404} \cdot d_{si}^{0.595}} \right) \right\}^2 \tag{3.30}$$

In Equation (3.30), d_{li} and d_{si} refer to the diameter of the larger and smaller sieves in each sequential pair, and f_i refers to the weight fraction retained on both sieves. The exponents shown were derived based on a log-linear relationship in grain size (Carrier, 2003). Combining terms from Equations (3.30), (3.29), and (3.27) yields the Kozeny-Carman formula expressed in terms of effective grain size:

$$k = \frac{e^3}{C_{K-C} \left(\dfrac{SF}{d_{eff}} \right)^2 (1+e)} = \frac{\theta^3}{C_{K-C} \left(\dfrac{SF}{d_{eff}} \right)^2 (1-\theta)^2} \tag{3.31}$$

The advantage of the Kozeny-Carman equation is that it incorporates the range of the grain sizes in the distribution, allowing it to be applied more effectively to finer- and coarser-grained sediments ranging from non-plastic, cohesionless silts to sand and gravel mixtures. The method is less reliable for very-poorly sorted soils, or soils with highly irregular shapes (Carrier 2003). The method is not suitable for estimating permeability of plastic soils (e.g., significant clay or organic content) or well-sorted cobble-sized gravel. The Kozeny-Carman approach is suitable for an initial estimate of permeability, or hydraulic conductivity, by incorporating the fluid properties as in Equation (3.24). Notwithstanding these limitations, the true advantage of the approach is that it enables hydrogeologists to efficiently estimate and correlate permeability with the facies trends using readily available grain-size data.

Table 3.1 presents calculated values of hydraulic conductivity for soils with effective grain-size diameters corresponding to the Udden-Wentworth grain classification as presented in **Table 2.1** using Equation (3.31). The values presented were calculated using a range of porosity between 25- and 45-percent, shape factors between 6 and 8.4 and values of density and dynamic viscosity corresponding to $10°$ C. The range of porosity and shape factors results in a computed range of conductivity for each effective diameter that typically spans a surprisingly narrow range of 3 to 4-fold between the minimum and maximum. This result indicates that using grain-size analysis and the Udden-Wentworth classification system would provide a relatively good correlation between permeability distribution and facies in the aquifer. This also underscores the primary deficiency associated with the USCS soil classification — the soil class divisions are too coarse to provide meaningful correlation of permeability, or to evaluate subtle facies changes that control it in aquifers.

The Udden-Wentworth classification is based on a geometric progression of diameters (e.g., fine sand is defined between 1/8 and 1/4 mm), and from **Table 3.1** we can see that natural variability in the mean diameter or sorting could result in up to an order of magnitude range in conductivity for each generic soil classification division.

Figure 3.9 presents the relationship between porosity, permeability, grain size and sorting for non-cohesive sediments. This figure shows that aquifer soils capable of transmitting meaningful quantities of water fall within a limited range of physical properties. Total porosity of water transmitting soils varies between 25 to 40 % while hydraulic conductivity varies across a range of only four orders of magnitude — much smaller than the range of soil properties we

TABLE 3.1

Representative hydraulic conductivities for Udden-Wentworth soil classifications based on the Kozeny-Carman formula as calculated using Equation (3.31) for water at 10°C. The range of values considers porosity values between 25- and 45-percent and shape factors between 6 and 8.4 and applies to well-sorted cohesionless soils. Note that the effective diameter used here corresponds to values that would be determined using sieve analysis and Equation (3.25).

Classification	Effective Diameter (mm)	Hydraulic Conductivity (cm/s)		
		Minimum	Median	Maximum
Coarse pebble	16	15	30	60
Pebble	4	1	2	4
granule	2	2×10^{-1}	4×10^{-1}	9×10^{-1}
Very-coarse sand	1	6×10^{-2}	1×10^{-1}	2×10^{-1}
Coarse sand	1/2	2×10^{-2}	3×10^{-2}	5×10^{-2}
Medium sand	1/4	4×10^{-3}	7×10^{-3}	1×10^{-2}
Fine sand	1/8	9×10^{-4}	2×10^{-3}	3×10^{-3}
Very fine sand	1/16	2×10^{-4}	4×10^{-4}	9×10^{-4}
Medium silt	1/32	6×10^{-5}	1×10^{-4}	2×10^{-4}

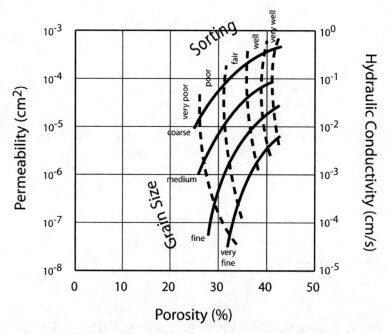

FIGURE 3.9 Conceptual relationship between porosity, permeability and particle sorting, for clay-free sand. Redrawn from Selly (1988).

encounter in natural aquifers. It is also clear that the relatively subtle effects of stratified bedding, or lateral channel succession trends ranging from fine sand to coarse sand could easily cause a 50-fold contrast in hydraulic conductivity. These examples point to the importance of accurately classifying soils using the Udden-Wentworth system and performing sieve analysis to confirm grain-size distributions in aquifers, because seemingly minor differences in grain size can have relatively large impacts on groundwater flow.

3.4 EQUIVALENT HYDRAULIC CONDUCTIVITY

The term **equivalent hydraulic** conductivity is used to describe the average hydraulic conductivity obtained for an aquifer, when the matrix is comprised of multiple permeabilities. The analysis of most hydraulic testing datasets yields equivalent hydraulic conductivities. Equivalent conductivities are particularly useful for water supply problems or containment system design, because the equivalent hydraulic conductivity gives a basis for estimating average behavior and total flow capacity. However, it is not very useful for remediation hydraulics because the average behavior can be dramatically different from individual behavior, especially when considering advective velocity. We show the approach here, not because we advocate its application in remediation hydraulics, but because it shows the contrast in flow that occurs in stratified systems and underscores the importance of applying Darcy's Law within each hydrostratigraphic unit.

Consider flow in the x-y plane of the three-stratum system, consisting of a sand aquifer situated between silty aquitard units, shown on **Figure 3.10**. The volumetric discharge, Q, through a unit thickness is the sum of the flow in the individual layers. Rewriting Equation (3.20) for the a general stratified system yields Equation (3.32):

$$q = \sum_{1}^{n} \frac{K_i \cdot d_i}{d} \cdot \frac{dh}{dl} = K_x \cdot \frac{dh}{dl} \qquad \left(\frac{cm^3}{cm^2 s} \right) \tag{3.32}$$

where the subscript **i** refers to the layer, d_i is the individual layer thickness, **d** is the total thickness of all layers, **dh/dl** is the head drop across the system and K_x is the equivalent flow in the horizontal direction, or parallel to layering. Simplifying the equation and solving for K_x yields the generalized solution for equivalent hydraulic conductivity parallel to layering in Equation (3.33):

$$K_x = \sum_{1}^{n} \frac{K_i \cdot d_i}{d} \quad \left(\frac{cm}{s} \right) \tag{3.33}$$

Assuming unit thickness for each of the three layers and substituting values from **Figure 3.10** into Equation (3.33) yields:

$$K_x = \frac{1.0 \times 10^{-5} \cdot 1.0 + 1.0 \times 10^{-2} \cdot 1.0 + 1.0 \times 10^{-5} \cdot 1.0}{3.0} = 3.3 \times 10^{-3} \left(\frac{cm}{s} \right)$$

This result indicates that the equivalent hydraulic conductivity for the three-layered system is reduced to 33-percent of the value for the high-K layer in the middle of the system. The important implication is revealed when comparing the flux in the entire system to the flux in the high-K layer. Assuming an average hydraulic gradient of 10^{-3} cm/cm and substituting the equivalent-K into the equation for Darcy flux yields 3.3 x 10^{-6} cm³/cm²s for the equivalent system shown at the bottom of **Figure 3.10**. Applying the K value for the sand, we get 1.0 x 10^{-5} cm³/cm²s. This flow is

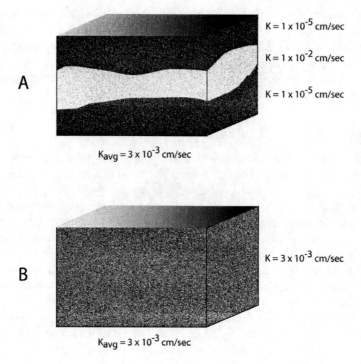

FIGURE 3.10 Two volume elements with equal average hydraulic conductivities. In the upper unit, the variance of conductivity is quite high, and solute transport rates will be dramatically underestimated by the average hydraulic parameters. In the lower unit, hydraulic conductivity is relatively uniformly distributed and solute transport rate estimates will not be adversely affected by use of the superficial representation of groundwater velocity.

concentrated in 1/3 of the total system thickness, which implies that nearly 100-percent of the flow occurs in the sand, and minimal flow is contributed from the low-K layers.

Now, consider flow perpendicular to the layers. Because conservation of flow holds, the specific discharge entering the system must equal the flow exiting the system. Therefore, the total head drop across the system must equal the sum of the head drop across each layer. Applying Darcy's Law from Equation (3.20) yields the general expression for equivalent conductivity perpendicular to x-y plane, K_z:

$$q \ = \ K_1 \cdot \frac{dh_1}{dl_1} \ = \ K_2 \cdot \frac{dh_2}{dl_2} \ = \ ... \ = \ K_n \cdot \frac{dh_n}{d\,l_n} = K_z \cdot \frac{dh_z}{dl} \tag{3.34}$$

Rearranging to solve for K_z:

$$K_z = \frac{qd}{dh} = \cdots \frac{qd}{\dfrac{qd_1}{K_1} + \dfrac{qd_2}{K_2} + \dfrac{qd_3}{K_3}} \ = \ \frac{d}{\displaystyle\sum_i^n \frac{d_i}{K_i}} \tag{3.35}$$

Substituting values from **Figure 3.10** yields:

$$K_z = \frac{3.0}{\dfrac{1}{1.0 \times 10^{-5}} + \dfrac{1}{1.0 \times 10^{-2}} + \dfrac{1}{1.0 \times 10^{-5}}} = 1.5 \times 10^{-5} \left(\frac{cm}{s} \right)$$

This result indicates that the vertical hydraulic conductivity across a stratified formation is approximately equal to the hydraulic conductivity of the least permeable unit within the aquifer — the low-K dominates in flow perpendicular to layers. A comparison of the horizontal and vertical permeabilities also illustrates the sharp contrast between permeability tensors — *the horizontal hydraulic conductivity is 667 times larger than the vertical.* The effect of the hydraulic conductivity contrast focuses groundwater flow within more permeable portions of the aquifer — the high-K dominates in flow parallel to stratified layers.

Natural aquifers that may be regarded as homogeneous and isotropic can have significant concentrations of flow, causing contaminant or reagent transport to proceed at velocities far in excess of the average value for the formation. **Figure 3.11** summarizes hydraulic conductivity measured at 5 cm intervals in two, 3-m soil cores collected from the Borden Aquifer (data estimated from Rivett, et al., 2001). This sector of the Borden Aquifer is characterized as a relatively homogeneous sand aquifer with a geometric mean hydraulic conductivity of 6.34×10^{-3} cm/sec. The discrete data illustrate a large amount of variability over short intervals, with hydraulic conductivity variations from 1×10^{-1} cm/s to 7×10^{-5} cm/s. **Figure 3.12** shows the frequency distributions for hydraulic conductivities in each of the cores.

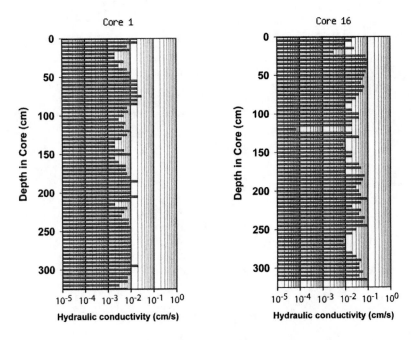

FIGURE 3.11 High-resolution vertical conductivity profiles developed from cores at the Borden aquifer (estimated from Rivett, et al., 2001). The geometric mean for hydraulic conductivities at their study site was 6.34×10^{-3} cm/sec.

The effective hydraulic conductivity for the Borden study area led to an estimated 8 cm/d average groundwater velocity. However, dissolved hydrophobic organic compounds placed in the formation traveled at more than 15 cm/d, indicating that the average velocity dramatically understates the maximum groundwater velocities and the associated chemical migration rates. Darcy's Law can be applied to the conductivities of the individual strata, to arrive at a much more accurate assessment of the range of velocities.

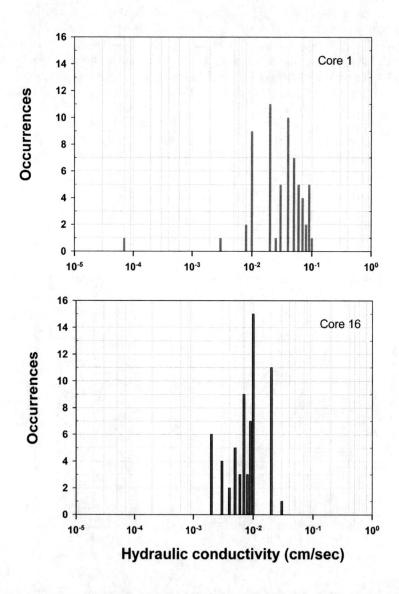

FIGURE 3.12 Hydraulic conductivity frequency distributions for subsamples collected at 5-cm intervals from 3-m Borden aquifer core sections. Approximated from Rivett, et al., 2001.

3.5 MOBILE AND IMMOBILE POROSITIES

From the detailed analysis of the "homogeneous" aquifer offered by the cores in **Figures 3.11** and **3.12**, we can see that there is a significant concentration of groundwater flow in a small fraction of the aquifer profile. Sites that we consider "heterogeneous" have an even greater concentration of flow in a small portion of the profiles. Because it is rarely possible to characterize sites at the level of detail shown in the Borden site cores, we need to find a simpler conceptual framework to discuss the concentration of groundwater flow and it implications for contaminant and reagent transport. We have created such a framework through the bifurcation of aquifer matrices into mobile and immobile porosities, θ_m and θ_i, respectively. The **mobile pore fraction**, or **mobile porosity,** corresponds to the segments of the aquifer with the highest permeability. This is distinct from the drainable fraction or specific yield values, that are typically much higher than the immobile porosity. The **immobile pore fraction**, or **immobile porosity**, corresponds to the segments of the aquifer with the lowest permeability. It is important to note that a large portion of the immobile porosity is likely to be drainable. Recognizing the fact that flow is concentrated in the mobile porosity allows us to focus our efforts on characterizing those segments of the aquifer, to better interpret the hydraulic data that we collect.

One of the obvious impacts of heterogeneous permeabilities is the concentration of flow in the more permeable segments of the formation. Standard aquifer testing protocols, which are discussed in detail in Chapter 11, are constructed specifically to obtain the average, or equivalent hydraulic conductivity across formations. This blends the high and low hydraulic conductivities and understates the actual flow velocities. When we bifurcate the porosities into mobile and immobile portions, we obtain a better estimate. We can use tracer studies to estimate the mobile porosity and, from that, the **mobile fraction velocity**, as shown in Equation (3.36).

$$V_{mobile} = \frac{V_{avg}}{\dfrac{\theta_m}{\theta_t}} \tag{3.36}$$

The actual groundwater velocity, represented by V_{mobile}, is significantly higher than V_{avg} for formations in which a small portion of the total porosity is in the mobile fraction.

Tracer testing demonstrates that, with the exception of very coarse grained aquifers, the mobile fraction is always less than the specific yield. **Table 3.2** presents mobile porosity estimates that we have obtained from tracer studies in sands and gravels, revealing a significant departure from the 20% value that is typically assumed. The mobile porosity of sand and gravel aquifers is often less than 10 percent. These observations should be considered when making estimates of aquifer behavior prior to field data collection. For the purpose of assessing plume migration rates, assuming mobile porosities between 0.02 and 0.10 would be more appropriate than using the common 0.20 value. The sand and gravel formations represented on this table yielded sufficient water that they are classified as aquifers. Injection rates and pressures were managed during injection to prevent soil fracturing.

Table 3.3 summarizes the subdivision of total porosity, θ_t, from the hydraulic and transport perspectives. The hydraulic perspective is what's necesssary to support development of reliable groundwater productivity estimates and the transport perspective is required to support more accurate estimates of solute transport velocities, time in transit between points along the flow path and the distance that a sustained contaminant source mass might reach over time.

TABLE 3.2
Summary of mobile porosity estimates determined by tracer tests.

Location	Aquifer	Aquifer Material	Mobile Porosity	Notes
Quebec, Canada	--	Poorly Sorted Sand & Gravel	8.5%	6.4 m^3 injection in 7.25 hours
Central Valley, California	--	Poorly Sorted Sand & Gravel	4% to 7%	575 m^3 injection over 30 days; arrival monitored in 7 wells
Northern Texas	Ogallala	Poorly Sorted Sand & Gravel	9%	1460 m^3 injection over 28 days
New Jersey	Passaic Formation	Fractured Sandstone	0.1% to 0.7%	24.6 m^3 injection over 2 days
Los Angeles, California	Gaspur Aquifer	Alluvial Formation	10.2%	17 m^3 injection over 8 hours
Northern New Jersey	--	Glacial Outwash	14.5%	7.57 m^3 in 3 days
Northern Missouri	--	Weathered Mudstone Regolith	7% to 10%	4.54 m^3 in 9 days
Sao Paulo, Brazil	--	Alluvial Formation	7%	18.9 m^3 injection over 2.5 days
Phoenix Arizona	--	Alluvial Formation	7%	2.27 m^3 in 8 hours
Savannah River Site, South Carolina[1]	Atlantic Coastal Plain	Silty Sand	5%	Model Calibration
Kaiserslautern, Germany	Trifels Formation	Fractured Sandstone	0.08% to 0.1%	Multiple injections and volumes between 0.1 m^3 and 5 m^3
West Texas	Rio Grande River Valley	Alluvium, Sand & Gravel	1.7%	18.9 m^3
Northern Texas	Ogallala	Alluvium, Poorly Sorted Sand & Gravel	0.3 to 1.7%	Dipole test, 61.3 m^3 in 22.8 hrs
Central Colorado	Cherry Creek	Alluvium, Sand & Gravel	11% to 18%	Two Injection tests, 4.9 m^3 and 7.6 m^3
Central Colorado	Denver Formation	Siltstone, Sandstone, Mudstone	1% to 5%	Monipole – Tracer Injected in monitoring well / Pumping well

[1]Hamm et al

TABLE 3.3
Summary of mobile porosity estimates determined by tracer tests.

Hydraulic Perspective		Transport Perspective
Total Porosity		Total Porosity
θ_t	$=$	θ_t
Drainable Fraction (Specific Yield)		Mobile Fraction
S_y	$\neq$	θ_m
Irreducible Fraction (Field Capacity)		Immobile Porosity
θ_{fc}	$\neq$	θ_i
$\theta_t = S_y + \theta_{fc}$		$\theta_t = \theta_m + \theta_i$

3.6 ANISOTROPY AND STRUCTURE

The depositional processes that lay down aquifer matrix materials create a non-random structure at many scales (patterned, rather than random heterogeneities and anisotropies). Because porous media are structured, intrinsic permeability must be viewed not as a scalar quantity, but rather as a tensor quantity as shown in Equation (3.37):

$$\hat{k} = \begin{bmatrix} k_{xx} & k_{xy} & k_{xz} \\ k_{yx} & k_{yy} & k_{yz} \\ k_{zx} & k_{zy} & k_{zz} \end{bmatrix} \tag{3.37}$$

The subscripts x, y and z denote the principal axes in Cartesian coordinates. This implies that hydraulic conductivity is also a tensor quantity. Hydraulic conductivity can be treated as a vector quantity when the principal directions of anisotropy are orthogonal and aligned with the principal axes, such that $k_{xy} = k_{yx} = k_{xz} = k_{zx} = k_{yz} = k_{zy} = 0$, leading to the vector representation as in Equation (3.38):

$$\vec{K} = \vec{K}_x + \vec{K}_y + \vec{K}_z \tag{3.38}$$

In simple terms, hydraulic conductivity could be treated as a vector when the bedding planes are aligned parallel with the horizontal x-axis, and the y- and z-axes are oriented orthogonal to bedding, such that the y-axis is in the plane of bedding and the z-axis is perpendicular to bedding.

If the three vector components of hydraulic conductivity are equal, then the system is considered isotropic, and hydraulic conductivity can be approximated as a scalar quantity in Equation (3.24).

However, nearly all real systems show some degree of directional dependence in hydraulic conductivity. Most sedimentary systems exhibit 3 to 10-fold higher hydraulic conductivity parallel to bedding than perpendicular to it, and stratified systems often show several orders-of-magnitude higher hydraulic conductivity in the bedding plane. When there are differences in vector components of hydraulic conductivity, or directional dependence, then the system is considered **anisotropic**. Normally, fractured bedrock systems are archetypically anisotropic, because the permeability is dominated by the orientation of the fractures. Anisotropy also arises in heterogeneous systems, such as the braided river or alluvial fan depositional environment, because the channel facies are organized more or less parallel to the direction of the channel, leading to a dominant orientation of the highest permeability structures in the system. The contrast in hydraulic conductivity between the high-permeability channels facies and overbank facies in these systems is often two- or more orders of magnitude, leading to concentration of flow in the channels that is out of proportion to their portion of the aquifer volume (e.g., 95 percent of the flow is concentrated in the 5 percent of the aquifer volume which is 1,000-fold more permeable than the average for the formation).

USING FLOW VECTORS

The key aquifer properties that control groundwater flow — groundwater elevation gradient and permeability — are anisotropic in most systems (although we often don't have sufficient data to map the tensors for these characteristics). Nonetheless, it can be helpful to view groundwater flow and solute transport from the vector, rather than scalar, perspective. Even in cases where we don't have the data needed for a full tensorial analysis, it is valuable to use available data to formulate a vector analysis of likely flow and transport pathways.

Figure 3.13 shows the cross-section of a site where contaminants were observed in a wetland (at 5 ug/L) and 40 feet below the aquifer surface at more than 1,000 ug/L. The hydraulic gradient was strongest along a path from the contaminant hot spot, at depth, to the overlying wetland. Because the dissolved-phase concentration at the wetland interface was quite low (< 10 ug/L), a question arose — Is horizontal plume migration occurring? To answer this question, a simple vector analysis was performed. The gradient in the x-direction (horizontal) was 0.0025 cm/cm and the maximum gradient was near-vertical, at 0.058 cm/cm, 20-fold stronger than the horizontal gradient. The aquifer in the area of concern lies in relict sand dunes and the formation comprises interbedded sands and silts on horizontal bedding planes. The equivalent hydraulic conductivities were estimated: for the horizontal direction, $K_x = 3.3 \times 10^{-3}$ cm/s and in the vertical direction, $K_z = 1.4 \times 10^{-5}$ cm/sec. Combining these factors to calculate the specific discharge component along each axis:

$$q_x = K_x \cdot \nabla_x = 1.9 \times 10^{-4} \quad \frac{cm^3}{cm^2 \cdot s}$$

$$q_z = K_z \cdot \nabla_z = 3.5 \times 10^{-8} \quad \frac{cm^2}{cm^2 \cdot s}$$

Although the strong elevation gradient suggested a large vertical flow component, the aquifer structure encouraged much greater flow, as indicated by the specific yield vectors, in the horizontal direction. This site is discussed further in Chapter 8.

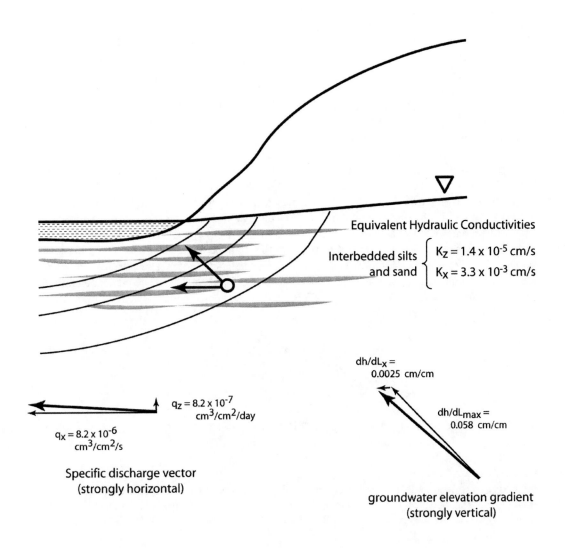

FIGURE 3.13 Construction of a flow vector from elevation gradient and equivalent hydraulic conductivities, aligned along alternative flow paths from the point indicated by the circle. Groundwater elevation gradients suggest a net upward movement of groundwater. However, when equivalent hydraulic conductivities are assessed, the net flow vector, represented as specific discharge, is strongly horizontal.

FLOW AT DISCONTINUITIES

Complex groundwater flow fields develop around permeability discontinuities that can deflect or concentrate and disperse specific discharge. **Figure 3.14** shows a lens of high permeability embedded in a larger volume of low-permeability material. Extending the structured flow analysis, above, to a 2-dimensional system, the higher-permeability lens consolidates flow at its upgradient end and disperses flow in the downgradient direction. Assuming the flow mapping orients the x-axis along the net flow direction, the flow consolidation would be apparent in the x-y and x-z planes. In heterogeneous porous media, flow consolidation and dispersal must occur throughout. High resolution model based mapping of flows in the synthetic aquifer system of Chapter 11 detect consolidation and dispersal, but elevation data cannot be collected in natural aquifers with sufficient resolution to detect these behaviors. Given currently available field instrumentation, many discontinuities can only be confirmed through tracer testing or contaminant plume mapping.

Hydrofracturing and in-filling the resulting fractures with high-permeability granular propping agent (e.g., coarse sand) can be used to create a stack of permeability discontinuities of the type shown in **Figure 3.14**. The formation access obtained through hydrofracturing is always accompanied by some amount of flow concentration. The formation access associated with fracturing can be very valuable and it is discussed further in Chapter 14.

When a discontinuity intersects a formation at an oblique angle, relative to the groundwater surface elevation gradient, the flow can be refracted, as depicted in **Figure 3.15**. If a tracer pulse were injected into the high-K zone, it would be smeared into the lower-K zone on the downgradient edge of the anisotropy, as shown on **Figure 3.16**. For a more complete discussion of flow refraction, refer to Freeze and Cherry (1979).

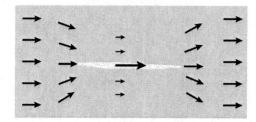

FIGURE 3.14 Groundwater flow, represented by specific discharge vectors, at a permeability discontinuity. In this example, a high-conductivity lens is embedded within a lower-conductivity porous medium. Groundwater flow is consolidated at the upgradient end of the discontinuity and dispersed at the downgradient end.

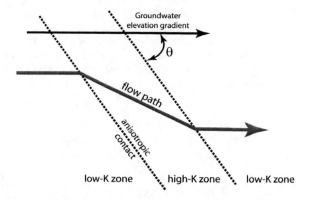

FIGURE 3.15 Refracted groundwater particle track across an anisotropic contact between high and low-conductivity segments of a heterogeneous aquifer.

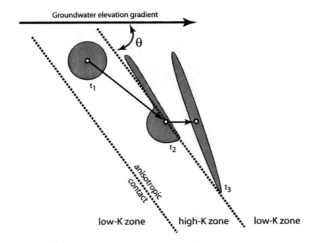

FIGURE 3.16 Refracted solute mass transport in groundwater flowing across an anisotropic contact between high and low-conductivity segments of a heterogeneous aquifer.

3.11 DARCY'S LAW IN RADIAL FLOW

Up to this point, our discussion of flow in porous media has been based on uniform, one-dimensional flow fields. **Figure 3.17** shows a conceptual volume element with one-dimensional flow. The rate of pressure drop along the volume element is constant and the cumulative pressure drop along a flow path is simply the rate of pressure loss times the path length.

Water moving to or from isolated vertical wells in an ideal porous medium follows a radial flow pattern. Unlike the one-dimensional flows we presume for natural aquifers, radial flows travel through a variable cross-sectional area, as depicted in **Figure 3.18**. The cross-sectional area, A, is a function of the radial distance from the well (r), and the thickness of the porous medium (h). Cross-sectional area, A, can be described as a function of r and h as follows:

$$A = 2 \cdot \pi \cdot r \cdot h \tag{3.39}$$

This value can be substituted into Equation (3.19). In this case, the designation for distance traveled is r, rather than L:

$$Q = K \cdot 2 \cdot \pi \cdot r \cdot h \cdot \frac{dP}{dr} \tag{3.40}$$

Rearranging,

$$\frac{dP}{dr} = \frac{Q}{K \cdot 2 \cdot \pi \cdot r \cdot h} \tag{3.41}$$

In radial flows that are generated by vertical wells, the area over which the flow is collected (for extraction) or dissipated (for injection) varies as a function of radial distance, as shown in **Figure 3.18**. As a result, the specific discharge decreases as the radius increases, for any fixed flow rate

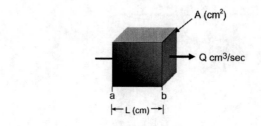

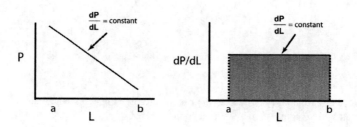

FIGURE 3.17 Relationship between total head [or pressure] (P) and path length (L), when cross-sectional area (A) is constant (one-dimensional flow). dP/dL is the slope of the pressure [head] path relationship. The total head loss [pressure drop] is the algebraic product of dP/dL and the path length, L.

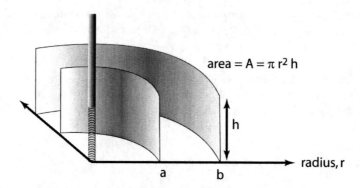

FIGURE 3.18 Cross-sectional area (A) for radial flow increases as a function of increasing radial distance from a cylindrical flow source or sink (illustrated for 1/4 of the arc).

and the rate of pressure drop becomes a function of 1/r. **Figure 3.19** is a graphical presentation of Equation (3.40) showing the dramatic reduction of the rate of pressure drop as a function of increasing radial distance, r, from the radial flow source (or sink).

Equation (3.40) can be integrated to provide the pressure drop exerted along a radial flow path, as determined by the flow rate and hydraulic conductivity. Rearranging,

$$dP = \frac{Q}{K \cdot 2 \cdot \pi \cdot r \cdot h} dr \qquad (3.42)$$

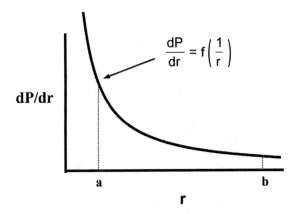

FIGURE 3.19 Graphical representation of the integration of the rate of pressure drop, dP/dr, over a path from "a" to "b", when the cross-sectional area, A, is a function of 1/r.

Moving constants outside the integrands,

$$\int_a^b dP = \frac{Q}{K \cdot 2 \cdot \pi \cdot h} \cdot \int_a^b \frac{1}{r} dr \tag{3.43}$$

Integrating both sides of the equation,

$$P_b - P_a = \frac{Q}{K \cdot 2 \cdot \pi \cdot h} \cdot \left[\ln(b) - \ln(a) \right] \tag{3.44}$$

Simplifying,

$$\Delta P_{a \rightarrow b} = \frac{Q}{K \cdot 2 \cdot \pi \cdot h} \cdot \ln\left(\frac{b}{a}\right) \tag{3.45}$$

Equation (3.45) is the formulation of Darcy's Law for radial (two-dimensional) flow in a porous medium. This equation describes flow in an unbounded porous medium that is uniform in all directions, and for which the supply of fluid is constant. It is a foundation for the more familiar solutions to flow problems that account for the boundary conditions and some of the structural heterogeneities encountered in real aquifers.

FLOW EQUATIONS

The assumptions of Equation (3.45) are not met in natural aquifers, so it is necessary to develop equations that account for boundary conditions encountered in realistic applications. The equations of groundwater flow are derived by combining Darcy's Law with the continuity equation. The continuity equation is based on conservation of mass (not volume) which states that with a fixed volume, the mass into a system minus the mass out the system equals the change in mass storage with time. Detailed derivations are provided in a number of references including Bear (1972), Domenico and Schwartz (1990) and Zheng and Bennett (2002) and the reader is referred to these sources for more information. There are no general solutions to these equations; however, many idealized solutions have been developed based upon simplified boundary conditions. These solutions have been applied to aquifer systems encountered in water supply development and used to accurately predict long-term water yields. The underlying assumptions for these flow equations are also probably not met, exactly, for most natural formations, but the equations have been proven to provide valuable estimates of average hydraulic properties, successfully manage groundwater resources and accurately estimate the zone-of-capture for groundwater containment systems.

The flow equations have been developed for multiple coordinate systems, such as Equation (3.40), and each has appropriate applications. The selection of coordinates for flow equations is usually a strategic decision to simplify problem solving by aligning a coordinate axis with a principal flow direction. A simple example is flow to a well; in Cartesian coordinates (x,y) flow is 2-dimensional, while in Cylindrical coordinates flow is 1-dimensional[5].

The equation for **transient flow in a homogeneous anisotropic aquifer** in cylindrical coordinates is as follows:

$$K_r \frac{\partial^2 H}{\partial r^2} + K_r \frac{1}{r} \frac{\partial H}{\partial r} + K_\theta \frac{\partial^2 H}{\partial \theta^2} + K_z \frac{\partial^2 H}{\partial z^2} = S \frac{\partial H}{\partial t} \qquad (3.46)$$

The first important solution of Equation (3.46) describes transient flow to or from a well in a homogeneous, isotropic, confined aquifer. The important assumptions are that the aquifer is infinite, the well is screened across the entire aquifer thickness and that water is elastically stored as a function of pressure. For these assumptions, there is no flow in the θ and z coordinate directions, which simplifies Equation (3.46) to:

$$K_r d \frac{\partial^2 H}{\partial r^2} + K_r d \frac{1}{r} \frac{\partial H}{\partial r} = S \frac{\partial H}{\partial t} \qquad (3.47)$$

or

$$\frac{\partial^2 H}{\partial r^2} + \frac{1}{r} \frac{\partial H}{\partial r} = \frac{S}{K_r d} \frac{\partial H}{\partial t} = \frac{S}{T} \frac{\partial H}{\partial t} \qquad (3.48)$$

The solution to Equation (3.48) was developed by Theis (1935). While obviously quite limited by the assumptions, the solution, known as the **Theis equation,** nonetheless led to the development of standard methods to estimate the *in situ* hydraulic conductivity of an aquifer.

$$h_0 - h = \frac{Q}{4\pi T} \int_u^\infty \frac{e^{-u}}{u} du = \frac{Q}{4\pi T} W(u) \qquad (3.49)$$

where $h_0 - h$ is drawdown, Q is the pumping rate, T is transmissivity, W(u) is the exponential integral commonly referred to as the well function and u is a related to radial distance (r), time (t) and aquifer properties (S, T).

---

[5] The groundwater flow equations can be developed in any coordinate systems with mutually orthogonal axes; lines drawn from a specific point (α, β, γ) are perpendicular with respect to each other for increasing α, β and γ.

$$u = \frac{r^2 S}{4Tt} \qquad (3.50)$$

Many researchers have contributed to the development of additional solutions to Equation (3.48) to describe a broader range in aquifer conditions including; leakage through confining beds, wells that partially penetrate aquifers, unconfined conditions and vertical anisotropy. The details of aquifer testing and the available techniques are not the focus of *Remediation Hydraulics* and the reader is directed to Kruseman & De Ridder (1990) and Butler (1997) for additional information.

At the remediation scale, the flow equations must be solved using numerical methods to allow for detailed representation of aquifer properties and application of site boundary conditions. There are many numerical models available to solve the groundwater flow equations with MODFLOW being one of the most widely used tools. Numerical models allow direct application of the groundwater flow equations to virtually any problem providing the ability to evaluate site specific conditions; however, this increase in precision must always be carefully assessed as it will not provide an increase in accuracy if geologic conditions have not been adequately characterized. Some of the well-studied forms and assumptions of the groundwater flow equations are summarized below:

Steady-state flow in a homogeneous aquifer: The groundwater flow equation for this class of problems reduces to the Laplace Equation.

$$K\frac{\partial^2 H}{\partial x^2} + K\frac{\partial^2 H}{\partial y^2} + K\frac{\partial^2 H}{\partial z^2} = 0$$

or

$$\frac{\partial^2 H}{\partial x^2} + \frac{\partial^2 H}{\partial y^2} + \frac{\partial^2 H}{\partial z^2} = 0 \qquad (3.51)$$

Transient flow in a homogeneous aquifer: Transient conditions introduce a storage term to the right hand side of Equation (3.51), where S is an aquifer storage term [feet-water/feet-head].

$$\frac{\partial^2 H}{\partial x^2} + \frac{\partial^2 H}{\partial y^2} + \frac{\partial^2 H}{\partial z^2} = \frac{S}{K}\frac{\partial H}{\partial t} \qquad (3.52)$$

Transient flow in a homogeneous anisotropic aquifer: The introduction of anisotropic conditions introduces the hydraulic conductivity as a constant in each coordinate direction.

$$K_x\frac{\partial^2 H}{\partial x^2} + K_y\frac{\partial^2 H}{\partial y^2} + K_z\frac{\partial^2 H}{\partial z^2} = S\frac{\partial H}{\partial t} \qquad (3.53)$$

For steady-state conditions the right hand side of Equation (3.52) is set to zero.

Transient flow in a homogeneous anisotropic aquifer: With heterogeneous conditions the hydraulic conductivity varies spatially and thus is differentiable.

$$\frac{\partial}{\partial x}\left(K_x\frac{\partial H}{\partial x}\right)+\frac{\partial}{\partial y}\left(K_y\frac{\partial H}{\partial y}\right)+\frac{\partial}{\partial z}\left(K_z\frac{\partial H}{\partial z}\right)=S\frac{\partial H}{\partial t} \tag{3.54}$$

For steady-state conditions the right hand side of Equation (3.52) is set to zero.

Transient Flow in the Vadose zone: The flow of water through soils with pores that are partially saturated with air and water can be described using Richards (1931) equation. Richards equation expresses hydraulic properties as a function of either water content (θ) or matrix potential (ψ=H-z) and assumes that the soil is always well vented, and air flow does not affect water flow.

$$\frac{\partial}{\partial x}\left(K_x\left(\psi\right)\frac{\partial H}{\partial x}\right)+\frac{\partial}{\partial y}\left(K_y\left(\psi\right)\frac{\partial H}{\partial y}\right)+\frac{\partial}{\partial z}\left(K_z\left(\psi\right)\frac{\partial H}{\partial z}\right)=S\frac{\partial H}{\partial t} \tag{3.55}$$

Where S is the specific moisture capacity

$$S=\frac{\partial\theta}{\partial\psi}$$

The solution of Equation (3.55) requires knowledge of the relationship between moisture content, matrix potential and hydraulic conductivity. The most common expressions relating these parameters are the Brooks-Corey or the van Genuchten equations.

For further study of the flow equations, and for discussion of some of the terms that we have not defined for the purposes of *Remediation Hydraulics* (e.g., transmissivity), we refer the reader to Freeze and Cherry (1979) and Driscoll (1986).

3.12 CONCENTRATION OF FLOW

The analysis of flow provided by Darcy's Law and the many calculation tools developed from it can be applied to an aquifer, as a whole (reflecting average conditions), or to an isolated unit that is relatively uniform and can be adequately characterized. In natural aquifer systems, our study tools are somewhat limited and it is quite challenging to obtain sufficient information to characterize any particular element of an aquifer. To help quantify and explain some of these limitations, we developed a synthetic aquifer domain that reflects the heterogeneities and anisotropies imprinted by natural depositional processes. The synthetic aquifer was 2-dimensional, measuring 300 x 400 feet. The cells were made 20 feet thick, to maintain water in all cells during pumping test simulations, which are discussed in greater detail in Chapter 11. Distinct hydraulic conductivity values were assigned for each square-foot block, yielding an aquifer domain of 120,000 point estimates of conductivity.

One of the difficult choices was the selection of a method to imprint hydraulic conductivities that would match the patterns we could expect from natural aquifer depositional processes (discussed in detail in Chapter 9). The challenge arises from the fact that natural anisotropies cannot be generated easily, *de novo*, by the stochastic approaches that have been described in the literature[6]. We elected to simulate a braided depositional environment, of the type shown in the upper panel of Figure 3.20. The simulation was built in four steps:

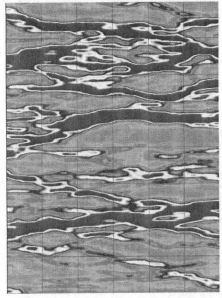

FIGURE 3.20 Building a synthetic aquifer to represent anisotropic hydraulic conductivity structure in aquifers. The upper panel shows a braided stream example (the Teklanika River in Denali National Park, Alaska). The lower panel shows an excerpt of a synthetic domain built to represent the braided stream environment.

---

[6] de Marsily et al. (2005) provided an excellent monograph on the difficulties associated with heterogeneous systems and we encourage readers to refer to that paper for a more complete explanation of modeling natural heterogeneous systems.

1. The first task was to secure a template of a braided stream. Aerial photos of braided rivers offered one possibility and **Plate 13** is such an example. Photos were also available from laboratory studies, such as those reported by Sapozhnikov and Foufoula-Georgiou (1997). We utilized an image from Bassler, et al. (1999, Figure 1), because it could be most easily digitized to represent a range of hydraulic conductivities.
2. The selected image was de-constructed from its native image format, creating a file of x and y coordinates, with a "z" value representing hydraulic conductivity that was created by calculating a single number from the red-green-blue color values in the original image.
3. The resulting file contained several million x-y elements and a 400 x 300 subset was carved out of the larger image, to serve as the model domain.
4. The z-values were scaled to represent a range of hydraulic conductivity values — for this work we examined flow behaviors at 100-fold, 1,000-fold and 10,000-fold conductivity ranges. In the most homogeneous natural aquifer systems, hydraulic conductivity contrasts of at least 1,000-fold are observed when detailed conductivity analyses are performed (Figure 3.11, for example).

    Plate 7 (left side) shows the 2-dimensional synthetic braided stream system that was developed from the digitization process. The anisotropies are organized by the depositional process and the connections between high-conductivity zones and the conductivity contrasts are aligned generally parallel the axis of the stream. Flow through the system was analyzed using MODFLOW (McDonald and Harbaugh, 1988; Harbaugh, et al., 2000), with no-flow side boundaries (the top and bottom of the domain shown on **Plate 7**) and a constant flux through the domain of 80 ft³/day, moving from left to right.

    The groundwater elevation field and flow vectors that developed from the model are shown in **Plate 7**. On the left, groundwater elevation contours are given at 0.5-ft intervals, based on the head reading at each of the 120,000 nodes in the domain. With this resolution, it is possible to observe small details of the flow field that are expressed in the elevation contours, but that expression is very small, despite significant concentration of flow in the higher-conductivity zones. At the

PLATE 13 Braided stream channels on the Delta River near Fort Greely, Alaska. *(See Plate section for color version of Plate 13.)*

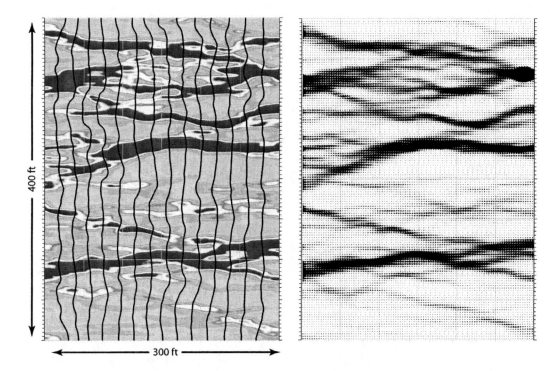

PLATE 7 Heterogeneous porous medium, synthesized to analyze the detectability of high-flow zones using water elevation contours. The relative hydraulic conductivities are shown on the left diagram, ranging over three orders of magnitude (dark blue is 1,000-fold higher conductivity than the light green shaded areas). The grid area is 300 x 400 feet and the water elevation contours, shown in black, are at 0.5-foot intervals. Groundwater flows are highly concentrated in this heterogeneous system, as shown by shading on diagram at the right. In this setting, roughly 70% of the flow occurred in 25% of the cross-sectional area, and more than 90% of the flow occurred in 50% of the cross-section. These flow concentrations are not discernable in the groundwater elevation contours. (*See Plate section for color version of Plate 7.*)

upgradient end of high-conductivity zones, the elevation contours show the collection of flow from larger areas of low hydraulic conductivity. At the downgradient end of a high-conductivity zone, the elevation contours show the dispersal of flow into larger areas of a less conductive matrix. The locations of high-conductivity zones can be inferred from these observations in the synthetic model. However, if the 300 x 400-ft domain had been sampled at a realistic frequency (say, for example, 10 wells in the 3-acre area), none of the groundwater elevation details would have been discernable. This suggests that significant concentrations of flow will be undetectable by groundwater elevation mapping.

The concentration of flow in the higher-conductivity zones is shown on the right side of **Plate 7**. This is a field of flow vectors, with the magnitude and direction of flow indicated by the size and orientation of the vectors. At this scale, the individual vectors can't be resolved, but the high-flow zones appear as darkened areas, clearly marking the areas of greatest flow. A transverse cross-section was constructed from the model output and the flow through each of the 400 nodes was tabulated. **Figure 3.21** shows the results of the tabulation for synthetic domains with 100-fold and 1,000-fold hydraulic conductivity contrasts (very small compared to the contrasts that are observed in most natural aquifers). The concentration of flow for this domain is indicated by divergence

Range of conductivity values:

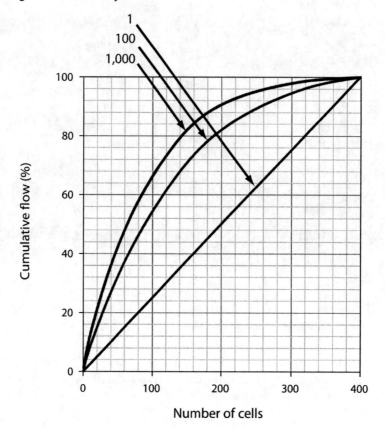

FIGURE 3.21 Concentration of flow in a transverse cross-section of the 2-dimensional synthetic braided stream aquifer presented in Figure 3.20 and Plate 7. For a 100-fold range of conductivities, 80 percent of the flow occurred in 50 percent of the cross-section; for 1,000-fold range of conductivities, 80 percent of the flow was concentrated in one-third of the cross-section. For natural, 3-dimensional systems, flow concentrations are expected to be much greater. Flow across a uniform system (range = 1) is provided for reference.

from the straight line that represents uniform flow distribution (equal flow through each node in the cross-section). As the range of hydraulic conductivities increases, so does the concentration of flow. In the 1,000-fold conductivity range, 90-percent of the flow was concentrated in 50-percent of the cross-section. The flow concentration is domain-specific, but the pattern is quite general. The concentration in this case is limited because the domain was only 2-dimensional. In natural aquifers, the 3-dimensional structures will cause much greater concentration of flow and Chapter 9 provides analytical methods to map the heterogeneities and to locate zones of concentrated groundwater flow.

4 Multi-Fluid and Non-Aqueous Flow

Water is the dominant wetting fluid for aquifer systems and groundwater fluid flow calculations are typically structured on that assumption. We encounter numerous fluids other than water in remediation engineering applications and it can be quite difficult to describe or predict their movement through aquifers. The simplest fluids to work with are aqueous solutions with fluid properties similar to groundwater. Tracer solutions, for example, often include salts that increase the fluid density, which can be a factor in their movement through aquifers. Oxidant solutions often have very high dissolved solids content and their densities and viscosity and the temperature increases associated with exothermic reactions all contribute to significant modifications to their fluid properties and temperature and density influences on hydraulic conductivity were examined in Chapter 3. When aquifers receive fluids that are immiscible with water, the prediction of fluid flows becomes much more challenging. These fluids may form two- or three-phase systems, with groundwater, a gas phase and a non-aqueous liquid phase all occupying the pore spaces of an aquifer. **Figure 4.1** illustrates the production of oxygen-dominated gas mixture associated with the application of Fenton's Reagent in aquifer cleanup. The disproportionation of hydrogen peroxide generates oxygen, the oxidation process generates carbon dioxide and the heat generated by the reactions vaporizes organic mass, all of which contribute to the very high rate of gas production

FIGURE 4.1 Oxygen bubbles erupting from the annular space between a monitoring well and its protective casing during a Fenton's Reagent application in a near-surface aquifer. Gases produced during these reactions can displace a significant fraction of the groundwater, reducing the hydraulic conductivity. Large gas eruptions can also generate safety hazards.

seen in **Figure 4.1**. These gases displace groundwater and substantially decrease the water-phase (hydraulic) conductivity of the aquifer.

At the opposite end of the spectrum of fluid behaviors, coal tar residues are often associated with former manufactured gas plants. These are typically denser than water, forming very high viscosity, non-aqueous fluids that may be mobile in aquifers. **Figure 4.2** shows coal tar oozing from a waste staging pile, immediately following its excavation from a former manufactured gas plant site in the Midwestern United States. The handling allowed coalescing of the tar and warming and the weight of soil that was added to the pile to suppress vapor formation combined to push the tar out of the pile, as seen in the figure.

Would coal tar in **Figure 4.2** be so mobile in the aquifer? How much groundwater flow reduction can we expect when we bubble gas into an aquifer in air sparging or generate uncontrolled gas eruptions with *in situ* oxidation? Those are questions we work to answer in the following examination of multi-fluid flow in porous media.

4.1 INTRINSIC PERMEABILITY AND SATURATED CONDUCTIVITY

The formulations for hydraulic conductivity that were derived in Chapter 3 were based on the premise that the aquifer matrix was saturated with a single fluid — water. Recall that the definition of saturated hydraulic conductivity, K_s, from Equation (3.25) is based on the intrinsic permeability, **k**, which is a property of the porous medium, and properties of the saturating fluid, density, ρ, and viscosity, μ, and the gravitational constant, **g**.

$$K = \frac{\rho \cdot g}{\mu} k \tag{4.1}$$

FIGURE 4.2 DNAPL tar excavated from a manufactured gas plant site. The tar became mobile when it was placed in a stockpile and clean fill sand was placed on top of the pile to suppress volatilization.

If we introduce a fluid other than groundwater into an aquifer and it fully saturates[1] the porous matrix, we can predict its behavior from Equation (4.1) by substituting appropriate values for density and viscosity. The intrinsic permeability can be estimated if we know the hydraulic conductivity of the porous matrix; otherwise, it would be necessary to test the porous matrix and obtain the needed value.

SATURATED NAPL FLOW

When the fluid of interest is non-aqueous liquid, we can describe the **NAPL-saturated conductivity**, K_{NAPL}, which can be calculated using the NAPL fluid properties, instead of those for water. These calculations should be based on site-specific measurements, as NAPL fluids typically consist of mixed liquids (e.g., solvents and oils or petroleum) and exhibit properties that can vary with weathering *in situ*. **Table 1.3** provided fluid properties for several NAPL field samples, including mixed chlorinated solvent waste, weathered fuel and manufactured gas plant tar. The field sample properties can be compared to neat trichloroethene and perchloroethene, for which data are also provided in **Table 1.3**. The fluid properties of weathered NAPL waste mixtures are strikingly distinct from those of neat solutions, making it clear that field sample collections are required to support relevant calculations of NAPL mobility in groundwater.

Intrinsic permeability is the second variable to be entered into Equation (4.1). Often, we have estimates of saturated hydraulic conductivity and measured NAPL density and viscosity values, but not measured intrinsic permeability values. There are two algebraically equivalent approaches commonly used to estimate the NAPL saturated conductivity under these circumstances: 1) solve for k using measured saturated hydraulic conductivity and published fluid properties of water, then substitute values for NAPL density and viscosity back into Equation (4.1), and 2) use NAPL-to-water ratios for density and viscosity to scale the saturated NAPL conductivity from a known hydraulic conductivity. In both cases, an estimate of hydraulic conductivity is the basis for calculating saturated NAPL conductivity.

WASTE SOLVENT PIT EXAMPLE

To illustrate these concepts, calculations are presented using data obtained from a site in the Midwestern United States. A large, mixed DNAPL source mass was developed from long-term disposal of solvent, paint and fuel wastes in pits that were back-filled more than 20 years ago. Hydraulic slug tests were performed on groundwater monitoring wells in the vicinity of the NAPL source mass to obtain estimates of the water-saturated hydraulic conductivity. The geometric mean of these tests was 1.2×10^{-3} cm/s, representing the average hydraulic conductivity for the zone of interest. Samples of the NAPL source mass were collected from two other wells, one in the center of the former waste disposal pit (sample A) and a second from the periphery of the source mass (sample B). Site-specific measurements of the DNAPL density, dynamic viscosity, and interfacial tension are compared to water in **Table 4.1**. As shown, the viscosity of the DNAPL samples ranged from 10.4 to 51.7 centipoise, which is approximately 10 to 52 times greater than water. The density of the NAPL ranged from 1.04 to 1.05 g/cm^3, as compared to 1.00 for pure water. Slug tests were performed on shallow wells at the site to estimate saturated hydraulic conductivity, yielding an arithmetic mean of 3.6×10^{-3} and a geometric mean of 1.2×10^{-3} cm/s. These wells were screened in an interbedded zone (e.g., interbedded sands, silts and clays that are above an aquifer) at approximately 25 to 30 feet below ground surface.

Substituting the fluid properties of water and the arithmetic and geometric mean hydraulic conductivity values into Equation (4.1) yields arithmetic-mean and geometric-mean permeability values of 3.7×10^{-4} and 1.2×10^{-4} cm^2, respectively. The NAPL-saturated conductivity values are

---

[1] It is important to note that it is very difficult to fully saturate a porous medium that has already been wetted by another fluid. This will be discussed further, in later sections.

TABLE 4.1.

Properties of NAPL and groundwater collected from a former waste disposal site in the Midwestern United States (Samples A , B and C), LNAPL from a former manufactured gas plant site in the Western United States (Sample D) and DNAPL from a former manufactured gas plant site in the Midwestern United Sates (Sample E). The field condition NAPLs are compared to selected non-aqueous-phase liquids commonly encountered during remedial technology applications. Both dense, non-aqueous liquid (DNAPL) and groundwater collections were made at the disposal site sampling locations.

Compound		Interfacial tension, σ_{OW} for NAPL and σ_{aw} for water phase at 20 °C (dynes·cm^{-1})	Density at 20°C (g/cm³)	Dynamic viscosity at 25 °C (cP)	Scaling factor (β_{ow})
PCE		44.4	1.62	0.844	
TCE		34.5	1.47	0.444	
Methylene chloride (MC)		28.3	1.33	0.413	
Toluene		35	0.867	0.560	
Octane		50.8	0.699	0.508	
water		72.7	0.998	1.002	
Sample A	DNAPL	11.6 (24°C)	1.05 (21°C)	10.4 (21°C)	6.29
Sample A	groundwater	---	1.002 (21°C)	1.007 (21°C)	1.00
Sample B	DNAPL	14.3 (24°C)	1.0399 (21°C)	51.73 (21°C)	5.10
Sample B	groundwater	---	1.002 (21°C)	0.9858 (21°C)	1.00
Sample C	DNAPL	13.6	1.014	20.3	
Sample C	groundwater	66.4	1.004	1.01	
Sample D	LNAPL	39.2	0.983	1,500	
Sample D	groundwater	51.9	1.003	1.015	
Sample E	DNAPL (tar)	15.5	1.013	1,500	
Sample E	groundwater		data not available		

obtained by substituting the NAPL values of density and viscosity into Equation (4.1). Results for the field NAPL samples A and B are summarized in **Table 4.2**.

Under the second approach, the saturated conductivity of the DNAPL can be calculated using Equation (4.2), as shown below:

$$K_{NAPL} = K_s \cdot \frac{\dfrac{\rho_{NAPL}}{\rho_w}}{\dfrac{\mu_{NAPL}}{\mu_w}} \qquad (4.2)$$

The calculation is illustrated using the geometric mean saturated groundwater conductivity of 1.2 x 10^{-3} cm/s and the fluid properties for sample A in **Table 4.1**:

$$K_{NAPL(MW-A)} = 1.2 \times 10^{-3} \cdot \frac{\dfrac{1.05}{1.00}}{\dfrac{10.4}{1.0}} \frac{cm}{s} \times \frac{\dfrac{g}{cm^3}}{\dfrac{g}{cm^3}} \frac{\dfrac{g}{cm^3}}{\dfrac{cP}{cP}}$$

$$K_{NAPL(MW-A)} = 1.2 \times 10^{-3} \cdot \frac{1.05}{10.4} = 1.2 \times 10^{-4} \quad \frac{cm}{s}$$

The results for field NAPL samples, A and B, are summarized in **Table 4.2**. The calculated NAPL-saturated conductivities are approximately 10- to 50-fold lower than the saturated hydraulic conductivity values. Because these calculations assumed a NAPL-wetted porous matrix, the resulting values represent the maximum possible conductivities for NAPL flow and the conductivities will be much lower in actual field conditions.

TABLE 4.2.

Comparison of Hydraulic and NAPL-Saturated Conductivities for the interbedded aquifer matrix at a former waste disposal site in the Midwestern United States.

	Hydraulic Conductivity (cm/s)	Sample A NAPL-saturated conductivity (cm/s)	Sample B NAPL-saturated conductivity (cm/s)
Arithmetic Mean	3.6×10^{-3}	4.8×10^{-4}	7.3×10^{-5}
Geometric Mean	1.2×10^{-3}	1.4×10^{-4}	2.4×10^{-5}

The viscosities of the two samples from the waste disposal DNAPL source were quite distinct, with the higher viscosity (sample B) generating a much lower NAPL fluid conductivity, which directly translates to lower mobility in the aquifer. Although groundwater samples at both locations had chlorinated solvent concentrations near aqueous-phase solubility limits, the DNAPL sample from the center of the source mass (sample A) comprised 17 percent chlorinated alkenes, while the sample from the periphery of the DNAPL source mass (sample B) comprised less than 0.1 percent chlorinated solvents. The difference may be attributable to weathering of the DNAPL source mass, through contact with groundwater. Over time, that contact resulted in leaching of the more highly water-soluble fraction, which also acts as a thinner for the DNAPL mixture. **Figure 4.3** shows a conceptual model of how this weathering process could affect DNAPL source mass, bounding the relatively mobile, solvent-rich core with a much less mobile, solvent poor shell.

COAL TAR EXAMPLE

The coal tar shown in **Figure 4.2** was excavated from a medium sand aquifer that has a hydraulic conductivity of approximately 1×10^{-3} cm/s. We can obtain an approximate value for the intrinsic permeability of the aquifer, using Equation (4.1) and the standard fluid properties of water, $\rho = 1.0$ g/cm^3 and $\mu = 1.0$ cP (cm·s/g). The value for g is 980 cm/s. Rearranging and completing the calculation:

$$k_{int} = K \cdot \frac{\mu}{\rho \cdot g} \quad \frac{cm}{s} \cdot \frac{g}{cm \cdot s} \cdot \frac{s^2}{cm} = cm^2$$

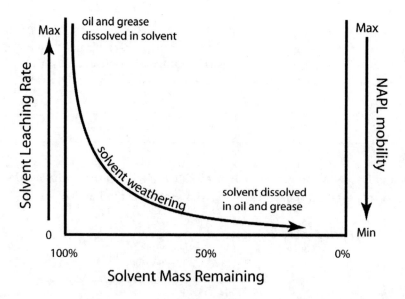

FIGURE 4.3 Conceptual model showing the influence of weathering processes on mixed DNAPLs below the groundwater surface. Dissolution of relatively more water-soluble fractions decrease the solvent content and the mixture transitions from "oil and grease dissolved in solvent" to "solvent dissolved in oil and grease." The impact of this transition is to dramatically decrease the mobility of the NAPL mass, through increased viscosity and decreased density, relative to solvent-rich mixtures.

the value for k_{int} is 1.02 x 10^{-6} cm². We can then enter the fluid properties of the coal tar, Sample E in **Table 4.1**: ρ = 1.013 g/cm³ and μ = 1,500 cP. Using Equation (4.1), we can calculate a coal tar fluid conductivity for the aquifer that has a hydraulic conductivity of 1 x 10^{-3} cm/s:

$$K_{coal\ tar} = \frac{1.013 \cdot 980}{1,500} \cdot 1.02 \times 10^{-6} = 6.8 \times 10^{-7} \quad cm/\sec$$

If the tar were driven by an elevation gradient of 1 percent (0.01 cm/cm) in an aquifer with a mobile porosity of 0.1, the tar could move at a pace of 6.8 x 10^{-8} cm/s, or 2.1 cm/year (based on Darcy's Law, Equation 3.19). This hypothetical velocity is quite small and it is not surprising that the tar that had been excavated in **Figure 4.2** had not moved appreciably in more than 50 years in the aquifer. This result suggests that the coal tar is almost immovable, but not quite. Further analysis is required to determine whether surface tensions of the tar or other non-aqueous liquids prevent their invasion of water-saturated aquifer pore space. These analyses will be introduced in the following sections.

4.2 RELATIVE PERMEABILITY AND CONDUCTIVITY IN TWO-FLUID SYSTEMS

The groundwater flow characterized in Chapter 3 and the saturated NAPL flow described above apply to porous media that are fully saturated by a single fluid. In natural aquifer systems, where water and air mingle above the water table, and at sites where hydrophobic organic compounds have spilled into the subsurface, the porous medium is simultaneously occupied by two or three fluids (air, water and NAPL) and flows become dramatically more complex. The behavior of NAPLs in aquifers is controlled by such small variations in aquifer structure that site-specific remediation-scale prediction of NAPL behavior is an unrealistic objective. There is, however, value to be derived from an understanding of the mechanisms of subsurface NAPL behavior. The literature on NAPL behavior in aquifers is well developed and we direct the reader to Cohen and Mercer (1993), and the many papers authored by David McWhorter, Beth Parker, John Cherry, Tom Sale and Bernard Keuper (e.g., Keuper and McWhorter, 1991; Parker, McWhorter and Cherry, 1997; Sale and McWhorter, 2001) to find a starting point in the literature. The following sections provide a cursory overview of two-fluid systems and, in particular, NAPL migration potential.

SATURATION AND CAPILLARY PRESSURE

In two-fluid systems, a wetting fluid is bound to the surface of the porous matrix particles and occupies a portion of the pore spaces and a non-wetting fluid occupies the remainder of the pore space. The aquifer void space occupied by a fluid is termed saturated and the portion of a matrix saturated by the wetting fluid is denoted by S_w, and the non-wetting fluid saturation is S_{n-w} or S_{nw}. Saturation is defined as the portion of the void-space or **total porosity**, θ_t, which is filled with each respective fluid. For systems with water as the wetting fluid (i.e., aquifers) the **volumetric water content**, θ_w, is related to saturation and porosity as follows (Corey, 1994):

$$\theta_w = S_w \cdot \theta_t \tag{4.3}$$

After a wetting fluid (water, for example) has fully saturated a porous matrix, a portion becomes relatively immobile. The wetting forces we described in Chapter 1 draw water into small-aperture voids, causing capillary rise[2]. The height of capillary rise, reflecting the strength of the lifting force, is greater for smaller-aperture conduits [refer to Equation (1.5)]. These forces also cause water to be retained in the soil matrix when we drain a fully saturated soil.

The pressure differential across an interface between immiscible fluids in a small-aperture void is termed the **capillary pressure**. The fluid pressure of the non-wetting (invading) fluid is balanced by the combined forces of the wetting fluid pressure and the surface tension of the wetting fluid contact with the pore wall, in a force balance that was shown earlier, in **Figure 1.4**. Because the force on the wetting fluid side of the interface gets a boost from surface tension, the pressure of the invading (non-wetting) fluid is greater than the pressure of the wetting phase, when the interface is at equilibrium (stationary).

Figure 4.4 illustrates a drainage curve, capillary pressure (P_c) versus S_r, for a fine-grained sand with a saturated hydraulic conductivity of 8×10^{-4} cm/s, as water is drained from the matrix. Residual water saturation values for the air-water system are strongly influenced by the grain-size distribution of the soils and commonly range from less than 5 percent for coarse sands to greater than 30 percent for clays based on the Carsel and Parrish database (Carsel and Parrish, 1988).

It is possible to measure the capillary pressure of the water-air interface in a soil sample and to follow the changes in capillary pressure that occur when the soil water content (saturation) is increased or decreased, as shown schematically in **Figure 4.5**. When the soil is fully saturated, the capillary pressure is zero. As the soil moisture decreases, the capillary pressure increases

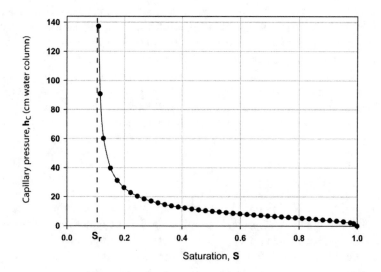

FIGURE 4.4 Synthetic capillary pressure curve, h_c, versus saturation, S, for a fine-grained sand with a saturated hydraulic conductivity of 7.9×10^{-4} cm/s, after Carsel and Parrish (1988). The residual saturation, S_r, of approximately 0.11 is determined by reading saturation at the point on the capillary pressure curve where the slope approaches vertical.

---

[2] The surface tension of water is higher than that of many aquifer solids, so water isn't a perfect wetting fluid for aquifer matrices (perfect wetting occurs when the solid surface draws the fluid into a thin layer that binds to the solid). None-theless, the energy released by the water wetting of soil matrix minerals is sufficient to lift water against the force of gravity.

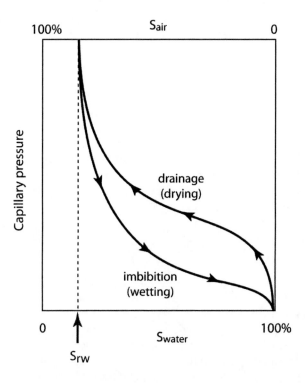

FIGURE 4.5 Path-dependent relationship between water saturation and capillary pressure for an air-water two-fluid system. The residual water (wetting phase) saturation is denoted as S_{rw}.

dramatically near the gravity-drained saturation level. The saturation level at which capillary pressure rises sharply is termed the **residual saturation**, represented by the variable $S_{r,w}$. It is approximately the amount of water that would be retained after gravity drainage. If the soil is dried to less than its residual saturation, its capillary pressure increases even more, to as high as 10 to 30 atmospheres.

From the curves shown on **Figure 4.5**, we can see that the relationship between capillary pressure and water saturation for a soil is path-dependent; that is, two different curves emerge — one for the wetting or imbibition process and a second path for the drying process (Corey, 1994). The path dependence of the relationship between capillary pressure and saturation is termed **hysteresis**. **Figure 4.4** illustrates the typical capillary pressure response during drainage and imbibition for an air-water (vadose zone) system.

The term **effective saturation, S_e**, is used to describe the portion of the saturated pore-space that contributes to drainable or flowing water/fluid in a two-phase system such as air-water or water-NAPL in Equation (4.4), where S is the saturation of the fluid as defined in Equation (4.3):

$$S_e = \frac{S - S_r}{1 - S_r} \tag{4.4}$$

For the example on **Figure 4.4**, the water residual saturation (wetting fluid), S_{r-w} corresponds to a water saturation, S, of approximately 11 percent. It is important to note that when $S_{e-w} = 0$, the saturation with respect to water is 11 percent. For the two-phase system of water and air in this example, the maximum total air saturation could be no greater than 0.89.

The effective water saturation can be determined using measurements of **capillary pressure head, h_c**, and the empirical van Genuchten relationship shown in Equation (4.5) (van Genuchten, 1980):

$$S_{e-w} = \left\lfloor 1 + (\alpha h_c)^N \right\rfloor^{-m}$$

(4.5)

The **van Genuchten parameters**, α (cm^{-1}) and **N** (cm), are characteristic properties of the porous medium, and **m** $= (1-1/N)$. The program SOILPARA was used to estimate the van Genuchten parameters for three soil types, sand, silt and silty-clay, using the database relationships developed by Carsel and Parrish (1988). SOILPARA was used for each case, to compute the retention curves as a function of saturation via Mualem's model and estimate the van Genuchten parameters, α, m and N. **Table 4.3** summarizes the van Genuchten parameters that were computed for the soil samples using the methods described above.

NAPL-Water Systems

Water-saturated aquifer matrices can be invaded by NAPL and, unlike the air-water system, NAPL-water fluid pairs readily form residual saturation in both the wetting and non-wetting phases. **Figure 4.6** shows capillary pressure – saturation curves for NAPL-water systems, for which the NAPL residual saturation ($S_{r,n-w}$) exceeds that of water. Under cyclical drainage and imbibition (displacement of NAPL by water), portions of the NAPL are trapped by the water with each cycle, further increasing the residual, or immobilized, volume of NAPL in the system.

Calculations associated with NAPL residuals are complex and we refer the reader to the literature cited above, for details on this topic. For the simple case of equilibrium drainage (e.g., gravity infiltration), the residual saturation of the NAPL would be similar to that for water, but scaled based on the ratio of the air-water interfacial tension, σ_{aw}, to the NAPL-water interfacial tension, σ_{ow} (Lenhard and Parker, 1987). The interfacial tension scaling factor for the NAPL-water system, β_{ow}, is defined in Equation (4.6):

TABLE 4.3.

The van Genuchten parameters used in calculating relative conductivity versus NAPL saturation for sand, silt and silty-clay samples, as estimated using the Carsel and Parrish database (Carsel and Parrish 1988) and SOILPARA (DAEM 2005).

	Hydraulic Conductivity (cm/s)	Residual Water Saturation	α	N	m	λ	γ
Sand	6×10^{-3}	0.05	0.145	2.68	0.63	1.68	2.19
Silt	7×10^{-5}	0.09	0.016	1.37	0.27	0.37	6.40
Silty clay	6×10^{-6}	0.16	0.005	1.09	0.08	0.90	23.21

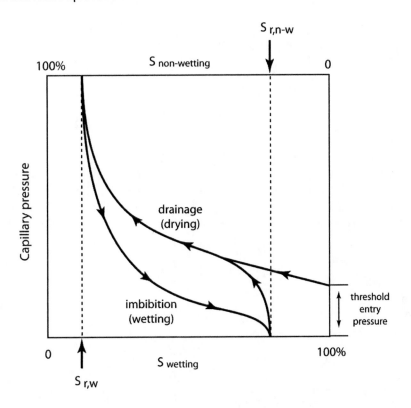

FIGURE 4.6 Path-dependent relationship between water saturation and capillary pressure for an NAPL-water two-fluid system. The residual water (wetting phase) saturation is denoted as $S_{r,w}$ and the residual NAPL saturation (non-wetting phase) is denoted as $S_{r,n-w}$.

$$\beta_{ow} = \frac{\sigma_{aw}}{\sigma_{ow}} \tag{4.6}$$

Using the scaling approach of Lenhard and Parker, Equation (4.5) becomes:

$$S_{e-nw} = \left[1 + (\alpha \beta_{ow} h_c)^N\right]^{-m} \tag{4.7}$$

Substituting NAPL-specific interfacial tensions for sample B (14.3 dynes/cm) and sample A (11.6 dynes/cm) of **Table 4.1** into Equation (4.6) yields β_{ow} values between 6.29 and 5.10. From calculations using Equation (4.7) we note that many NAPLs can have residual saturations greater than for water in the same aquifer.

When dissimilar fluids occupy a porous medium, each of their fluid conductivities must be adjusted to account for partial saturation — pores containing water reduce the area available for NAPL flow and *vice versa*. Corey (1994) and Van Genuchten (1980) developed empirical relationships that describe the relative water and NAPL permeabilities as a function of their respective saturations, S_w and S_{n-w}.

For the water-NAPL system, water is the wetting fluid and NAPL is the non-wetting fluid. For the wetting fluid, water, the **relative permeability, k_{r-w}**, can be computed using the van Genuchten equation (van Genuchten, 1980):

$$k_{r-w} = k \cdot S_{e-w}^{1/2} \left[1 - \left(1 - S_{e-w}^{1/m} \right)^m \right]^2 \tag{4.8}$$

where S_{e-w} is the effective saturation of water phase and the exponent "m" is a medium constant derived from the P_c versus saturation curve. Using the relationship between permeability and conductivity in Equation (4.8) allows direct computation of the **effective conductivity of water**, K_{rw}, by replacing the intrinsic permeability, k, with the saturated hydraulic conductivity, K_w.

The **relative permeability for the non-wetting fluid, k_{nw}**, is also defined by van Genuchten (1980) as follows:

$$k_{nw} = k \left(1 - S_{e-nw} \right)^2 \cdot \left(1 - S_{e-nw}^{\gamma} \right) \tag{4.9}$$

where

$$\gamma = \frac{2 + \lambda}{\lambda} \tag{4.10}$$

and

$$\lambda = m \cdot N \tag{4.11}$$

The constants m and N are as previously defined. The NAPL-phase relative conductivity, K_{r-nw}, can be computed by substituting the saturated NAPL-phase conductivity, K_{nw}, into Equation (4.9) for the intrinsic permeability parameters.

The methods of calculating relative permeability and conductivity discussed above are relatively simple to implement and provide a more realistic means of evaluating the potential mobility of DNAPL in a water-NAPL system through a range of water-NAPL saturation conditions. The following series of examples was developed using the NAPL characteristics from the waste disposal pit example introduced in Section 4.1, to illustrate the influence of NAPL saturation on NAPL mobility using relative conductivity as a metric. More rigorous multi-phase flow analysis can be completed using numerical codes such as STOMP (Nichols, Aimo et al., 2005) or Tough2 (Pruess, Oldenburg et al., 1999).

The relative hydraulic conductivity was computed for soil types observed in the waste disposal pit area, using the following steps:

1) Apply the scaling approach by Leonard and Parker in Equation (4.7) to calculate the retention curves for the water-NAPL system;
2) Apply the van Genuchten relations in Equations (4.8) and (4.9) to calculate the relative permeability and conductivity for water and NAPL in the three soil types.
3) Plot computed relative conductivities for water, K_{r-w}, and NAPL, K_{r-nw}, as a function of effective saturation.

The results of the relative conductivity calculations are shown graphically for the sand, silt and silty-clay soils on **Figures 4.7** through **4.9,** respectively. As shown on each of the three figures, the maximum relative conductivity for water and NAPL occurs at 100 percent water or NAPL

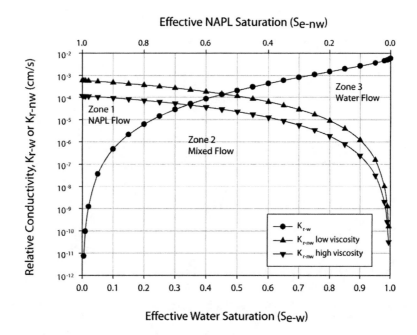

FIGURE 4.7 Computed relative conductivity for sand with saturated hydraulic conductivity of 6.0 x 10⁻³ cm/s and residual water saturation of 0.05. $K_{r\text{-}w}$ is the wetting (water)-phase relative conductivity. $K_{r\text{-}nw}$ is the non-wetting (NAPL) phase relative conductivity, where the low-viscosity fluid corresponds to the sample A from Table 4.1 and the high-viscosity fluid corresponds to sample B in Table 4.1.

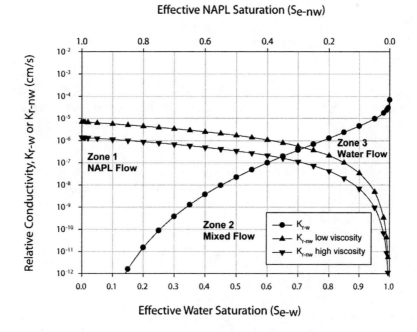

FIGURE 4.8 Computed relative conductivity for silt with saturated hydraulic conductivity of 7.0 x 10⁻⁵ cm/s and residual water saturation of 0.09. $K_{r\text{-}w}$ is the wetting (water)-phase relative conductivity. $K_{r\text{-}nw}$ is the non-wetting (NAPL) phase relative conductivity, where the low-viscosity fluid corresponds to the sample A from Table 4.1 and the high-viscosity fluid corresponds to sample B in Table 4.1.

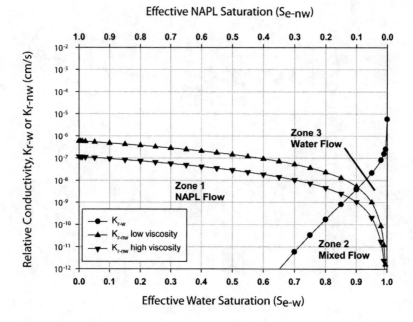

FIGURE 4.9 Computed relative conductivity for silty clay with saturated hydraulic conductivity of 6.0×10^{-6} cm/s and residual water saturation of 0.16. K_{r-w} is the wetting (water)-phase relative conductivity. K_{r-nw} is the non-wetting (NAPL) phase relative conductivity, where the low-viscosity fluid corresponds to the sample A from Table 4.1 and the high-viscosity fluid corresponds to sample B in Table 4.1.

saturation. These values correspond to the expected result using Equation (4.1). The NAPL-phase conductivity was evaluated using the fluid properties from Sample A (low viscosity NAPL) and Sample B (high viscosity NAPL) to illustrate the influence of NAPL viscosity on computed conductivity. The relative conductivity curves shown for the water- and NAPL-phase depend on effective saturation and decrease according to the relationships defined in Equations (4.8) and (4.9). Each of the relative conductivity graphs in **Figures 4.7** through **4.9** are annotated using a "flow zone" convention developed by Newell, et al. (1995) to evaluate relative mobility of the fluids in the system.

Zone 1 — **"NAPL Flow"** is the region delimited by the NAPL-phase relative-conductivity curve above and the water-phase relative-conductivity curve to the right. In this region, the NAPL-phase relative conductivity is greater than water due to high NAPL saturations, and fluid flow is dominated by NAPL.

Zone 2 — **"Mixed Flow"** is the region where the water-phase and NAPL-phase relative conductivities are similar and the proportion of fluid flow varies as a function of effective saturation for the fluids.

Zone 3 — **"Water Flow"** is the region where the water-phase relative conductivity is much greater than NAPL-phase relative conductivity due to high effective water saturation. Fluid flow is dominated by water flow in this region.

When the contrast in relative conductivities is greater than 100-fold, the low-conductivity fluid will be "hydraulically immobile" even though the effective saturation levels might be greater than the residual saturation levels.

The results shown for the sand (**Figure 4.7**) indicate that water-flow conditions would predominate (e.g., K_{r-w} is $>> K_{r-nw}$) up to $S_{e-w} = 0.4$ to 0.5. At NAPL saturations greater than 0.8, NAPL-flow would predominate. In contrast to the silt and clay materials shown on **Figures 4.8** and **4.9**, the sand shows modest decreases in relative conductivity values that are conducive to mixed-fluid and NAPL flow, particularly for higher degrees of saturation and lower viscosity NAPL.

The results for the silt and silty-clay (**Figures 4.7** and **4.8**) indicate that the contrast in viscosity of the fluids leads to very low saturated NAPL conductivity values. Based on the relationship between effective saturation and relative conductivity for the NAPL in the silt, separate phase NAPL flow would not be expected except at effective NAPL saturations above 0.4 to 0.5. As shown in **Figure 4.8**, the slope of the NAPL-phase relative conductivity curves is modest for S_{e-nw} between 0.4 and 1, which leads to less than 1 order-of-magnitude decrease in conductivity over this range of saturation; however, the magnitude of NAPL-phase conductivity is less than 10^{-5} cm/s for the silt. For the silty-clay in **Figure 4.9**, the NAPL-phase mobility is greater than the water at effective NAPL saturations of 0.2 to 0.3. As shown in **Figure 4.9**, the slope of the NAPL-phase relative conductivity curves is modest to $S_{e-nw} = 0.4$, but increases to $S_{e-nw} = 0.2$, which leads to almost 2 orders-of-magnitude decrease in NAPL conductivity. At low NAPL saturations, the NAPL-phase conductivity is exceedingly low and separate phase flow would be negligible.

4.3 ENTRY PRESSURES AND VERTICAL NON-AQUEOUS LIQUID MOBILITY

Non-aqueous-phase liquids are, by definition, immiscible with water. To form persistent non-aqueous fluid bodies within an aquifer, a fluid must meet two basic criteria:

1) Their solubilities must be exceedingly small so that the water surrounding the separate-phase fluid is at, or near, the solubility limit for all the fluid's components (otherwise, the NAPL would be lost entirely to the aqueous phase), and
2) The interfacial tension at the NAPL-water interface is sufficient to prevent easy mixing of the fluid masses.

Our field staff regularly distinguish NAPLs in monitoring wells when the non-aqueous fluids generate an interfacial tension greater than 10 dynes/cm, in contact with groundwater. **Table 4.1** shows fundamental fluid properties, including organic-water interfacial tension σ_{ow}, for NAPL and groundwater collected at several contaminated sites in the United States, for several pure organic liquids of interest and for pure water.

In most aquifers, including those that are highly contaminated, the porous matrix is wetted by water. That means that water adheres to the particle surfaces and bridges the spaces between particles, giving rise to a capillary pressure. To force its way into a water-wetted aquifer matrix, any non-aqueous-phase fluid must develop a pressure that exceeds the capillary water pressure. The threshold pressure is called the fluid entry pressure and it is denoted (for an organic fluid entering a water-wetted pore) as P_c^{ow}. The fluid entry pressure increases for smaller pore apertures because capillary effects become more difficult to break through.

Many of the publications on NAPL behavior in aquifers describe the movement of NAPL into fractures or macropores of specific aperture and build equations from a first-principles analysis (e.g., Kueper and McWhorter, 1991). **Figure 4.10** shows the surface of a wetting fluid occupying a small-aperture gap between two parallel surfaces, such as we might find in a fractured aquifer matrix or in a well screen. Kueper and McWhorter provided the calculation given by Equation (4.12) for the fluid entry pressure between two parallel surfaces. The contact angle, θ, and the pore space aperture, e, are depicted in **Figure 4.10**.

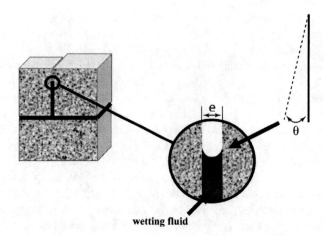

FIGURE 4.10 Illustration of the contact angle, θ, for a wetting fluid filling the space between two parallel surfaces with a gap dimension of "e". Water is assumed to be a wetting fluid and the theoretical contact is $0°$. The pore aperture between parallel surfaces is designated by the variable e and for tubular pore spaces the aperture dimension is represented by the tube diameter, designated by the variable ϕ.

$$P_{entry} = \frac{2 \cdot \sigma_{ow} \cdot \cos\theta}{e} \quad (dynes \cdot cm^{-2}) \tag{4.12}$$

For the fluid entry pressure in a circular aperture of diameter ϕ, Kueper and McWhorter (1991) provided Equation (4.13).

$$P_{entry} = \frac{4 \cdot \sigma_{ow} \cdot \cos\theta}{\phi} \quad (dynes \cdot cm^{-2}) \tag{4.13}$$

The result for both equations is fluid entry pressure required to displace the wetting fluid, water, from a pore of aperture dimension, e, or diameter, ϕ. The contact angle for water as a wetting fluid is theoretically $0°$, so the $\cos\theta$ is assumed to be 1.0 for these calculations. Refer to Chapter 1 for further discussion of wetting processes.

Equations (4.12) and (4.13) are applicable to situations in which we know or can reasonably speculate on the pore aperture dimensions (such as fractured clays or bedrock and well screens). For porous media such as sand, silt and clay, there is a very wide range of pore apertures and we can't reasonably speculate on the aperture dimension that will control NAPL entry pressures. To handle porous media, McWhorter (1996) described an experimentally-based calculation of fluid entry pressures that depend on the fluid properties and the hydraulic conductivity of the porous medium. Equation (4.14) combines McWhorter's equations and provides the entry pressure in equivalent water column (cm H_2O).

$$P_{entry}^{ow} = \frac{\sigma_{ow}}{\sigma_{aw}} \cdot 1.34 \cdot K^{-0.43} \quad cm\,H_2O \tag{4.14}$$

In this equation, σ_{ow} is the interfacial tension between the organic fluid and water, and σ_{aw} is the interfacial tension of the air-groundwater interface. The air-water interfacial tension for pure water is approximately 73 dynes/cm at 15 °C, a typical aquifer temperature for middle latitudes. However, microbial activity in highly contaminated groundwater is often affected by biosurfactants and we have observed significant reductions of air-water interfacial tensions in these systems. It is important to measure the groundwater-air interfacial tension, rather than to presume the pure water value as a default. In recent analyses of groundwater associated with major DNAPL source zones, the air-groundwater interfacial tensions were approximately 66 dynes/cm. The value K in Equation (4.14) is the hydraulic conductivity in cm·sec^{-1}.

Equation (4.14) can be used to generate entry pressures for a range of hydraulic conductivities. **Table 4.4** provides entry pressure estimates for NAPL samples A and B from **Table 4.1**, for hydraulic conductivities ranging from 1×10^{-6} to 1×10^{-2} cm·sec^{-1}. From these data, we can see that NAPL entry pressures for these NAPLs range from 8.9 cm H_2O for high-conductivity strata (k = 1×10^{-3} cm/sec) to 170 cm H_2O for low-conductivity soils that we encounter at many sites. The weathering of the mixed NAPL mass appears to impact entry pressures in two ways: the microbial communities appear to decrease the air-groundwater interfacial tension (up to 10 percent in many cases) and, more importantly, the organic-water interfacial tension decreases to low levels.

The pressure that develops at the base of a continuous NAPL mass provides the principal driving force to generate fluid entry pressure. It is a function of the pool height, the fluid density and the buoyant force of the surrounding water, which subtracts from the NAPL's downward force. In some aquifers, there is also a vertical gradient in the piezometric surface (dh/dz), which adds

TABLE 4.4.
Pore entry pressures for field DNAPL samples reported in Table 4.1, collected from a waste disposal site in the Midwestern United States. Interfacial tension values (σ_{ow}) were determined using both groundwater and NAPL collected from each monitoring well. Calculated using Equation (4.14).

Hydraulic conductivity, k cm·sec^{-1}	P^{ow}_{entry} (cm H_2O) Sample A NAPL σ_{ow} = 11.6 dynes·cm^{-1}	P^{ow}_{entry} (cm H_2O) Sample B NAPL σ_{ow} = 14.3 dynes·cm^{-1}
1×10^{-6}	81.0	99.8
5×10^{-6}	40.5	50.0
1×10^{-5}	30.1	37.1
5×10^{-5}	15.1	18.6
1×10^{-4}	11.2	13.8
5×10^{-4}	5.6	6.9
1×10^{-3}	4.2	5.1
5×10^{-3}	2.1	2.6
1×10^{-2}	1.5	1.9

(for a downward gradient) or subtracts (for an upward gradient) from the pressure a NAPL is able to exert at a water-filled pore aperture. Equation (4.15) provides a calculation of the fluid pressure at the base of a NAPL mass, as a function of the vertical thickness of the mass (h_{NAPL}), the densities of the NAPL and water (ρ_{NAPL} and ρ_{water}, respectively) and the piezometric head change over the NAPL thickness (Δh_{water}). **Figure 4.11** summarizes the factors that enter the calculation of available pressure at the base of a NAPL pool.

$$P_{NAPL\ base}\left(cm\ H_2O\right)=\frac{h_{NAPL}\cdot g\cdot\left(\rho_{NAPL}-\rho_{water}\right)}{\rho_{water}\cdot g}+\Delta h_{water} \tag{4.15}$$

where g is the acceleration due to gravity. The piezometric elevation change over the NAPL thickness is given in Equation (4.16), in which $h_{water\ A}$ is the shallower of the two observations.

$$\Delta h_{water}=h_{water_A}-h_{water_B}\left(cm\ H_2O\right) \tag{4.16}$$

We normally don't have piezometric elevation measurements that correspond to the top and base of any particular NAPL pool. In practice, we estimate the Δh_{water} from the vertical gradient (dh/dz) in the aquifer of interest as a function of NAPL pool height, as shown in Equation (4.17).

$$\Delta h_{water}=h_{NAPL}\cdot\frac{dh_{water}}{dz} \tag{4.17}$$

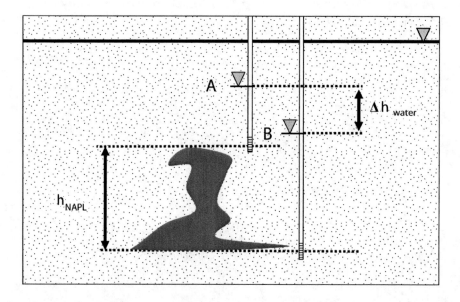

FIGURE 4.11 Graphical depiction of variables that are used to calculate pore entry pressure at the base of a contiguous non-aqueous liquid (NAPL) mass within an aquifer.

Substituting,

$$P_{NAPL\,base}\left(cm\;H_2O\right)=\frac{h_{NAPL}\cdot g\cdot\left(\rho_{NAPL}-\rho_{water}\right)}{\rho_{water}\cdot g}+h_{NAPL}\cdot\frac{dh_{water}}{dz} \tag{4.18}$$

and simplifying,

$$P_{NAPL\,base}\left(cm\;H_2O\right)=h_{NAPL}\cdot\left(\frac{\rho_{NAPL}-\rho_{water}}{\rho_{water}}+\frac{dh_{water}}{dz}\right) \tag{4.19}$$

To determine the pool height that can overcome the porous matrix entry pressure, set the NAPL base pressure equal to the entry pressure and solve for h_{NAPL}.

$$h_{NAPL}=\frac{P_{entry}^{ow}}{\left(\dfrac{\rho_{NAPL}-\rho_{water}}{\rho_{water}}+\dfrac{dh_{water}}{dz}\right)}\quad(cm) \tag{4.20}$$

Using Equation (4.20), we can re-visit the mobility of coal tar from Sample E in Table 4.1. Recall that we estimated the potential migration rate at 2.1 cm/year in a moderately permeable aquifer matrix. The entry pressure for that material is calculated from Equation (4.14) as follows:

$$P_{entry}^{ow}=\frac{15.5}{65}\cdot1.34\cdot0.001^{-0.43}=6.2\;cm$$

Then, the pool height required to push into a water-filled pore space is calculated from Equation (4.20), assuming a 0.01 cm/cm hydraulic gradient:

$$h_{NAPL}=\frac{6.2}{\left(\dfrac{1.013-0.998}{0.998}-0.01\right)}=1,232\;cm$$

The density of the material is so close to that of groundwater that a very large pool thickness is required to overcome the threshold entry pressure. The material originally was placed in the subsurface in a tar well — it wouldn't have migrated into its current position without a very large pool height buildup.

The pool height needed to push the DNAPL from Sample C in **Table 4.1**, vertically through aquifer soils with a range of hydraulic conductivity, was calculated and the results of these calculations are given in **Figure 4.12**. We also calculated the pool height required for the manufactured gas plant LNAPL, Sample D from **Table 4.1** and the results are given in **Figure 4.13**. For comparison, **Figure 4.14** provides pool height required to achieve pore entry threshold for pure perchloroethene in a pure-water aquifer.

The pool height graphs can be read to get an estimate of the contiguous NAPL pool height that would be required to push into a water saturated porous medium with various hydraulic conductivities. The water gradient provides a boost to the pool height for each curve. Realistic

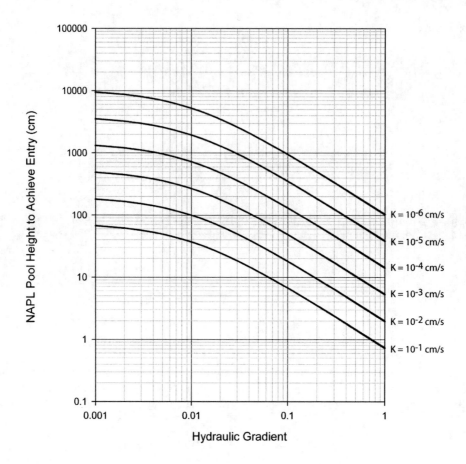

FIGURE 4.12 NAPL pool height required to achieve threshold entry pressure as a function of groundwater elevation gradient across the NAPL pool. Calculated for a field NAPL (Sample C, Table 4.1), with the following parameters: σ_{ow} = 13.6; σ_{aw} = 66.4; P_{NAPL} = 1.014; P_w = 1.004. Each curve represents the noted hydraulic conductivity.

natural gradients are on the far left of each curve and the gradients were extended to 1.0, to include a range of water elevation gradients that occur near groundwater extraction wells, which can obviously induce NAPL movement in aquifers where the NAPL would otherwise be stalled without sufficient pool height to achieve the entry pressure threshold. Comparison of the field-condition NAPLs (**Figures 4.12** and **4.13**) to pure perchloroethene DNAPL (**Figure 4.14**) shows how DNAPL-oil-grease mixtures and weathering can dramatically alter the NAPL fluid properties, reducing mobility of the separate-phase liquids in aquifers. Many of the NAPLs we encounter at field sites are much less mobile than the neat solvents typically used for experimental purposes.

NAPL Entry Into Well Screens

We also need to determine how easily a NAPL can enter a monitoring well screen. For that, we use the entry pressure calculation given in Equation (4.12), because the well screen gaps are equivalent to a parallel-walled fracture. For the "10-slot" screen used in monitoring wells at many sites, the

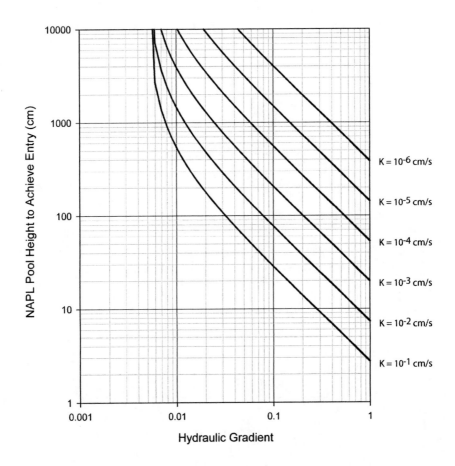

FIGURE 4.13 NAPL pool height required to achieve threshold entry pressure as a function of groundwater elevation gradient across the NAPL pool. Calculated for a field NAPL (Sample D, Table 4.1), with the following parameters: $\sigma_{ow} = 39.2$; $\sigma_{aw} = 51.9$; $P_{NAPL} = 0.983$; $P_w = 1.003$. Each curve represents the noted hydraulic conductivity.

gap is 0.010 inches, or 0.0254 cm. From Equation (4.12), we calculate the entry pressure for 10-slot screen as follows (we assume that water is an ideal wetting fluid and the angle, θ, equals 0°):

$$P_{entry}^{ow} = \frac{2 \cdot \sigma_{ow} \cdot \cos\theta}{0.0254} \quad (dynes \cdot cm^{-2}) \tag{4.21}$$

For the NAPL observed in sample A on **Table 4.1**, the observed interfacial tension was 11.6 dynes/cm, and the entry pressure for the 10-slot screen was, therefore, 913 dynes/cm². The entry pressure can be converted from dynes/cm² to cm H₂O as follows:

$$P_{entry}^{ow}(cm\ H_2O) = \frac{P_{entry}}{\rho_{water} \cdot g} \left(\frac{dynes \cdot cm^{-2}}{\frac{g}{cm^3} \cdot \frac{cm}{s^2}} \right) \tag{4.22}$$

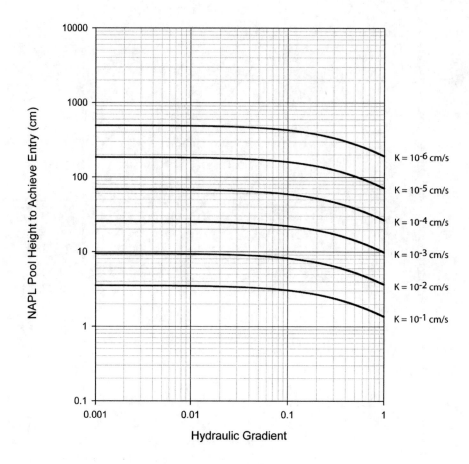

FIGURE 4.14 NAPL pool height required to achieve threshold entry pressure as a function of groundwater elevation gradient across the NAPL pool. Calculated for pure perchloroethene (PCE, Table 4.1), with the following parameters: $\sigma_{ow} = 44.4$; $\sigma_{aw} = 7274$; $P_{NAPL} = 1.62$; $P_w = 0.998$. Each curve represents the noted hydraulic conductivity.

For sample A, the density of water is 1.002, the conversion factor is 0.0107 cm H_2O per dynes/cm² and the entry pressure of 913 dynes/cm² equals 0.97807 cm H_2O. With this conversion, Equation (4.20) can be applied to determine the effect of vertical groundwater elevation gradients on the pool height required to overcome the fluid entry pressure for a 10-slot monitoring well.

For the NAPL encountered in sample A of **Table 4.1**, the NAPL pool would have to be 20 cm high to exceed the entry pressure for a 10-slot screen, if the vertical gradient was 0. This calculation provides an estimate of the *maximum* pool height required to achieve fluid entry pressure, which occurs when the NAPL entry point is at the base of the screened interval. In cases where the NAPL entry point is above the base of the screen, the vertical gradient may be larger than we're calculating in Equation (4.20). Consequently, the pool heights required to cause NAPL to enter a 10-slot well screen calculated with these assumptions are conservatively high.

4.4 IMPACT OF SITE ACTIVITIES ON VERTICAL NAPL MOBILITY

The vertical mobility of dense non-aqueous liquids is very sensitive to hydraulic gradients and other aquifer characteristics that can be easily (and accidentally) modified by restoration activities associated with DNAPL source zone cleanup.

Drilling — Drilling into the subsurface at DNAPL source zones is a very high-risk activity. As we can see be Equations (4.14) and (4.20), it is possible for a large pool of DNAPL to accumulate at an aquifer surface, without penetrating the groundwater. Drilling can short-circuit the entry pressure requirement and generate an immediate cascade of DNAPL penetration along the drill hole. Drilling through DNAPL source zones, particularly into the underlying groundwater, should be carefully controlled or avoided, altogether.

Vertical gradients — Vertical groundwater elevation gradients, whether naturally-occurring or induced by groundwater injection or extraction, can be critically important in mobilizing or stabilizing DNAPL. The vertical gradient is a key component of Equation (4.20), and its impact can be seen in Figures 4.11 and 4.12, where we calculated DNAPL pool height that would generate vertical migration over a range of hydraulic conductivities and vertical gradients (both upward and downward).

Surfactants — Aquifer bacteria can generate biosurfactants that significantly decrease the interfacial tension between groundwater and NAPL. That, in turn, significantly decreases the pool height needed to overcome the fluid entry pressure threshold. The production of surfactants is associated with highly productive bacterial communities and can be observed in association with enhanced reductive dechlorination systems.

Injection impacts on aquifer matrix structure — The impact of fluid injections depend on the soil type and where the injections occur:
- Injection of fluid directly into NAPL-bearing aquifer soils generates an expansion of the pore apertures for any soil type, effectively increasing its susceptibility to vertical NAPL migration.
- For injections performed below the NAPL-bearing zone, two mechanisms will decrease the aquifer's susceptibility to downward NAPL migration: 1) injection pressures exert compressive force on the aquifer matrix, decreasing the pore aperture dimensions, and 2) injection below the NAPL zone decreases the magnitude of downward water elevation gradients or reverses the gradients to upward.
- Injections above the NAPL-bearing zone compress the formation, but strengthen downward vertical water elevation gradients.

Groundwater extraction — As with injection processes, the impact of groundwater extraction depends on the depth of extraction relative to the DNAPL depth. Extraction that occurs below a NAPL-bearing zone strengthens the vertical downward gradient, enhancing downward DNAPL migration potential. If groundwater extraction were occurring immediately above a NAPL-bearing zone, downward vertical gradients would be reduced or even reversed, limiting the downward DNAPL migration potential.

Construction induced compression and vibration — Movement of heavy equipment, excavation activity and vibrations from drilling (especially sonic rigs) can induce NAPL mobilization by literally squeezing the material from its original lodgement in the aquifer matrix. Vibration can coagulate DNAPL mass that has been separated over time, to form a DNAPL residual. We have observed 1,1,1-trichloroethane form several liters of recoverable DNAPL in a monitoring well during sonic drilling for an adjacent well installation. The well collecting the DNAPL had been in place for more than 2 years and had not given any indication of DNAPL and dissolved-phase trichloroethene concentrations had been far below 0.1 percent of its aqueous solubility.

4.5 GAS INDUCED GROUNDWATER CONDUCTIVITY DECREASES

Many aquifer restoration techniques generate gases that occupy, at least temporarily, a portion of the aquifer porosity. Technologies that use or generate gases include: air sparging, aggressive *in situ* oxidation and enhanced reductive dechlorination, and these methods can exert an impact on the mobility of groundwater. Suthersan and Payne (2005) provided an approximation for the reduction of groundwater permeability k_w that is caused by gas displacement of groundwater in an aquifer:

$$k_w = k_{int} \left(\frac{S_w}{n} \right)^3 \quad cm^2 \tag{4.23}$$

where k_{int} is the intrinsic permeability (discussed above), S_w is the groundwater saturation and n is the drainable porosity of the formation. When the porous medium is fully saturated with water, $k_w = k_{int}$. The gas-phase permeability is estimated by:

$$k_{gas} = k_{int} \cdot \left(\frac{n - S_w}{n} \right)^3 \tag{4.24}$$

where k_{gas} is the gas-phase permeability. Figure 4.15 shows the relative permeabilities for water and gas for effective water saturation values from 0 to 100 percent. Notice that a relatively small displacement of water by gas generates a significant decrease in water permeability, as indicated by the dashed lines on the figure.

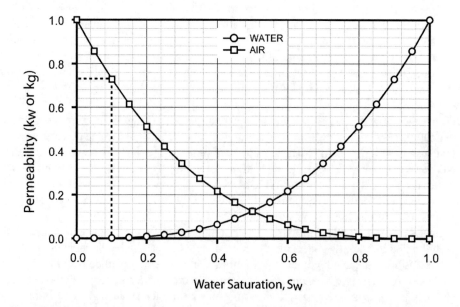

FIGURE 4.15 Permeabilities of gas and water phases in a two-fluid system. The dashed line indicates that a 10-percent displacement of groundwater by gas causes a 25-percent reduction of the water permeability.

5 Solute Dispersion in Porous Media

Dispersion denotes the macroscopic tendency of a solute plume to spread and to be diluted as groundwater moves away from a source location. It is often invoked as a concentration attenuation mechanism for contaminants in groundwater modeling and conceptual site models and the standard view of dispersion envisions two components: molecular diffusivity and hydrodynamic dispersivity. Both dispersivity and diffusivity are superficial characteristics, in which the behavior we observe is an average of processes occurring at much smaller scales. In the case of molecular-level diffusivity, we observe system behavior at scales several orders of magnitude larger than that at which the process occurs and the superficial characterization reliably represents the underlying process. Hydrodynamic dispersivity is a more conceptual process, based on the expectation that solutes will spread as a result of pore-scale bifurcation of groundwater flows and bulk water mixing. The conventional view in hydrogeology has been that hydrodynamic dispersivity is the dominant contributor to overall plume spreading and diffusivity makes only a small contribution at the large scales of contaminant plumes. However, detailed solute concentration mappings, whether for plumes arising from persistent sources or from short-term tracer pulses, typically do not follow patterns predicted by standard dispersion models.

We present an analysis in this chapter that suggests: 1) the effects of pore-scale hydrodynamic dispersivity (random walk effects) are probably quite limited, 2) that aquifer matrix depositional features may cause splitting of concentrated flows, but laminarity of groundwater flow limits dilution due to bulk mixing and 3) that diffusivity, acting through small-scale stratigraphic heterogeneities in porous media, can account for the longitudinal dispersion that is commonly observed in groundwater contaminant plumes and tracer pulse migration. In this approach, there are three fundamental modes of solute spreading:

Time-dependent — Repeated collisions with water molecules generate microscopic spreading of solute molecules that, over time, are observable at larger scales. This process, aqueous-phase *diffusion*, is entirely independent of the bulk movement of water through the porous medium and is a significant contributor to macroscopic solute dispersion in aquifers.

Flow-dependent — Groundwater moves through the pore-scale structure of porous media in networks of channels that split and recombine repeatedly along the flow path. Each cycle of splitting and recombination offers an opportunity for spreading of solute particles in the bulk water mass, either along the flow path (longitudinal dispersivity) or perpendicular to the flow path (transverse dispersivity). This form of solute spreading is entirely dependent on the movement of water through the porous medium and is termed *hydrodynamic dispersion*. Hydrodynamic dispersion is accompanied by decreasing solute concentration.

Formation dependent — Aquifer matrix depositional processes form higher-flow pathways that may bifurcate or anastomose (split and recombine), spreading the total cross-section of the contaminated groundwater front, without decreasing observed concentrations in the flow conduit. Although the cross-sectional average decreases in this case, it is not a meaningful measure of solute behavior.

A cohort of solute molecules in an aquifer can be characterized by its total mass (or number of molecules), its 3-dimensional center of mass and its bounding volume — the volume of groundwater that is just sufficient to envelop the entire cohort. The movement of a cohort of solute molecules in flowing groundwater is accompanied by some amount of increase in the bounding volume and a concurrent decrease in the average concentration within the bounding volume. Random motion processes force a cohort of solute molecules to occupy an always-increasing bounding envelope as it traverses the flow path, and the flow path itself may split and recombine frequently in natural porous media. **Figure 5.1** illustrates the bounding envelope concept as applied to solute pulses and persistent solute sources in naturally heterogeneous porous media. The illustration shows the concentration of solute in mobile portions of the porous medium, reflecting conditions expected during a solute pulse or at the leading edge of a plume arising from a persistent source mass.

If we could narrow our focus to a perfectly homogeneous porous medium, solute spreading would be caused by the combined effects of two random motion processes: molecular diffusion and bulk hydrodynamic dispersion. These are the processes that have been most often considered as the mechanisms that underlie contaminant plume development. However, there is mounting evidence that transverse dispersivities are small and that hydrodynamic dispersion is more realistically viewed as a deterministic result of solute migration through three-dimensionally heterogeneous porous matrices (i.e., real aquifer formations), in the manner depicted in **Figure 5.1**. There is also evidence that the impact of diffusion can be very significant, especially in the case of solute pulse migration through heterogeneous media. In the following sections, we examine each of these processes and draw several conclusions:

- The classic view of dispersivity leads to an overestimate of solute transverse spreading at points downgradient from a persistent source or solute pulse.

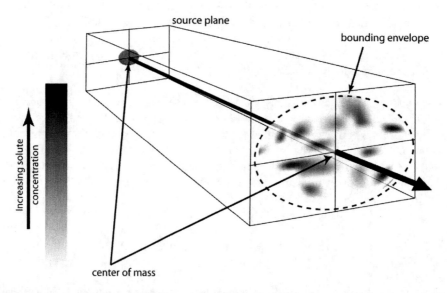

FIGURE 5.1 Conceptual diagram showing expansion of the cross-sectional distribution for a solute (dark areas), within a bounding envelope that increases along the flow axis. Solute migration is primarily along high-permeability pathways embedded within heterogeneous porous media.

- The flow through any particular site is most appropriately viewed as predominantly deterministic, not stochastic.
- The heterogeneous conductivity patterns that are primarily responsible for solute spreading are built into porous matrices at the time of deposition. Knowledge of the depositional environment for an aquifer provides important guidance for the stochastic modeling process.
- Although aqueous-phase diffusion cannot effectively drive solute migration over large distances, the cumulative impact of diffusive solute movement over short distances in heterogeneous media can have a striking impact on the distribution of solute pulses.
- The longitudinal dispersivity observed in real aquifers is attributable to multiple groundwater velocities, even in "homogeneous" aquifers, and diffusive interchange between mobile and immobile pore spaces in the aquifer matrices.

Figure 5.2 illustrates the combined impact of multiple velocities and diffusion, showing the modeled distribution of a tracer pulse injected into a "homogeneous" sand aquifer with the permeability characteristics of the Borden aquifer (Core 1 in **Figure 3.11**). We observe these complex patterns of solute movement, with channeling along multiple transport paths, limited transverse spreading and diffusive interaction between mobile and immobile pore spaces, at all the field sites where we have applied tracer studies or detailed contaminant distribution mapping. In the following sections, we will discuss pore-scale mechanisms of solute spreading, then extend the discussion to formation-scale solute spreading and its impact on transport in **Chapter 7**.

5.1 DEFINITIONS

The groundwater literature can be very confusing in its treatment of these processes, in particular due to the convention that dispersivity and diffusivity are distinct processes, but dispersion refers

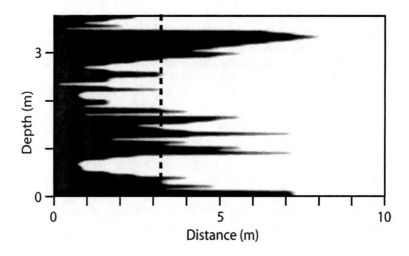

FIGURE 5.2 Simulated impact of natural aquifer heterogeneity on solute migration. Shading represents the progress of a bromide tracer, 5.5 days after the start of a constant source, uniformly distributed over the aquifer thickness. The dashed line indicates the average groundwater travel over the 5.5-day interval. The hydraulic conductivity structure was taken from the Borden aquifer (Core 1), shown in Figure 3.11.

to the effects of both processes. The following definitions reflect the general usage of terms and will be used in *Remediation Hydraulics*:

> *Dispersion* — the spreading of solute molecules over increasing volumes of water, independent of the mechanism generating the spreading. Dispersion decreases solute concentrations and includes the effects of molecular diffusion and hydrodynamic dispersion.
> *Diffusion* — molecular-scale spreading of solutes in aqueous phase, generated by the random motions of water and solute molecules.
> *Diffusivity* — the rate of molecular diffusion per unit time.
> *Dispersivity* — the rate of solute spreading per unit length of groundwater travel, attributable to bulk mixing of the water mass. It does not include solute spreading due to diffusivity, the movement of molecules within the water mass. Dispersivity can occur in three dimensions:
> > *Longitudinal dispersivity* — the spreading of a solute concentration front or tracer pulse along the flow axis (on the x-axis). The variable α_x is used to denote longitudinal dispersivity.
> > *Transverse horizontal dispersivity* — the lateral spreading of solute that generates a plume that broadens with distance from the source location (on the y-axis). The variable α_y is used to denote transverse horizontal diffusivity.
> > *Transverse vertical dispersivity* — the vertical spreading of solute that generates a vertical thickening plume, increasing with distance from the source zone (on the z-axis). The variable α_z denotes transverse vertical diffusivity.
>
> *Hydrodynamic dispersion* — the spreading of solutes attributable to dispersivity, the bulk water mixing that occurs when groundwater moves through porous media. It is also known by the term **bulk hydrodynamic dispersion**.

5.2 ADVECTION-DISPERSION

The bounding envelope for a solute pulse or persistent source tends to increase in size with increasing distance from the point of entry into the groundwater flow. It has been a common practice to lump processes that contribute to the increasing bounding envelope under the general term dispersion and to represent dispersivity as a scale-dependent random walk.

Dispersion along the x-axis (longitudinal dispersion) is calculated from diffusivity and dispersivity values, as follows:

$$D_x = \alpha_x \cdot v_x + D^* \quad \left(\frac{cm^2}{s} \right) \qquad (5.1)$$

where α_x is the longitudinal dispersivity, v_x is the groundwater velocity along the x-axis, and D^* is the aqueous-phase molecular diffusivity, corrected for porous media. The dispersion term can then be combined with the groundwater flow term, as in the one-dimensional advection-dispersion equation (5.2), shown here for dispersion along the flow path:

$$\frac{\partial C}{\partial t} = D_x \frac{\partial^2 C}{\partial x^2} - v_x \frac{\partial C}{\partial x} \qquad (5.2)$$

A solution to Equation (5.2) was given by Ogata (1970):

$$C = \frac{C_0}{2}\left[erfc\left(\frac{X - v_x t}{2\sqrt{D_x t}}\right) + \exp\left(\frac{v_x X}{D_x}\right) erfc\left(\frac{X + v_x t}{2\sqrt{D_x t}}\right)\right] \tag{5.3}$$

where: C_0 is the source concentration, X is the distance from the source, along the flow path, t is the time since the source was initiated, v_x is the groundwater velocity in the x-direction and D_x is the dispersion coefficient in the x-direction.

Domenico (Domenico 1987) provided further analytical equations that described aquifer concentrations of a solute that would be generated downgradient from a persistent source. The underlying logic of this and other dispersive models is that, without dispersion, the groundwater concentration at a point downgradient from a persistent solute source would match the source concentration, at steady state. However, it was believed, steady-state concentrations in aquifers are everywhere less than the persistent source concentrations[1]. Therefore, various dispersive processes must cause the observed dilution, which increases with distance from the source. Equation (5.4) shows the concentration of a persistent source that generates a plume at x = 0 with cross-sectional dimensions Y horizontal and Z vertical. The concentration at any point and time along the plume is determined by the dispersivities in each dimension (α_x, α_y and α_z).

$$C(x,y,x,t) = \left(\frac{C_0}{8}\right) erfc\left[\frac{(x - vt)}{2\left(\alpha_x vt\right)^{1/2}}\right] \cdot$$

$$\left\{ erf\left[\frac{\left(y + \frac{Y}{2}\right)}{2\left(\alpha_y x\right)^{1/2}}\right] - erf\left[\frac{\left(y - \frac{Y}{2}\right)}{2\left(\alpha_y x\right)^{1/2}}\right]\right\} \cdot \tag{5.4}$$

$$\left\{ erf\left[\frac{z + \frac{Z}{2}}{2\left(\alpha_z x\right)^{1/2}}\right] - erf\left[\frac{\left(z - \frac{Z}{2}\right)}{2\left(\alpha_z x\right)^{1/2}}\right]\right\}$$

Each form of the advection-dispersion equation generates a Gaussian solute concentration distribution, as shown in **Figure 5.3**, which was developed from Equation (5.4) with the parameters shown on the figure.

The advection-dispersion concept provided groundwater modelers with a simple basis for projecting solute concentration distributions over time and space, developed from persistent solute sources. However, current field experience indicates that we should reconsider the advection-dispersion concept — observed transverse dispersivities, for example, are very small compared to most model assumptions.

5.3 RECONSIDERATION OF THE DISPERSIVITY TERM

Figure 5.4 shows an example of a random walk process, balls falling through a lattice of pegs in a manner similar to the pinball game, Pachinko. As each ball falls through the lattice, there is an equal chance of going to the right or left at each junction. This is less complicated than the

---

[1] We now understand, through the use of high-resolution contaminant mapping, that uniform spreading is not typically observed and that stream tubes of high-concentration solute often travel great distances from persistent sources.

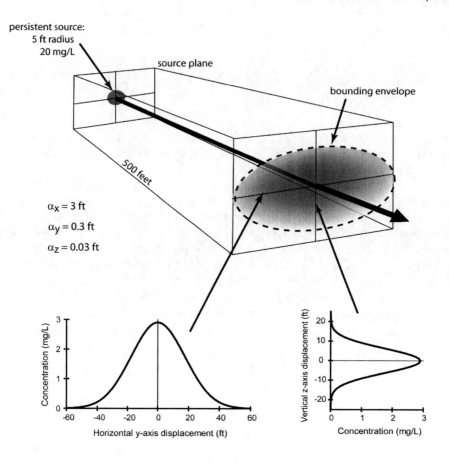

FIGURE 5.3 Calculated distribution of steady-state solute concentrations in a transect located 500 feet from a persistent source. The calculation shows the impact of dispersivity, without sorption or decomposition effects.

flow of water through a porous matrix, but it shows how a random walk process might induce bulk hydrodynamic dispersion with fluid flow in aquifers. The lattice metaphor suggests there is a significant lateral displacement potential for balls (or solute molecules) entering a binary choice matrix. However, a few calculations show that a binary choice process cannot generate significant lateral dispersion.

The lattice in **Figure 5.4** can be viewed as an instance of Pascal's triangle, for which there are 2^n possible routes that lead to n+1 outlets from the *nth* layer of the lattice. For a ball to move significantly away from the center of the lattice requires a disproportionate number of right or left moves. To arrive at the lateral extremes, a ball must repeatedly move either right or left, which is highly improbable.

We prepared a simulation of the lattice at work, in which we dropped 10,000 balls into the entry point and allowed them to pass through 1,000 steps of binary left-right choices, with an unbiased, equal probability of moving left or right at each step. The contents of each outlet bin at the base of the triangle were counted after each batch of 10,000 balls passed through the lattice. **Figure 5.5** shows the outcomes for a single trial[2] and the average of outcomes for 100 trials. As expected from the probability patterns developed for a 4-step lattice in **Figure 5.4**, outlets near the centerline

---

[2] A batch of 10,000 balls passing through the lattice, one at a time, constituted a *trial*.

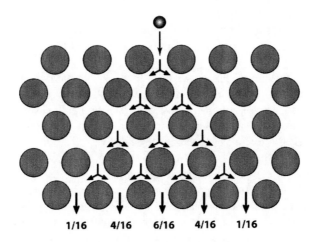

FIGURE 5.4 A simple random motion process: balls are dropped into a lattice of pegs and at each intersection, there is an equal chance of the ball proceeding to the right or left. There are 16 possible pathways through this lattice and the probability of reaching any one outlet is the number of pathways that lead to that outlet, divided by the total number of possible paths.

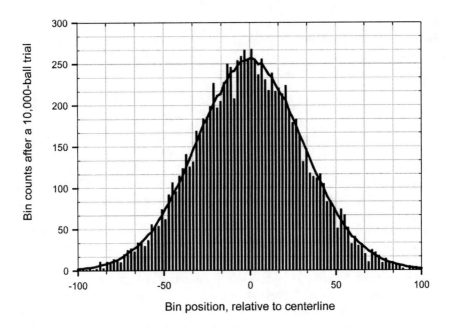

FIGURE 5.5 Results of a single trial (bars) and the averge of 100 trials (solid line) for a 1,000-step binary lattice. In each trial, 10,000 balls were dropped into an unbiased, binary-choice lattice, in which the ball moved randomly to the right or left at each step.

received the most balls. Also, even though it was possible for a ball to move 1,000 steps off the centerline (1,000 "rights" or 1,000 "lefts" out of 1,000 choices), only four balls were observed to arrive more than 150 units from the centerline in the 1,000,000 runs through the lattice (100 trials of 10,000 balls, each). Two balls arrived 153 units to the right and two were observed 155 units to the left.

That leads to an important conclusion we need to take from the random walk lattice — even though there is a visual impression of a 45-degree spreading, encouraged by depictions such as those in **Figure 5.4**, random migration to a location far off the centerline is quite improbable. For solute transport, those few molecules that stray far off the centerline through random walk processes are effectively diluted to less than detection.

The probability that a particle will reach any particular outlet from Pascal's Triangle is determined by the number of routes leading to that outlet, divided by the total number of routes leading to that layer (2^n). For example, a particle entering the lattice can reach any one of 5 outlets from the 4th layer and it can follow any one of 16 possible routes to get there. There are 6 routes leading to the center path, 4 routes to each of the next-to-center outlets, and only one route that reaches either of the outermost pathways (there are zero routes available that could carry a particle farther from the center line). These probabilities are termed the Bernoulli coefficients and we can calculate the probability for other outcomes using Equation (5.5). (Schwarzenbach, Gschwend et al., 2003)

$$p(n,m) = \frac{n!}{2^n \times \left[\frac{1}{2}(n-m)\right]! \times \left[\frac{1}{2}(n+m)\right]!} \tag{5.5}$$

From Equation (5.5), we can calculate the probabilities for each possible outcome, m, after n steps through a lattice as shown in **Figure 5.4**.

For large values of n, it's impractical to calculate Equation (5.5), due to the factorials. DeMoivre and Laplace (Schwarzenbach, Gschwend et al., 2003) developed an approximation for the Bernoulli coefficients that can be used for large n:

$$p(n,m) = \left(\frac{2}{\pi \cdot n}\right)^{\frac{1}{2}} \cdot e^{\left(-\frac{m^2}{2 \cdot n}\right)} \tag{5.6}$$

We applied Equation (5.6) to track the distribution of balls through 1,000,000 binary choice steps and the results are shown in **Figure 5.6**. After 600,000 steps, 99.99 percent of particles entering along the centerline will exit the lattice in the first 3,000 outlet channels to each side of the centerline. This shows that there are limits on the spreading induced by a random walk process due to the low probability of persistently moving in the same direction (right or left in this example) at successive choice points.

To transition from the binary choices to a geometric interpretation, we need to determine how frequently flowing groundwater encounters choice points. In the anastomosing channel structure we observed in porous matrix castings (**Section 2.7**), there were typically one to three channel-to-channel contacts (binary choice points) in the length of a single matrix particle. That means the relationship between choice points (channel intersections) encountered and Cartesian travel distance is a function of the matrix particle dimensions and packing structure.

Conceptually, we can reduce a two-dimensional particle matrix to a series of binary choice points. If the dominant particle size were 1 mm and there were an average of two channel intersections per particle, there would be 2,000 choices associated with every meter of travel through the aquifer matrix and 1,000 channel outlets per meter on a transverse cross-section of the aquifer matrix. 1,000,000 channel contacts would be encountered in 500 meters of travel along the flow axis and from **Figure 5.6**, we see that 99.99 percent of particles entering at the centerline will exit within the first 3,800 outlet channels, or less than ± 3.8 meters from the centerline in this hypothetical 2-dimensional matrix.

If the particle dimensions in **Figure 5.6** were 0.01 mm (10 microns), there would be 200,000 channel intersection choice points per meter of travel and 100,000 outlets per meter on a transverse cross-section. At this smaller grain size, 1,000,000 binary choice points will be encountered in 5 meters of travel along the flow axis. The 3,800-channel displacement for 99.99 percent of particles entering the lattice corresponds to a spread of ± 38 mm from the centerline, in 5 meters of travel.

These hypothetical two-dimensional examples show that a random walk process can only generate limited transverse dispersion (perpendicular to the flow path). This outcome is consistent with observations in laboratory studies and in real aquifers with very constant flow conditions, where researchers have noted that transverse dispersivity is very limited (in some cases, less than 3 degrees of arc) (Rivett, Feenstra et al., 2001; Cirpka, Olsson et al., 2006; Theodoropoulou, 2007), for example.

Another outcome of the simplified two-dimensional system is that all transit times through the matrix are identical and there is no longitudinal dispersion (spreading of the transit times). That is a consequence of the uniform particle dimensions and regular packing structure used to build the example. In more realistic porous media simulations, the possible paths to any point would not be of equal length and some variability would be introduced in arrival times at the outlet points (longitudinal dispersion). However, the magnitude of longitudinal dispersion induced by a random

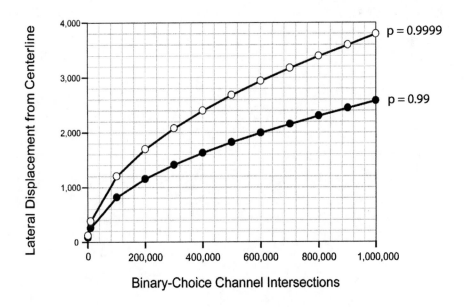

FIGURE 5.6 Lateral displacement from the flow axis, measured in channel outlets, as a function of the number of binary-choice channel intersections encountered along the flow axis. The solid circles show the divergence of 99 percent of mass and the open circles indicate 99.99 percent of mass fall within the indicated number of channel outlets from the centerline. For example, after 600,000 binary-choice channel intersections have been traversed, 99.99 percent of the mass will remain within 3,000 outlets from the center outlet.

walk through a more heterogeneous matrix is also likely to be small, compared to the longitudinal dispersion observed in the field.

In the next sections, we will examine the contribution of pore-scale and fine-scale diffusion processes to the development of longitudinal dispersion in solute transport. We will see that diffusion effects are quite large in heterogeneous porous media and probably exceed contributions from the random walk dispersivity induced by groundwater flow through anastomosing aquifer flow channels.

5.4 FUNDAMENTALS OF MOLECULAR DIFFUSION

Molecular diffusion is a mass transfer mechanism that has a very important role in contaminant, reagent and tracer movement through aquifers. Its impact is most significant at short distances (up to a few centimeters) and is negligible at distances greater than a few tens of centimeters. Molecular diffusion is often lumped with bulk hydrodynamic mixing under a superficial aquifer parameter, dispersivity.

When a population of solute molecules is placed into a solvent, its members move randomly in all available directions, at a velocity that is characteristic of the solvent-solute pair. The random movement is driven by Brownian motion, the continuous jostling that results from collisions between molecules in a liquid. A solution temperature increase is accompanied by increasing collision frequency and an accompanying increase in the velocity of solute molecules.

Although the movement of any one solute molecule is absolutely random and unpredictable, the net movement of the entire population of solute molecules is highly predictable. The predictable net movement of the solute population is an emergent behavior that we know as molecular diffusion — the statistical outcome of the random motions of large populations of molecules.

Figure 5.7 shows the fate of a cohort of solute molecules (bromide anion, Br-, for example) entering a large solvent volume. Initially, the cohort of solute molecules occupies a small portion of the solvent volume, yielding a high solute concentration at the entry point. Each of these molecules is moving randomly within the confines of the solvent volume. The predictable net effect of

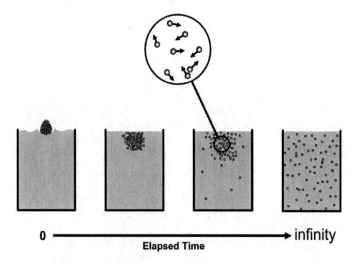

0 ⟶ infinity

Elapsed Time

FIGURE 5.7 Diffusion of a cohort of molecules added as a concentrated droplet to a solvent. Random movement of the solute molecules generates a net movement from areas of high concentration to areas of lower concentration. The movement of each molecule is independent of the movement of all the others. When the elapsed time is large, the solute molecules become evenly (although randomly) distributed throughout the available solvent volume.

all the unpredictable individual motions is to spread the solute molecules over a larger solvent volume, decreasing the solute concentration at the entry point and increasing it elsewhere. Over an extended time, the random motions distribute the solute molecules evenly over the available solvent volume.

We often speak of concentration gradients in solutions as though there's a chemical driving force directing the movement of solute molecules. In fact, diffusive movement of solute molecules is a strictly passive statistical behavior that emerges from the random motions of large populations of completely independent solute and solvent molecules. Collisions between particles of the solvent and solute populations generate the random motions that, with time, disperse aggregations of solute molecules.

Fick's First Law of diffusion predicts the diffusive flux of solute particles across an imaginary plane segment within a solution, as a function of the rate of concentration change with distance (i.e., the concentration gradient) perpendicular to the plane. Fick's First Law, Equation (5.7), shows that the solute flux in the x-direction, J_x, is proportional to the concentration gradient, $\partial C/\partial x$, perpendicular to the plane segment of area A. The constant of proportionality, D, is the diffusion coefficient, which has units of length^2/time. **Table 5.1** provides values of D for several compounds of interest.

TABLE 5.1

Values of aqueous-phase molecular diffusivity for ions and compounds of interest in remediation hydrogeology. Thermal diffusivity in water is provided for comparison.

Compound	Diffusivity (cm^2/sec)
Chemical diffusivities	
Bromide anion	1.9×10^{-5}
Chloroform	1.0×10^{-5}
cis-dichloroethene	1.1×10^{-5}
Perchloroethene	8.2×10^{-6}
1,1,1-Trichloroethane	8.8×10^{-6}
Trichloroethene	9.1×10^{-6}
Thermal diffusivities	
Heat in saturated sand ($\theta_{tot} = 0.40$)	7.4×10^{-3}
Heat in saturated clay ($\theta_{tot} = 0.40$)	5.1×10^{-3}
Heat in water	1.4×10^{-3}

$$J_x\left(\frac{mol}{sec}\right) = D \cdot \frac{\partial C}{\partial x} \cdot A \quad \left(\frac{cm^2}{sec}\frac{mol}{cm^3}\frac{1}{cm}cm^2\right) \tag{5.7}$$

Fick's First Law describes an instantaneous mass flux rate along a gradient of specified strength. The movement of solute particles in finite systems changes the magnitude of the concentration gradient as elapsed time increases. Consequently, Fick's First Law is useful for determining instantaneous flux in finite (real) systems, but it would be necessary to re-calculate the flux very frequently to keep up with the changing gradient of an aquifer (finite) system.

The diffusive movement of solutes in aquifers and the resulting changes in concentration gradients require that we examine non-steady-state gradient conditions, which are described by Fick's Second Law [Equation (5.8)]:

$$\frac{dC}{dt}\left(\frac{mol}{cm^3 \cdot sec}\right) = D \cdot \frac{\partial^2 C}{\partial x^2} \quad \left(\frac{cm^2}{sec}\frac{mol}{cm^3}\frac{1}{cm^2}\right) \tag{5.8}$$

Crank (1975) provided solutions to Fick's Second Law that correspond to numerous initial solute concentration distributions and solvent volume shapes and dimensions. The simplest case [Equation (5.9)] describes diffusion across the interface between two quasi-infinite fluid masses of uniform initial concentrations: $C = C_0$ in one fluid mass and $C = 0$ on the other side. The interface is located at $x = 0$ and diffusive mass transfer across the interface begins at $t = 0$.

$$C(x,t) = \frac{1}{2}C_0 erfc\left(\frac{x}{2\sqrt{D \cdot t}}\right) \tag{5.9}$$

Equation (5.9) can be modified to account for non-zero initial concentrations, A and B, on each side of the interface, as follows,

$$C(x,t) = C_{0,A} + \left(\frac{C_{0,B} - C_{0,A}}{2}\right) \cdot erfc\left(\frac{x}{2\sqrt{D \cdot t}}\right) \tag{5.10}$$

where $C_{0,B} > C_{0,A}$. Inspecting Equation (5.10), the error function complement (erfc) of 0 equals 0, so the concentration at the interface (where $x = 0$) is the average of $C_{0,A}$ and $C_{0,A}$, for all values of time. You may also notice that as time approaches infinity, the concentration approaches the average of $C_{0,A}$ and $C_{0,B}$, for all values of x. However, the solution to Fick's Second Law provided by Equation (5.10) is unbounded in the x-direction. Consequently, concentrations approach the average very slowly for positions in the gradient that are not extremely close to the solution interface.

Figure 5.8 shows the breakdown of the concentration gradient across the interface between two quasi-infinite fluid masses, A and B, described above. On the left is a fluid with an initial bromide ion concentration of 200 mg/L (Solution B) and on the right is a fluid mass of similar dimension with an initial concentration of 2 mg/L (Solution A). The four lines depict concentration as a function of distance at 0, 1, 10 and 100 days after the start of diffusion. The solute concentration pattern depicted in **Figure 5.8** might be encountered at the contact between an uncontaminated low-permeability matrix and an adjacent porous channel that has just been flooded with a high-concentration contaminant pulse. After 100 days of diffusive bromide migration, the concentration at a point 40 cm into the Solution A has only increased from 2 mg/L to 4.6 mg/L. Calculations developed from the unbounded solution to Fick's Second Law [e.g., Equation (5.10)] suggest that diffusion processes will be relatively unimportant in contaminant or tracer distribution within porous media. However, there is a significant difference in the outcome if we shift from an unbounded to a bounded system.

Another solute distribution we're likely to encounter in an aquifer is a pulse of high concentration solute that initially extends over a short distance in a channel, then diffuses along the channel axis over time. Equation (5.11) describes the diffusion from a pulse along a quasi-infinite (unbounded) channel,

$$C = \frac{1}{2} \cdot C_0 \left(erf\left(\frac{h-x}{2 \cdot \sqrt{D \cdot t}} \right) + erf\left(\frac{h+x}{2 \cdot \sqrt{D \cdot t}} \right) \right) \tag{5.11}$$

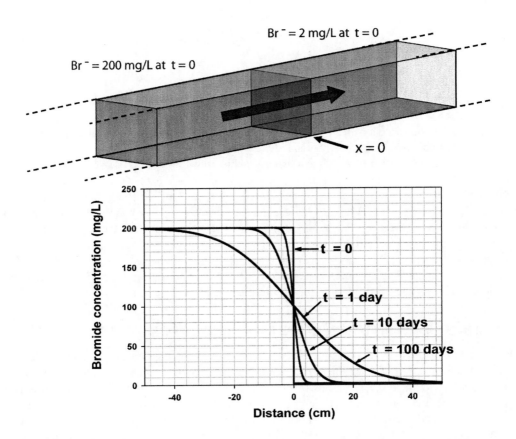

FIGURE 5.8 Relaxation of a concentration gradient in a quasi-infinite medium. At t = 0, fluid on the left side has a uniform concentration of 200 mg/L bromide and on the right side there is a uniform 2 mg/L bromide. The diffusion process begins at t = 0. After 100 days, the bromide concentration at 40 centimeters from the interface has only increased from 2 mg/L to 4.6 mg/L.

in which the initial solute concentration, C_0, is centered at $x = 0$ and extends from $x = -h$ to $x = +h$. The solute is unbounded, meaning it is free to diffuse over unlimited distance along the channel axis.

In the unbounded case, the initial (finite) solute pulse is diffusing into an infinite solvent volume and the solute concentration tends toward zero at large time values. For a bounded case (finite solvent volume), the solute concentration reaches a non-zero uniform concentration at large time values. In a physical interpretation of the bounded system (Crank, 1975), solute migration is reflected at the end wall. Equation (5.12) shows a solution to the one-dimensional bounded-volume diffusion case, using an infinite series.

$$C = \frac{1}{2} \cdot C_0 \cdot \sum_{n=-\infty}^{\infty} \left(erf\left(\frac{h + 2 \cdot n \cdot L - x}{2 \cdot \sqrt{D \cdot t}} \right) + erf\left(\frac{h - 2 \cdot n \cdot L + x}{2 \cdot \sqrt{D \cdot t}} \right) \right) \tag{5.12}$$

Crank (1975) pointed out that Equation (5.12) converges rapidly. We tested the convergence rate for a 10-cm wide, 2,000 mg/L tracer pulse, centered in a 40-cm long volume. The solutions varied by less than one part in 10,000 for values of n ranging from ±1 to ±500. Equation (5.13) provides a suitable substitute for Equation (5.12) for values of x and t that we typically wish to examine in the practice of remediation hydraulics.

$$C = \frac{1}{2} \cdot C_0 \cdot \sum_{n=-5}^{5} \left(erf\left(\frac{h + 2 \cdot n \cdot L - x}{2 \cdot \sqrt{D \cdot t}} \right) + erf\left(\frac{h - 2 \cdot n \cdot L + x}{2 \cdot \sqrt{D \cdot t}} \right) \right) \tag{5.13}$$

Equation (5.13) can be solved easily by MathCad or similar software, and will provide a close approximation to the infinite series of Equation (5.12).

Figure 5.9 shows concentration distributions that develop from an initial tracer pulse in bounded and unbounded systems. The bounded system in this example extends from $x = -20$ to $x = +20$ cm, representative of small-scale stratigraphic and fine-scale processes. A 2,000-mg/L tracer pulse is inserted into each segment, extending from $x = -5$ to $x = +5$ cm. At 1 and 10 days elapsed time, the bounded and unbounded solute distributions are indistinguishable. Between 10 and 100 days elapsed time, the diffusive migration reaches the edge of containment in the bounded system, and its concentration distribution begins to flatten at the average concentration (500 mg/L in this case). Because solute in the unbounded system moves freely to distances beyond the 40-cm segment, its average concentration decreased to much lower levels than we observe in the bounded segment. The bounded case is most representative of diffusive migration into and out of the small dead-end channels and slow-moving groundwater that we wish to examine at the pore scale.

5.5 MOLECULAR DIFFUSION IN HETEROGENEOUS POROUS MEDIA

When molecular diffusion is superimposed on a moving solvent mass, solutions to Fick's Second Law become somewhat challenging and if the solvent is moving through the complex geometry of a porous medium, direct solutions to Fick's Second Law are unattainable. That leaves two alternatives: 1) develop a superficial descriptor of diffusive mass transport that can be applied in concert with an advective solvent mass transport model (e.g., a mass transfer coefficient like that we can apply in the MT3D model), or 2) develop an estimate of the diffusive solute movement using Fick's First Law in an iterative calculation, over small time and distance increments.

The iterative solution is quite accurate and scientifically satisfying, because we can examine diffusive solute movement at the aquifer pore scale and gain important insights into solute behaviors that we observe at larger scales. However, it is impractical to develop iterative solutions to all

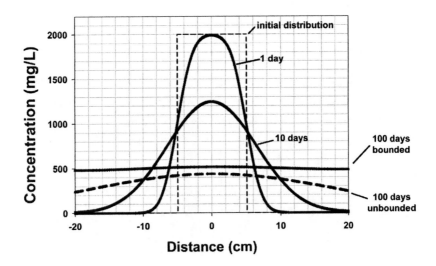

FIGURE 5.9 Comparison of the unbounded solution to Fick's Second Law to the bounded solution, for a 2,000-mg/L bromide source that initially occupies a 20-cm span. The unbounded solution was made with the assumption that bromide was free to diffuse into an unlimited mass of water initially at 0 mg/L, throughout. The bounded solution was developed with the assumption that the bromide source was in a container 40 cm in total length, with the source at its center. The dashed line indicates the initial bromide concentration for both bounded and unbounded cases.

problems of interest. We'll use the iterative calculation tool to study details of the diffusive process at the aquifer pore scale to guide our development and application of superficial mass transfer process estimators that can be applied more broadly.

Figure 5.10 shows the calculation strategy for an iterative solution to a mass transfer process. To study the diffusion process in flowing groundwater, we created a three-dimensional matrix of cells, each a 1-mm cube. The mass content of each cell was tabulated at 1-second intervals, accounting for mass transfers that would occur across cell boundaries in 1 second. The mass transfers for any time slice are accumulated in a temporary matrix and the net of all transfers is applied in a system-wide mass update that occurs every second. We assume that the contents of each cell is well mixed and that the mass transfers occur over 1-mm distances (cell center to cell center). For this discussion, we calculated the effects of advection and diffusion, with a central line of cells carrying advective flow, flanked by various configurations of static-water cells, mimicking dead-end pores that interacted with the flowing cells by diffusion, alone.

To test the impact of the well mixed cell assumption, we ran an advection-only calculation with no static water contact, simulating a 100-mm wide tracer pulse after five days of travel. The initial tracer concentration was 2,000 mg/L. We compared that result to a direct calculation, using Equation (5.11), Crank's equation for unbounded diffusion from an extended source (Crank 1975) and the two results are shown in **Figure 5.11**. Tracer mass recovery was acceptable for both approaches (minus 8 parts in 1 million for the direct calculation and plus 365 parts in 1 million for the iterative simulation). The assumption that each cell is well mixed generated a 28% increase in the effective diffusion coefficient. This is a common problem for all numerical modeling approaches and the magnitude of the error is determined by the cell dimension — relatively small at the 1-mm cell dimension, increasing significantly for larger cell sizes. For the following discussion, the modest

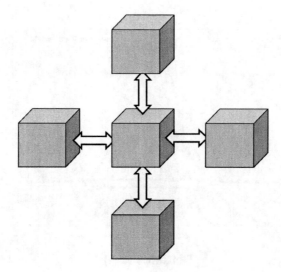

FIGURE 5.10 Iterative model concept, showing possible interactions between cells in a two-dimensional system. At each time step, all cell-to-cell interactions are tallied independently. Then, after all interactions that occur during a time step have been calculated, the status of each cell is updated to reflect the net change that occurred during the time step.

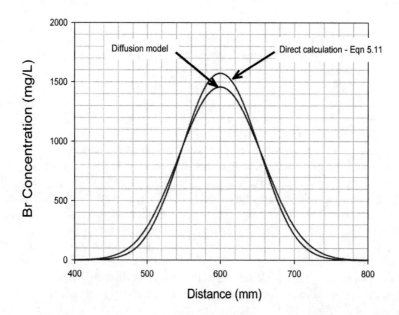

FIGURE 5.11 Comparison of bromide tracer pulse concentration estimates after five days of advective groundwater movement, using the iterative simulation approach and direct calculation by Equation (4.8). The areas under both curves showed good mass recovery; for direct calculation, there was a mass loss of 8 parts in 1 million and for the iterative solution there was a mass gain of 365 parts in 1 million. The diffusion model results in a slightly higher effective diffusivity, due to the assumption that the 1-mm cells are well mixed.

acceleration of bromide diffusion is tolerable. However, the mixing assumption can make a great difference in effective diffusivity for the large cell sizes that are commonly used in mass transport modeling, as discussed in Chapter 6.

One of the important capacities of the iterative simulation is to provide insights into problems that are too geometrically complex to solve by direct calculation. We are especially interested in diffusion that occurs at the pore-scale, between fast-flowing groundwater and the adjoining slow-flowing or static water masses. Most of groundwater flow is expected to occur in the larger of the pore apertures, but the static pore volume is intimately connected to the high-flow zones. Because the distances between flowing and static spaces are very small (sub-millimeter), diffusive solute movement can be a significant mass transfer process, contrary to its effect at larger scales (decimeters, for example).

Based on our iterative calculations, two characteristics of aquifer pore space structure appear to control diffusive solute migration patterns: 1) the mobile-to-immobile porosity ratio, and 2) the relative mass transfer geometry. These characteristics are shown diagrammatically in **Figure 5.12.** In the upper group of panels (A - C), each structure has an equal volume of mobile and immobile pore space, creating a mobile-to-immobile ratio of 1.0 ($\theta_i/\theta_m = 1.0$, or $\theta_m/\theta_t = 0.5$). In part A, the mobile pore space is in near-continuous contact with the immobile spaces and depth of the immobile space is limited. The pore geometry of Part A would be relatively favorable for mass transfer, what we're terming a *high-mass-transfer geometry*. Part C shows a pore geometry with the same mobile-to-immobile pore volume ratio as in Parts A and B, but the geometry is much less favorable for mass transfer: the contact between the mobile and immobile volumes is much less frequent and the depth of the immobile space is much greater, reducing the rate at which solutes perfuse the immobile pore spaces through diffusion. The lower two panels (D – F) of **Figure 5.12** show systems with pore volume ratios of 1 mobile to 2 immobile, or a 33.3% mobile to total ratio. As in the upper panels, the configurations show a range of relative mass transfer geometries. In **Table 3.2**, we provided field data that showed the mobile porosity in all aquifers we have tested is well below 30 percent of the total porosity. This suggests there is a potential for significant diffusive interaction between mobile and immobile groundwater masses and that pore-scale modeling can provide useful insights into diffusion-driven solute behaviors.

We ran a series of iterative simulations with diffusion and advective flows as the only mechanisms affecting the distribution of solute in the system. In each of these tests, there was a central advective flow path formed by 1-mm cubic cells. Each test started with a 2,000 mg/L tracer pulse placed in the advective cells from 50 to 150 mm and the advective water velocity was set at 100 mm/day[3]. The tests were run for 30-day trials, during which time advective water movement was 3,000 mm and tracer pulse migrations varied according to the mobile-to-total porosity ratios and the mass transfer geometries.

There are two approaches for tracking tracer pulses and other solute migrations in groundwater: 1) profiles, in which solute concentrations are mapped as a function of distance along a transect, and 2) breakthrough curves, in which solute concentrations are plotted as a function of time for a fixed observation point along the migration path. In field work, it isn't economically feasible to generate a high-resolution profile for most sites (due to drilling costs), so we normally work with breakthrough curves to obtain a high-resolution measure of solute migration. In simulations, we take advantage of our ability to generate high-resolution profiles because it's easier to visualize the solute mass in the mobile phase at any point in time.

Figure 5.13 compares tracer pulse profiles along the flow axis at several times after injection, for a system with no immobile porosity ($\theta_m/\theta_t = 1.0$) and a system with equal mobile and immobile porosities ($\theta_m/\theta_t = 0.5$). In both cases, the injected tracer pulse occupied a 300-mm span (50 to 350 positions on the flow axis). In the system with 100 percent mobile porosity (**Figure 5.13 A**), the

---

[3] 100 mm/day (approximately 0.3 ft/day) is a moderate groundwater velocity. It is somewhat higher than the superficial velocity estimate for the Borden aquifer (80 mm/day) reported by Rivett, et al.

tracer pulse migrated at the groundwater velocity (100 mm/day) and there was a slow decrease in peak concentration, caused by diffusive movement of solute ahead and behind the center of tracer mass ($V_w \pm V_{diffusion}$). The system with 50 percent mobile porosity shows the importance of diffusive solute migration at this scale, for which two outcomes can be seen in **Figure 5.13 B**: 1) There was a rapid loss of solute from the mobile pore space into the adjoining immobile pore space. Diffusive equilibration of the mobile and immobile volumes diluted the mobile phase solute; 2) the velocity of the solute peak was slowed to 50 percent of the groundwater velocity. Attenuation of the peak velocity is caused by solute losses from the leading edge of the pulse to "clean" immobile pore space, and repatriation of solute from high-concentration immobile space to "clean" mobile pore space at the trailing edge of the solute pulse.

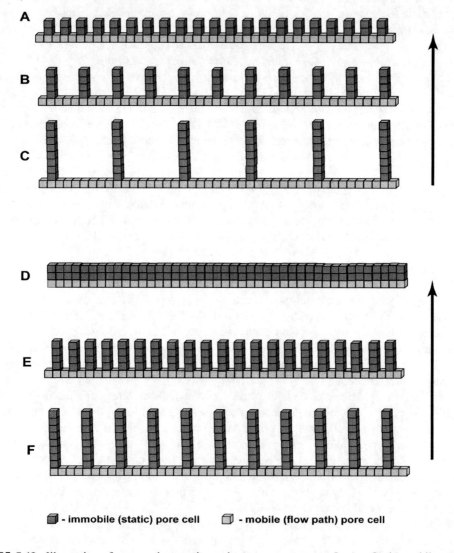

▨ - immobile (static) pore cell   ▨ - mobile (flow path) pore cell

FIGURE 5.12 Illustration of pore scale porosity and geometry concepts. In A - C, the mobile porosity is 50 percent of total porosity, while the geometries vary significantly. D - F shows three geometries that provide a 33-percent mobile porosity. Arrows indicate increasing relative mass transfer geometry within each grouping. Both the mass transfer geometries and the mobile-to-total porosity ratio have a significant impact on scale pore-processes that control solute mass transport.

After the mobile and immobile pore spaces first reach equilibrium, the area beneath the concentration profile was reduced to exactly half that of the tracer pulse. Effectively, only 50 percent of the solute mass was "visible" in the mobile pore space at any time after equilibration. Breakthrough curves show a dramatically different visualization of solute migration. All the injected solute molecules eventually pass any observation point on the flow axis in the mobile pore space. Consequently, the area under any breakthrough curve located at a point downgradient from the initial pulse will be determined by the initial tracer pulse width and concentration, independent of the mobile to total porosity ratio (the pore scale geometry has other, very striking, influences on the shape of tracer breakthrough curves).

At the center of the initial tracer pulse (the injection well) there are additional useful analyses available to us. Here, progress of the mobile-immobile pore space equilibration can be seen, as shown in **Figure 5.14**. If the tracer pulse injection spans several days of groundwater movement, the relative injection well concentration (C/C_0) approaches the mobile-to-total porosity ratio. From this, we suggest that tracer injection wells can be utilized to provide an estimate of the mobile-to-

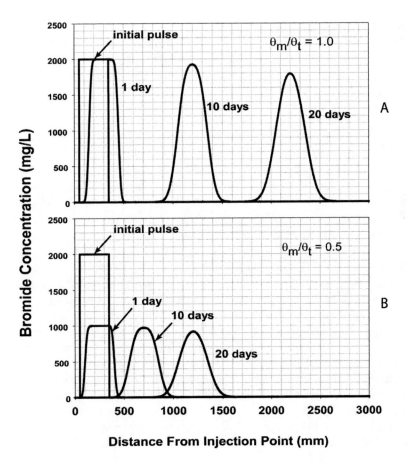

FIGURE 5.13 Effects of immobile pore space on tracer pulse migration. In each case, molecular diffusion is acting on a 2,000-mg/L tracer pulse in a system with a 100 mm/day groundwater velocity. In part A, there is no immobile pore space ($\theta_m/\theta_t = 1.0$) and in part B, there are equal volumes of mobile and immobile porosity ($\theta_m/\theta_t = 0.5$). Diffusion causes both sets of curves to assume a Gaussian distribution; in part B, the peak heights, areas beneath the curves and peak velocities have been attenuated in proportion to the ratio of mobile to total porosity.

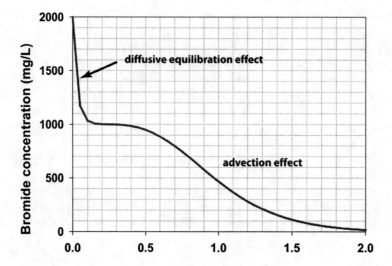

FIGURE 5.14 Breakthrough curve at the center of an injected tracer pulse. At this location, there is a rapid decrease in tracer concentration, reflecting equilibration of solute concentrations between the mobile and immobile pore volumes. In this example, the effects of advective water movement can be seen after 0.4 days.

total porosity ratio. We will make use of this approach in **Chapter 12**, where we provide details of tracer study design and analysis.

    Figure 5.15 provides a comparison of tracer breakthrough curves for high-mass-transfer geometries in mobile-to-total porosity ratios ranging from 1.0 (no immobile porosity) to 33 percent θ_m/θ_t. The groundwater velocity for all these tests was 100 mm/day and the breakthrough curves represent observations from a point 400 mm downgradient of the center of the initial tracer pulse; water originally at the center of the tracer pulse arrived at exactly 4 days for each of the test runs. **Figure 5.15-A** shows the effect of diffusion along the flow axis, lowering the peak concentration and spreading the solute ahead and behind the initial 100-mm pulse width; the peak and the center of mass (indicated by the dashed line for each scenario) arrived at 4 days, consistent with simple advective flow. In **Figure 5.15-B**, there was one immobile cell located adjacent to every third mobile cell, a 75 percent mobile-to-total porosity ratio. The center of mass arrived at approximately 5.33 days, traveling slower than the water velocity. In the succeeding panels, C and D, the center of mass arrival was further delayed in proportion to the decreasing mobile-to-total porosity ratios. Each of the tracer breakthrough curves in **Figure 5.15** would be considered Gaussian, or Fickian, because concentrations appear to be normally distributed around the average arrival time.

    These results are consistent with a diffusion-driven retardation effect on solute transport, observed in each of our pore-scale simulations, as well as larger-scale simulations that we introduce in **Chapter 7**. The **retardation coefficient** can be calculated as follows:

$$Rf_\theta = 1 + \frac{\theta_i}{\theta_m} \qquad\qquad (5.14)$$

The solute center of mass velocity is retarded, relative to the mobile fraction velocity, by diffusive transfers between the mobile and immobile porosities. The retardation factor is applied as follows to calculate the **center of mass velocity**:

$$V_{CofM} = \frac{V_{mobile}}{Rf_\theta} = \frac{V_{mobile}}{1 + \dfrac{\theta_i}{\theta_m}}$$

(5.15)

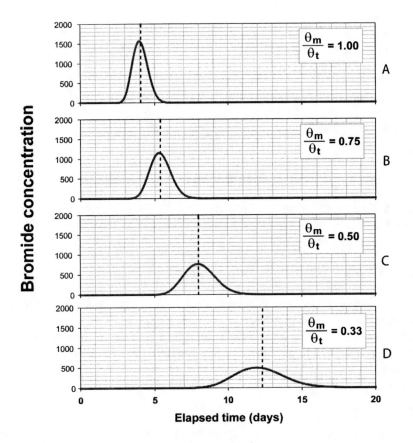

FIGURE 5.15 Comparison of tracer breakthrough curves, Part 1: High mass transfer geometry. These curves are calculated for a location 4 days downgradient from the center of mass of a 100-mm wide, 2,000-mg/L tracer injection. The dashed lines indicated arrival of the center of mass for each tracer pulse. When the entire pore space participates in advective flow ($\theta_m/\theta_t = 1.0$), the tracer pulse center of mass travels at the groundwater velocity. When immobile pore space lies adjacent to the mobile pore space ($\theta_m/\theta_t < 1.0$), diffusive mass interchange slows the center of mass, in exact proportion to the ratio of mobile to total porosities. When the pore space geometry supports mass transfer, the concentration peak arrives simultaneously with the center of mass.

The center of mass velocity coincides with the average groundwater velocity. Rearranging Equation 3.36 from Chapter 3,

$$V_{avg} = \frac{V_{mobile}}{1 + \dfrac{\theta_i}{\theta_m}}$$

(5.16)

In **Chapter 7**, we will compare Rf_{θ} to the retardation effects of equilibrium and kinetic sorption. The magnitude of Rf_{θ} exceeds that of the sorption retardation effects for many natural aquifer settings, because natural aquifer organic carbon content is typically quite low.

When we shift the system to low-mass-transfer geometries, as in **Figure 5.16**, the velocity of the center of mass is unaffected, but the peak velocities move faster, nearly at the advective velocity (compare the peak arrival times for B − D with the 100-percent mobile porosity in A). Diffusive solute migration into immobile pore spaces has spread the tracer breakthrough curves dramatically: first, the lower mass-transfer geometries reduce the rate of solute loss from the tracer pulse, increasing the peak velocity and concentration, relative to high-mass-transfer geometries, and, second, slowing the return of solute to the mobile pore space, causing an asymmetrical tailing of the breakthrough curve. We sometimes refer to the asymmetry as *snail-trailing*. The tracer breakthrough curves in these examples are strongly log-normal and are often referred to as exhibiting non-Fickian or anomalous behavior. There are numerous modeling studies that suggest mechanisms other than diffusion for log-normal behavior, most likely because diffusion effects are often not expected to be significant at large scales with groundwater velocities that far exceed diffusion velocities (i.e., essentially all sites). However, these simulations show clearly that diffusion effects arising out of pore scale interactions may have a significant impact on macroscopic tracer, contaminant and reagent behaviors. We will extend the discussion of diffusion effects into the fine and stratigraphic scales in **Chapter 7**.

When we examine breakthrough curves traveling over longer distances, we see that the initially strong log-normality appears to diminish, although a significant snail-trailing still remains. **Figure 5.17** shows a maturing tracer breakthrough curve for a system with a relatively low mass transfer geometry. Simulation results were logged at five distances from the center of the initial tracer pulse, from 100 to 900 mm downgradient. The initial sharp peak attenuated significantly over the 9-day period. The peak velocity is also decreasing, although it is difficult to discern visually from this type of graph. In **Figure 5.18**, we have compiled peak velocities for two systems with 50 percent mobile porosity: one with a high mass transfer geometry and another with a relatively low mass-transfer geometry (the same system as shown in **Figure 5.17**). From this graph, we can see that the peak velocity is initially near the groundwater velocity and slows, approaching the velocity predicted by Equation (5.14). This pattern is consistent for a wide range of porosity ratios and mass transfer geometries. **Figure 5.19**, for example, shows the same analysis for systems with 33 percent mobile porosity, over a range of mass transfer geometries.

There are now four velocities that may be considered to describe groundwater and solute flow velocities:

- Average groundwater velocity, V_{avg}, the value provided by steady-state pumping tests and averaging analysis of gradients and aquifer matrix properties. Also known as the superficial groundwater velocity.
- Mobile fraction groundwater velocity, V_{mobile}. This is the actual groundwater velocity through the formation and represents the fundamental solutes transporting velocity.
- Solute center of mass velocity, V_{CofM}, which coincides with the average groundwater velocity.

- Solute peak velocity, V_{peak}. This velocity varies along a flow path and its maximum value is determined by the mass transfer geometry of the formation and the diffusivity of the solute. Maximum values for V_{peak} approach those of the mobile pore fraction when the mass transfer geometry discourages diffusive interchanges between the mobile and immobile porosities.

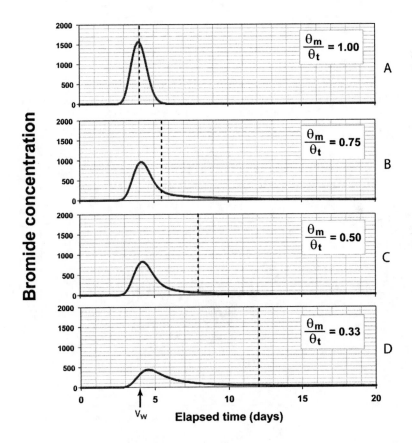

FIGURE 5.16 Comparison of tracer breakthrough curves, Part 2: Low mass transfer geometry. These curves are calculated for a location 4 days downgradient from the center of mass of a 100-mm wide, 2,000-mg/L tracer injection. The dashed lines indicated arrival of the center of mass for each tracer pulse and the arrow indicates the arrival of water from the center point of the tracer pulse. When the entire pore space participates in advective flow ($\theta_m/\theta_t < 1.0$), the tracer pulse center of mass travels at the groundwater velocity. When immobile pore space lies adjacent to the mobile pore space ($\theta_m/\theta_t < 1.0$), diffusive mass interchange slows the center of mass, in exact proportion to the ratio of mobile to total porosities. When the pore space geometry does not support rapid mass transfer between the mobile and immobile pore space, the concentration peak arrives earlier than the center of mass.

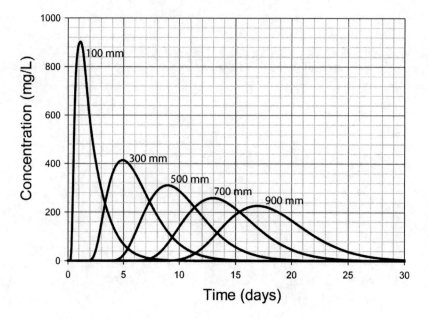

FIGURE 5.17 Breakthrough curves for a low-mass-transfer geometry in a pore structure with 50 percent mobile porosity and 100 mm/day groundwater velocity. The apparent velocity of the peak decreased with distance travelled. Numbers indicate the tracer pulse travel distance for each breakthrough curve.

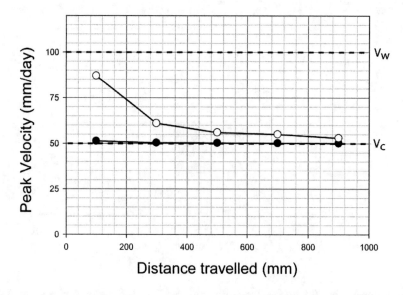

FIGURE 5.18 Changing peak velocities in low and high-mass-transfer geometries. In simulations with high-mass-transfer geometry (solid circles), the peak velocity rapidly matches the center of mass velocity, so the peak and center of mass arrival is effectively simultaneous at all travel distances. In simulations with low-mass-transfer geometries (open circles), the peak velocity is initially much faster than the center of mass velocity, causing significant separation of arrival times for the peak and center of mass. The peak velocity decrease causes the initial time separation to "lock in" for the low-mass-transfer geometry. The upper dashed line indicates the groundwater velocity, V_w, and the lower dashed line indicates the center of mass velocity, V_c, for a 50% mobile porosity system.

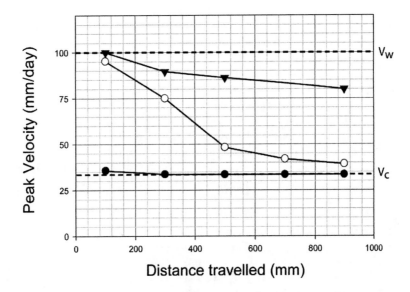

FIGURE 5.19 Changing peak velocities in a 33% mobile porosity system with varying mass-transfer geometries. In simulations with high-mass-transfer geometry (solid circles), the peak velocity rapidly matched the center of mass velocity, so the peak and center of mass arrival was effectively simultaneous at all travel distances. In simulations with low and very low mass-transfer geometries (open circles and triangles, respectively), the peak velocity was initially much faster than the center of mass velocity, causing significant separation of arrival times for the peak and center of mass. The peak velocity decrease caused the initial time separation to "lock in" for the low-mass-transfer geometry. The upper dashed line indicates the groundwater velocity, V_w, and the lower dashed line indicates the center of mass velocity, V_c, for a 33% mobile porosity system.

5.6 AN APPLICATIONS BASED OUTLOOK ON DISPERSIVITY AND DIFFUSIVITY

The advection-dispersion equation and its underlying mechanisms have drawn the attention of theory oriented hydrogeologists for many years, generating a substantial technical literature. We have provided only a cursory survey of the dispersivity arguments, recapitulating much of our own very energetic deliberations on the topic. Our evaluation of these issues was settled by three groups of observations:

- High-resolution field and laboratory studies (our own and many drawn from the literature) consistently find very small values for transverse dispersivity.
- Random processes appear incapable of generating significant transverse dispersivity, as presented above, and
- Diffusive interaction between mobile and immobile porosities provides a very good explanation of the longitudinal dispersivities we observe in reagent pulse and contaminant transport studies.

There are two remediation-relevant outcomes that arise from these conclusions:

1. We cannot rely on transverse dispersivity to spread injected reagents over a wide swath. As depicted in **Figure 5.20**, diffusive interaction between mobile and immobile porosities "smears" each injected reagent pulse. The small amount of transverse dispersivity that may occur generates only a small widening of the injected reagent pulse. Whatever transverse coverage we wish to achieve from an injection must be established through the initial injection – we cannot rely on passive spreading to provide broad coverage. This outcome is examined thoroughly in Chapter 13.

2. Diffusive mass transfer between mobile and immobile pore spaces places a large amount of solute into static storage in contaminated aquifers and in aquifers treated by reagent injection. This generates the "anomalous" (non-Gaussian) behaviors that have perturbed supporters of the advection-dispersion equation and causes the contaminant rebound commonly observed following groundwater restoration efforts. These issues are discussed further in Chapter 7.

Each time we add a high-resolution contaminant or tracer mapping to our experience base, the more clearly we see the tendency for solute bounding volumes to remain small. The formation-dependent element of control over solute migration appears to dominate transverse spreading behaviors and the flow-independent (diffusive) element appears to dominate spreading behavior along the flow axis. These concepts will be developed more fully in succeeding chapters. The challenge regarding formation-dependent and flow-independent transport mechanisms is that we have to give up the convenience and comfort of the predictive calculations offered by the advection-dispersion approach.

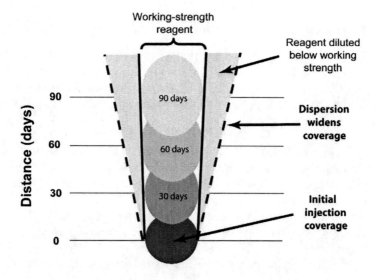

FIGURE 5.20 Limitations on the impact of dispersion in the distribution of reagents in porous media. Although transverse dispersivity carries injected reagent laterally from the flowline, dilution reduces concentrations to less than working strength.

6 Reactive and Sorptive Processes

Advective groundwater flow and aqueous-phase diffusion combine to facilitate contact between dissolved chemical compounds entrained in flowing groundwater and a large fraction of the aquifer's matrix material. Aquifer matrices comprise a wide array of materials, including igneous, metamorphic and sedimentary rock and mineral grains, fragments of plant and animal matter and colloidal masses of partially-decomposed organic matter. Each of these classes of matrix material can interact with some, or all, of the dissolved chemical species entrained in groundwater flow. Sorptive processes may slow the solute mass transport, relative to groundwater movement, and chemical reactions that occur when entrained solutes contact the aquifer matrix may consume or precipitate significant amounts of the mass in transit, or may add dissolved aquifer matrix mass to the transport through dissolution. We provide a brief overview of these processes and direct the reader to Suthersan and Payne (2005) for a more complete description of reactions and sorptive processes associated with *in situ* remediation engineering.

6.1 SORPTIVE PROCESSES

The sorptive interactions of greatest interest in remediation hydrogeology occur between hydrophobic organic compounds dissolved in groundwater and solid or colloidal organic matter of the aquifer matrix, or between dissolved cations and points of negative charge concentration on aquifer minerals (cation exchange). The sorption reactions are classified as **adsorption** or **absorption**, based on whether the organic matter behaves like a solvent (absorption) or, alternatively, provides specific binding sites that immobilize (adsorb) the dissolved molecules. The energy released in absorption is relatively low and the process is best characterized as fully reversible equilibrium partitioning of the solute between the aqueous and organic matter phases. There is an implicit assumption that absorbed solute molecules move freely in the organic matter phase. Adsorption is the more energetic reaction, an exothermic binding reaction between a solute molecule and a binding site on the organic matter. Adsorbed solute molecules are effectively pinned in place and do not move freely through the binding matrix.

ADSORPTION

The adsorptive binding process for any pairing of binding matrix and solute is characterized by the number of available binding sites on the matrix and the energy released upon binding. Because binding occurs at specific sites on the sorbing matrix, the adsorptive binding capacity is limited. Further, although the binding reaction for any matrix-solute pairing is exothermic, the reaction is not entirely irreversible (although adsorption is sometimes referred to as irreversible). A small fraction of the molecular population may have sufficient energy to break the binding, generating a very small rate of desorption, relative to adsorption. Alternatively, a different solute, with a stronger binding reaction, may displace a previous, weaker binding (cation exchange is an example of displacement of a weaker pairing).

The equations that describe adsorption pairings are termed isotherms, and two such equation forms are used in remediation hydrogeology. The Freundlich isotherm relates the aquifer soil (C_{soil}) and groundwater (C_{aq}) concentrations with a distribution coefficient, K_d, and is non-linear, as shown in Equation (6.1).

$$C_{soil} = K_d \cdot C_{aq}^b \quad \frac{mg}{kg} \tag{6.1}$$

The second is the Langmuir adsorption isotherm [Equation (6.2)], a special case of the Freundlich isotherm, in which the exponent, b, is 1.0.

$$C_{soil} = K_d \cdot C_{aq} \quad \frac{mg}{kg} \tag{6.2}$$

For both isotherms, the distribution coefficient, K_d, has units of L/kg and the aqueous phase units are mg/L.

Adsorption reactions may occur between dissolved hydrophobic organic molecules and inorganic aquifer matrix minerals or organic matter broadly classed as "hard carbon" forms. Materials that fall into the hard carbon class include coal fragments and pieces of wood char and soot from forest fires, materials that are unlikely to be measured in total organic carbon analyses. At this time, there is no commercial analysis that effectively captures the refractory carbon fractions. Luthy, Aiken et al. (1997) provide an extensive assessment of adsorbing and partitioning materials encountered in aquifers and sediments.

ABSORPTION

Equilibrium partitioning of hydrophobic organic compounds between groundwater and sorbing aquifer organic matter, absorption, is the sorption process most commonly considered in contaminant fate and transport calculations — the laboratory analysis of "soft" carbon sorbents in aquifer materials is commercially available and equilibrium partitioning provides a more conservative assessment of contaminant availability in the aqueous phase. Equation (6.3), describing equilibrium partitioning, was drawn from the U.S. EPA Soil Screening Guidance (EPA 1996),

$$C_{sorbed} = K_{oc} \cdot f_{oc} \cdot C_{aq} \tag{6.3}$$

in which K_{oc} is the organic carbon partition coefficient and f_{oc} is the organic carbon fraction of the aquifer matrix (a value that is estimated from total organic carbon laboratory analysis). Note that this series of calculations is developed assuming that the groundwater concentration, C_{aq}, and organic carbon fraction have been measured and that the organic carbon partition coefficient has been taken from published values, such as those in **Table 6.1**.

The sorbed mass (mg/m³) can be estimated from Equation (6.4), in which ρ_{bulk} is the aquifer matrix dry bulk density (assume 1,800 kg/m³ as a typical value).

$$Mass_{sorbed} = C_{aq} \times K_{OC} \times f_{OC} \times \rho_{bulk} \quad \frac{mg}{L} \times \frac{L}{kg_{org}} \times \frac{kg_{org}}{kg_{soil}} \times \frac{kg_{soil}}{m_{soil}^3} \tag{6.4}$$

TABLE 6.1
Values for soil organic carbon/water partition coefficients for common chlorinated alkenes, from US EPA (1996a), Table 39. All values were measured, unless otherwise noted.

Compound	K_{oc} (L/kg)
Chlorinated Alkenes	
Cis-1,2-dichloroethene	36[a]
1,1,1-trichloroethane	139
Trichloroethene	94
Perchloroethene	265
Vinyl chloride	19[a]
Aromatics	
Benzene	62
Ethylbenzene	204
Toluene	140
o-xylene	241
m-xylene	196
p-xylene	311
naphthalene	1,191

a Estimated value

The mass in the aqueous phase is simply the groundwater concentration, multiplied by the total saturated porosity, θ_w.

$$Mass_{aqueous} = C_{aq} \times \theta_w \qquad \frac{mg}{L} \times \frac{L}{m^3_{soil}} \tag{6.5}$$

It is also possible to express the expected groundwater concentration from a total aquifer matrix analysis (C_{total}), as given in Equation (6.6).

$$C_{aq} = \frac{C_{total}}{\left[(K_{OC} \cdot f_{OC}) + \dfrac{\theta_W}{\rho_{bulk}} \right]} \qquad \frac{mg}{L} \tag{6.6}$$

The sorbed and aqueous masses can be summed and the sorbed expressed as a fraction of the total contaminant mass, over a range of aquifer matrix carbon fraction values. **Figure 6.1** shows the sorbed fraction for several compounds of interest in groundwater restoration. This figure shows that compounds with high organic carbon partitioning values reside predominantly in the sorbed phase, whenever organic carbon is available in the aquifer matrix.

The soft carbon mass that supports equilibrium partitioning includes humic substances, natural biogenic oil droplets and partially decomposed plant and animal tissues. Organic carbon fraction values in the uppermost 1 m of a soil formation are typically greater than 1 percent ($f_{oc} > 0.01$). In hydric soil groups, the surficial value is likely to exceed 10 percent and in desert areas, the surface value may be somewhat lower than 1 percent. These materials are at least partially degradable in oxygenated habitats and our aquifer matrix sample analyses typically show very low organic carbon fraction values compared to those near the soil surface, often less than 100 mg/kg total organic carbon, or $f_{oc}<0.0001$. For example, Rivett, Feenstra et al. (2001) reported f_{oc} values of 0.00021 for their study area in the Borden aquifer. We have observed the use of default assumptions for organic carbon fraction that exceed the values we typically encounter by 10-fold, or more. This aquifer parameter is too critical to presume without sample collections, and the default assumption for aerobic aquifers should be very low ($f_{oc} = 0.0005$, or lower) to be conservative.

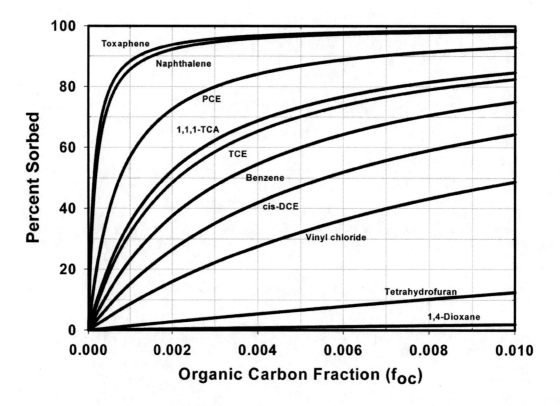

FIGURE 6.1 Sorbed (non-aqueous) fraction of selected hydrophobic organic compounds expressed as a function of aquifer organic carbon content.

Because the migration of any solute molecule stops during its time spent in the soft carbon phase, its velocity will be retarded, relative to the *average* groundwater velocity. The retardation has been approximated by:

$$Rf_{oc} = 1 + \frac{\rho_{bulk}}{\theta_{total}} \cdot K_{oc} \cdot f_{oc} \qquad (6.7)$$

where ρ_{soil} is the soil dry bulk density (1.8 g/cm³). The bulk density must be expressed in g/cm³ for the Rf_{oc} to generate "correct" results, even though other units for bulk density (kg/m³, in particular) might appear dimensionally consistent. The net **solute velocity** is given by:

$$V^*_{solute} = \frac{V_{avg}}{\left(1 + \dfrac{\rho_{soil}}{\theta_{total}} \cdot K_{oc} \cdot f_{oc}\right)} \left(\frac{cm}{s}\right) \qquad (6.8)$$

The term V* is used to distinguish sorption-retarded from porosity-retarded velocity estimates. Both diffusion (porosity) and sorption-driven velocity retardation occur as a result of time spent off the mobile flow path. *Both of the retardation factors apply to individual solute molecules, not to the net solute flow through the mobile path* — this is a very important distinction. The retardation of individual molecules has a noticeable effect on overall concentrations at the leading edge of a solute pulse. However, when solutes in the mobile flow path reach equilibrium (or near equilibrium) distribution, between dissolved phase concentrations in the mobile and immobile porosities, or between dissolved phase concentration and soft carbon mass in the mobile flow path, *there is no further net retardation of mass flux*. The retardation factors should not be applied to mass flux estimates — they are only valid for estimating the rate of migration of the leading edge of a solute pulse into a clean aquifer, or a clean water front entering a contaminated aquifer.

Calculation example — Perchloroethene is a common groundwater contaminant and its organic carbon partition coefficient is 265 L/kg (refer to **Table 6.1**). If the formation dry bulk density is 1.8 g/cm³, the organic carbon content is 100 mg/kg (f_{oc} = 0.0001) and the average groundwater velocity is 30 cm/day (values commonly observed in productive aquifer formations throughout the world), the perchloroethene velocity estimate would be calculated as follows:

$$V^*_{solute} = \frac{30}{1 + \dfrac{1.8}{0.3} \cdot 265 \cdot 0.0001} = 26 \frac{cm}{day} \qquad (6.9)$$

If the organic carbon content increased to 1,000 mg/kg (f_{oc} = 0.001), the estimated perchloroethene velocity would decrease to 12 cm/day. For 100 mg/kg organic carbon and a K_{oc} of 94 L/kg (corresponding to trichloroethene) the velocity would increase to 28 cm/day, representing a very small retardation factor. It is clear from these comparisons and from **Figure 6.1** that the organic carbon content is a dominant determinant in the expected retardation of solute velocities for these and other hydrophobic organic solutes.

The standard estimation for solute velocity, given in Equations (6.7) and (6.8), is based on average velocities, overlooking the role of mobile porosity in groundwater flow. If we inject a hydrophobic organic compound into an aquifer with mobile and immobile porosities, the observed retardation coefficient probably won't make sense, because *solute pulses can move faster than the average groundwater velocity*. Such an observation conflicts with the retarded velocity calculated from the average groundwater velocity - it would require a negative retardation coefficient. In fact, this is exactly what was observed at the Borden Aquifer, in studies reported by Rivett, et al. (2001). They reported on the contamination plume that developed from an emplaced DNAPL source comprising chloroform, trichloroethene and perchloroethene. The average groundwater velocity in their study area was 8.5 cm/day and the chloroform and trichloroethene peaks traveled at approximately 15 cm/day. The estimated retardation factors Rf_{oc} for the site were: chloroform = 1.06, trichloroethene = 1.10 and perchloroethene = 1.29. Information was not developed to estimate the mobile fraction at the study site, so it wasn't possible to develop an expected value for actual contaminant velocity (which is controlled by groundwater velocity in the mobile pore fraction). However, we can work with the available data to develop an estimate of the mobile pore fraction. First, we know that the observed velocity represents a retardation of the mobile velocity, hence:

$$V_{observed} = \frac{V_{mobile}}{Rf_{oc}}$$
(6.10)

Based on the report of the Rivett group, $V_{observed}$ = 15 cm/day for chloroform and the site-specific Rf_{oc} for chloroform was 1.06. The estimated mobile fraction velocity, V_{mobile}, was 16 cm/day. Recalling Equation (3.36),

$$V_{mobile} = \frac{V_{avg}}{\dfrac{\theta_m}{\theta_t}}$$
(6.11)

and we can, therefore, conclude from this analysis that the mobile porosity was approximately 50 percent of the total porosity.

That is not the end of the analysis, however. We have not considered the contribution of porosity-driven retardation on $V_{observed}$. Recalling Equation (5.14),

$$Rf_\theta = 1 + \frac{\theta_i}{\theta_m}$$
(6.12)

For the 50-percent mobile porosity estimate, Rf_θ would be at least 2.0 (assuming mass transfer occurred between mobile and immobile pore fractions), much larger than Rf_{oc} in this case. We can't do any more reconstructive analysis without adding assumptions, although we can be assured that the actual value for V_{mobile} is higher than 16 cm/day, possibly significantly higher.

Summarizing the important issues related to sorption processes:

- Sorption of dissolved hydrophobic organic compounds from groundwater into organic matter in the aquifer matrix provides off-line storage of solute (contaminant) mass;
- Adsorption of hydrophobic solutes by hard carbon matter is only marginally reversible - the rate of adsorption is much greater than the rate of desorption. The storage capacity in this compartment is strictly limited and cannot be estimated from commercially-available laboratory analysis.

- Equilibrium partitioning (absorption) occurs between hydrophobic solutes and soft carbon matter and is fully reversible. The storage in this compartment is estimated by the total organic carbon (mg/kg TOC) analysis, which is commercially available.
- When solute molecules are absorbed, they drop out of the flow path and their migration rate slows, relative to the groundwater velocity. Increasing the organic carbon fraction increases this retardation.
- Although the velocity of the leading edge of a contaminant plume is slowed, relative to the groundwater velocity, after a plume reaches steady state, there is no further net retardation. This is important in correctly estimating mass flux — do not apply retardation factors to steady-state mass flux estimates.
- Even though retardation affects solute velocities, hydrophobic solutes often travel faster than the average groundwater velocity, because it is the velocity in the mobile porosity — not the average velocity — that determines the mass transport velocity.

MULTI-COMPARTMENT SORPTION MODELS

The effects of adsorption and absorption can be combined into a multi-compartment model, such as that developed by Chen, et al. (Chen, Kan et al. 2002)(2002). In the Chen dual equilibrium model, the soil concentration, q, is the sum of values predicted by equilibrium partitioning (q^{1st}) and isotherm adsorption (q^{2nd}), as indicated by Equation (6.13).

$$q = q^{1st} + q^{2nd} \quad \frac{mg}{kg} \tag{6.13}$$

The equilibrium partitioning value is calculated as above and is given by Equation (6.14).

$$q^{1st} = K_{oc}^{1st} \cdot f_{oc} \cdot C_{aq} \quad \frac{mg}{kg} \tag{6.14}$$

Chen et al. (Chen, Kan et al. 2002)(2002) used Equation (6.15) to approximate the adsorbed mass.

$$q^{2nd} = \frac{K_{oc}^{2nd} \cdot f_{oc} \cdot f \cdot q_{max}^{2nd} \cdot C_{aq}}{f \cdot q_{max}^{2nd} + K_{oc}^{2nd} \cdot q_{max}^{2nd} \cdot C_{aq}} \quad \frac{mg}{kg} \tag{6.15}$$

They provided a "universal" empirical value for the second organic carbon partition coefficient, which drives the isotherm portion of the process:

$$K_{oc}^{2nd} = 10^{5.92 \pm 0.16} \quad \frac{L}{kg} \tag{6.16}$$

From that value, Equation (6.17) provides an estimate of the adsorbed-phase mass.

$$q_{max}^{2nd} = f_{oc} \cdot \left(\frac{K_{oc}^{1st}}{0.63} \cdot C_{sat} \right)^{0.534} \quad \frac{mg}{kg} \tag{6.17}$$

Equation (6.18) combines the two mass estimates.

$$q = K_{oc}^{1st} \cdot f_{oc} \cdot C_{aq} + \frac{K_{oc}^{2nd} \cdot f_{oc} \cdot q_{max}^{2nd} \cdot C_{aq}}{q_{max}^{2nd} + K_{oc}^{2nd} \cdot f_{oc} \cdot C_{aq}} \quad \frac{mg}{kg} \tag{6.18}$$

Figure 6.2 shows the effects of the dual-compartment model for perchloroethene, relying on the empirical value for K_{OC}^{2nd} given by Chen, et al. (2002) and the value for K_{OC}^{1st} provided in **Table 6.1**. In this model, adsorption is the dominant mechanism at concentrations up to 0.001 mg/L perchloroethene and is a significant factor through concentrations up to 10 mg/L. The equilibrium partitioning is dominant only between 10 and 110 mg/L, a small fraction of the range of concentrations typically encountered for perchloroethene. The relationship predicted by equilibrium partitioning represents the expected groundwater concentrations much more conservatively, as indicated by the line labeled "D" on **Figure 6.2**.

The dual equilibrium desorption and other multi-compartment solute behavior models are built on an implicit assumption of homogeneous, isotropic aquifer matrices. **Figure 6.3** provides an updated understanding of the various compartments, in a more realistic, multi-porosity aquifer setting. The slow-release component of the dual equilibrium model calculated by Chen, Kan et al. (2002) from field observations could be, at least partly, a result of simple diffusive interaction between mobile and immobile porosities. We now understand that a solute behavior model built on sorption processes, alone, cannot express all the important interactions between solutes and

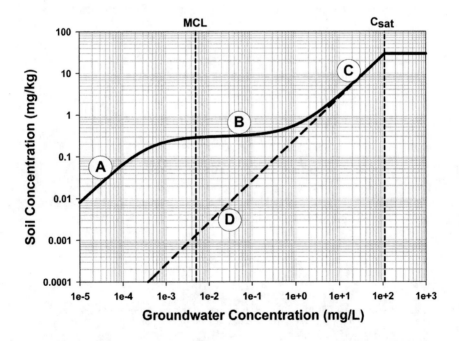

FIGURE 6.2 Dual-compartment sorption model for perchloroethene, showing Langmuir isotherm behavior (A), equilibrium partitioning (C) and transition from isotherm to partitioning (B). The line labelled (D) shows the predicted relationship from equilibrium partitioning, alone. C_{sat} is the estimated solubility limit and MCL indicates the current drinking water limit for perchloroethene.

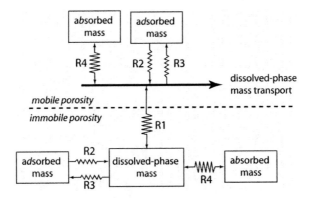

FIGURE 6.3 A multi-resistance conceptual model that combines the effects of sorption processes and diffusion on solute mass transport in a multi-porosity matrix. R1 represents diffusion as a resistance to solute movement between mobile and immobile porosities. R2 and R3 represent kinetic adsorption and desorption, respectively (R2 >> R3). R4 represents reversible (equilibrium) sorption.

aquifers. Moreover, the typically low values we observe for organic carbon fraction in many aquifers suggests that diffusive interactions between mobile and immobile porosities may be the more dominant factor in solute distribution and transport behavior.

6.2 PRECIPITATION

Precipitation reactions may be used as a remedial tool, to sequester dissolved inorganic contaminants such as heavy metals, or may occur as an indirect product of various chemical or biological treatment processes. For remediation hydrogeology, precipitate formation has the potential to decrease aquifer permeability and mass transfer between mobile and immobile porosities and that brings these reactions to our attention.

Figure 6.4 provides an interesting start for a discussion of precipitation reactions in the subsurface. This is a photograph of a laboratory soil column that was dry-packed with a fine sand. The column was perfused with an alcohol (to stimulate microbial productivity and develop an iron-reducing environment) and ferrous sulfate. Iron sulfides precipitated along the high-permeability flow pathways that developed as a result of the packing method. The dark coloration of the sulfide minerals shows where the precipitates were deposited and calls our attention to the concentration of flow in a small portion of the matrix volume. As we discussed in Chapter 3 (where we introduced mobile and immobile porosity concepts) and will reinforce in Chapter 12 (with the results of field tracer studies), groundwater flow in aquifers is concentrated in only a portion of the total matrix porosity, much like the flow was concentrated in the laboratory column of **Figure 6.4**.

There are several possible effects of precipitation in the mobile pore spaces:

Reduced hydraulic conductivity — The occupation of pore space by precipitates causes a reduction in the hydraulic conductivity;

Increased flow velocities — For permeable reactive barriers that are constructed to be much more conductive than the surrounding aquifer, local flow velocities may increase as a result of the reduced cross-section, reducing contact time with the formation;

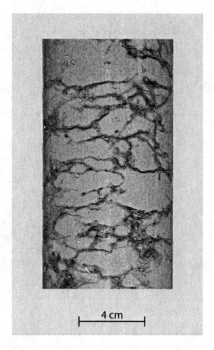

FIGURE 6.4 Reactive iron precipitation (dark areas) in a laboratory soil column. The column was packed by spooning fine sand into a dry column, causing uneven compaction and the formation of strongly preferred pathways through the matrix. The injected reagent followed the more permeable pathways and precipitation quickly occluded them.

Reduced mass transfer — Precipitates may block the interfacial contact between mobile
   and immobile porosity zones, reducing the mass transfer rates.

Researchers have focused on two remedial technologies that are susceptible to precipitation – *in situ* chemical oxidation by permanganate and permeable reactive barriers that use zero-valent iron as a reactant. In each of these cases, mineral precipitates form as a result of the chemical reactions, potentially blocking the flow of groundwater. Other *in situ* reactive zone strategies are designed to cause chemical precipitation of contaminants, especially metals, and these are susceptible to precipitation limitations, as well.

Permeable reactive barriers are typically built from mixtures of sand and zero-valent iron, and are built perpendicular to groundwater flow. Their effectiveness depends on maintaining contact between the reactive iron and contaminants in the groundwater flowing through the barrier. Dissolved minerals that are in equilibrium with the geochemistry upgradient of the barrier can be supersaturated with respect to the geochemistry within the barrier and are likely to precipitate. If the volume of precipitate formation is large, groundwater velocities can increase (reducing contact time) and the reactive surfaces can be passsivated. Li, et al (2005) provided an estimate of the hydraulic conductivity reduction associated with reductions in matrix porosity associated with mineral precipitation:

$$K' = \frac{K_0 \left[\dfrac{n_0 - \Delta n}{n_0} \right]^3}{\left[\dfrac{1 - n_0 + \Delta n}{1 - n_0} \right]^2} \tag{6.19}$$

in which K_0 and n_0 are the pre-existing hydraulic conductivity and total porosity (θ_t), respectively, and Δn is the reduction in total porosity generated by mineral precipitation. Li, et al (2005) observed that passive reactive barrier longevities are satisfactory, based on current field experience, although the rate of permeability decline predicted by Equation (6.19) increases as the porosity reduction, Δn, increases. Because permeable reactive barriers are typically constructed to be much more permeable than the surrounding aquifer, the reduction in permeability due to precipitation is manageable.

Remedial technologies that rely on the formation of mineral precipitates in natural aquifer formations are likely to be much more susceptible to blockage of the formation by precipitation. We used Equation (6.19) to generate a series of curves showing the reduction in hydraulic conductivity that would be expected to result from loss of porosity, for a range of mobile porosities. The results of these calculations are shown in **Figure 6.5**. For a permeable reactive barrier that might be constructed with a mobile porosity, θ_m, of 0.30, a precipitation loading of 5 percent of the aquifer volume only reduces the hydraulic conductivity by 50%. That is a manageable reduction, if the reactive barrier is sufficiently high in conductivity, relative to the surrounding formation.

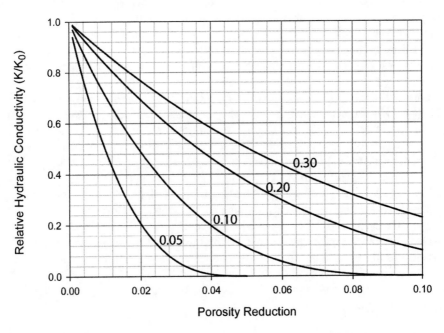

FIGURE 6.5 Reduction of hydraulic conductivity that results from reduction of porosity due to mineral precipitation, for initial mobile porosities ranging from 0.05 to 0.30. The calculations are made from Equation (6.19) and assume that the mobile porosity is the dominant contributor to observed hydraulic conductivity. Porosity reductions are in absolute units (0.10 represents a 100-percent reduction of porosity for the 0.10 initial mobile porosity).

If the original mobile porosity is in the range we typically observe in natural aquifers (0.02 to 0.10), the hydraulic conductivity will be much more sensitive to precipitation reactions. If the aquifer mobile porosity, θ_m, is 0.10, a precipitation load of 0.05 would reduce the hydraulic conductivity 10-fold. It is important to note that chemical precipitation reactions normally occur quite rapidly and as a result, they are most likely to occur in the mobile pore spaces — the reactions occur before significant amounts of reagents can diffuse into the immobile porosity. Precipitation is likely to be focused in the mobile pore spaces, as was observed in the laboratory column in **Figure 6.4**.

It is possible to calculate the pore space that could be occupied by mineral precipitates, if we know their molar volumes. However, for many of the reagents applied to aquifers, there are multiple precipitate species that might form. Moreover, the initial amorphous precipitates can have a higher volume than the mature crystals that form over time. There are a few strategies that could be employed to determine potential porosity derived limitations on precipitation reactions:

- Estimate molar volumes of known precipitates and calculate the pore volume that would be occupied at the design loading, using Equation (6.19);
- Conduct bench-scale studies to observed precipitation blockage — it is difficult to pack columns to accurately represent natural aquifer matrices, as the experience related in **Figure 6.4** attests;
- Conduct field trials — this is the more certain method of determining the potential for formation blockage. Field trials faithfully replicate the aquifer chemistry, which is difficult to match at the bench scale. Mineral precipitates can be identified and quantified through post-treatment aquifer matrix sample collections, eliminating speculation regarding which minerals were formed.

Many *in situ* reactive zone precipitation projects are intended to treat relatively low concentration target compounds and the mass of precipitate formation would be quite small. The *in situ* precipitation of 1 mg/L of hexavalent chromium, for example, generates such a small mass of precipitate that there is no need to consider the potential for blockage. If a quasi-permanent reactive barrier is under consideration, or if the target is a DNAPL source mass, with the potential to form tens of kilograms of precipitate per cubic meter of aquifer, there is clearly a need to consider the impacts of precipitation.

Several studies have focused specifically on the blockage of aquifer matrices by permanganate oxidation (e.g., Seigrist, et al., 2002; Li and Schwarz, 2004a, 2004b, and 2004c). The consensus of these studies was that the manganese dioxide, formed as a by-product of permanganate oxidation, was a source of permeability loss and that the rate of loss was directly related to the permanganate solution strength and the total mass of reagent consumed (related to the target compound mass and the loss of reagent to side-reactions with aquifer minerals and non-target organics). In the studies reported by Seigrist, et al. (2002) an important source of particulate matter was *ex situ*, from pre-reacted permanganate (the system was recirculation-based and trichloroethene from the extraction side reacted with added reagent before reaching the aquifer matrix). The permanganate precipitates were likely to be manageable in the treatment of dissolved-phase chlorinated solvent mass, but the buildups associated with attempts to treat large DNAPL source mass are likely to be quite large and significant reductions in permeability are to be expected. Li and Schwarz (2004b, 2004c) suggest that MnO_4 precipitates can be removed by organic acid flushing, although we are not aware that this has been attempted in the field.

6.3 GAS-GENERATING REACTIONS

Many of the chemical and biological processes we deploy in the subsurface generate large quantities of gaseous by-products and these gases may have a very large impact on the hydraulic conductivity of the treated formation. Enhanced reductive dechlorination generates carbon dioxide and methane gases, *in situ* chemical oxidation generates carbon dioxide and, in the case of Fenton's Reagent, large quantities of oxygen (refer to **Figure 4.1**). Although the impact of gases on hydraulic conductivity is often quite large, they are normally temporary, in contrast to the loss of conductivity associated with mineral precipitation reactions, discussed above.

Large quantities of gas can be formed from relatively small masses of dissolved or liquid reagent. Recalling the ideal gas law,

$$P \cdot V = n \cdot R \cdot T \quad atm \cdot L = mol \cdot \frac{L \cdot atm}{mol \cdot {}^\circ K} \cdot {}^\circ K \tag{6.20}$$

where P is the pressure in atmospheres, V is the gaseous volume, n is the number of moles, R is the universal gas constant (8.206×10^{-2} L·atm/mol·K) and T is the system temperature in Kelvin (°C + 273). At 10°C and 2 atm pressure (equivalent to 9.8 m below an aquifer surface), the volume of 1 mol of gas is 12 L. That volume expands to 24 L as the gas migrates to the aquifer surface, where the total system pressure is approximately 1 atm.

As an example, we will calculate the oxygen gas volume that could be generated by a Fenton's Reagent treatment that is injecting 30 percent peroxide at 4 L/min into an aquifer, 9.8 m below the aquifer surface. We will assume that half of the peroxide is lost to the disproportionation (oxygen formation), and the balance is productively consumed in oxidation reactions that generate oxidized products, including carbon dioxide. This calculation will focus on the oxygen producing reaction:

$$2H_2O_2 \xrightarrow{\ metal\ } 2H_2O + O_2\,(g)$$

A 30-percent hydrogen peroxide solution contains 300 g peroxide per liter and the molecular weight of peroxide is 34 g/mol; the injection rate of 4 L/min is placing 35 moles of hydrogen peroxide, per minute, into the aquifer and the molar yield is 50 percent, or 17.5 moles per minute of oxygen, created *in situ* by reaction of hydrogen peroxide with aquifer matrix minerals (and other disproportionation triggers in the oxidation *milieu*). From Equation (6.20), this translates to a gas generation rate of 210 L/min that expands to 420 L/min (15 cfm) at the aquifer surface. A one gallon-per-minute (4 L/min) peroxide injection can generate oxygen at a rate matching a high-flow aquifer sparge well.

When we calculate the gas generated by conversion of dissolved or liquid reagents for any of the gas generating technologies (oxidation, reduction or biological) they all generate large volumes that displace groundwater from the aquifer matrix. In Chapter 4 (**Figure 4.15** and associated equations), we showed the calculation of permeability reductions caused by air displacement of groundwater and it's clear that these processes can have a very large, temporary impact on groundwater transport.

6.4 MOBILIZATION OF COLLOIDS

At most contamination sites, colloid-size particles are immobile, as long as our monitoring wells are properly developed and our sampling methods are not too violent. The colloid particle fraction is generally defined as falling between 0.45-micron and 0.10-micron filtrations and this size range may include extremely small mineral particles, freshly precipitated reagents (amorphous crystalline

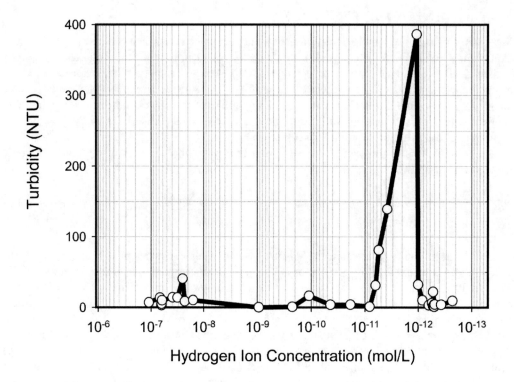

FIGURE 6.6 Effect of pH (shown as hydrogen ion concentration) on the mobility of colloids, reflected in measurements of turbidity in low-flow sampling. The site is in the Midwestern United States and contains unreacted calcium oxide residues in some areas. Turbidity increased as the H+ molarity declines (pH increased), reflecting decreased electrostatic retention of colloids on aquifer minerals. When the H+ falls below 10^{-12} M, the ionic strength increase associated with increasing hydroxyl anion causes the colloids to salt out.

forms that will mature into larger particles over time) and organic matter such as humic substances and kerogen. These particles are normally bound to soil particles, but they can be mobilized at high pH, where the surface charge of soil minerals becomes suppressed. **Figure 6.6** shows monitoring well turbidity measurements from a site in the Midwestern United States, where some portions of the aquifer were at high pH due to the disposal of calcium oxide containing material. Turbidities were generally low, until the hydrogen ion concentration fell to 10^{-11}M (pH 11), where a rise in turbidity was noted. The effect increased to pH 12, above which the turbidities dramatically decreased (the high ionic strength of the groundwater caused coagulation of the colloids).

Sites that have high pH impacts, and sites where pH adjustments are part of the remedial strategy, are susceptible to colloid mobilization. In some cases, colloid movement can facilitate transport of sorbed contaminants that would otherwise be immobile (PCBs and metals, for example) and consideration should be given to managing these potential impacts as part of remedy planning.

7 Contaminant and Reagent Mass Transport

> "I consider it certain that we need a new conceptual model, containing the known heterogeneities of natural aquifers, to explain the phenomenon of transport in groundwater."
>
> *Charles V. Theis[1]*

The long-standing conceptual model of groundwater flow, built on Darcian averaging over representative elementary volumes, provides the groundwater development industry with unquestionable predictive value. Conversely, the advective-dispersive conceptual model of contaminant and reagent mass transport, built on the averaging Darcian framework, has not provided such a reliable representation of mass transport processes for the scales at which we track contaminant migration and apply remedial processes. In contaminant and reagent mass transport, the distinctions between Darcian hydrogeology and the heterogeneous domain of remediation hydraulics become clearly evident and critically important. As suggested 40 years ago by Professor Theis, heterogeneities in aquifer structure play a dominant role in the migration and storage of solutes in aquifers and to obtain acceptable predictive power, our conceptual model of contaminant and mass transport must account for heterogeneous aquifer structure. In this chapter, we examine physical and chemical mechanisms that govern solute migration between flowing and essentially static groundwater, the mobile and immobile porosity. Although we can describe the distributive mechanisms and build conceptual site models that are sensitive to aquifer heterogeneities, there is no way to escape the requirement that we map aquifer structure as a foundation for understanding contaminant distribution patterns at each site.

There are four premises of remediation hydraulics, laid down in earlier chapters, that we now apply in earnest:

1) Classical aquifer characterization measurements are superficial averages of aquifer conditions at large scales and, when we examine aquifer structure in detail, we observe various complex patterns of fast and slow-moving groundwater. Darcy's Law can be applied to smaller aquifer segments very effectively (**Chapter 3**).
2) Transverse dispersivity is very limited in natural aquifers and the tendency of solutes is to flow in tightly-focused pathways, following patterns of higher hydraulic conductivity, embedded at the time of deposition (**Chapter 5**).
3) Diffusive interaction between solutes in fast-moving (mobile) and slow-moving or static (immobile) groundwater slow the overall migration rate of solutes and spreads the solute mass along the flow axis. Diffusion that continues over extended periods can push a large solute mass into slow-moving segments of an aquifer (**Chapter 5**).

---

[1] Theis, C.V. 1967. Aquifers and models. Proc. Symp. on Groundwater Hydrology, Amer. Water Resources Assoc., p 138.

4) Sorption of hydrophobic organic compounds provides additional mass storage in the aquifer matrix, although the magnitude of this storage may be smaller than the dissolved-phase mass stored in immobile porosity (**Chapter 6**).

Groundwater contaminant plume and injected reagent behaviors that we observe in the field are consistent with a conceptual model of contaminant mass transport built on these premises. Nonetheless, aquifer heterogeneities and anisotropies defeat attempts to model contaminant distribution without first developing a detailed aquifer structure mapping, validated and refined through tracer studies.

7.1 AQUIFER MASS STORAGE CAPACITIES

At this point, we have identified the storage compartments that account for the bulk of contaminant or reagent that may reside in an aquifer matrix. As a preface to discussions of mass transfer and plume propagation, it is useful to quantify and visualize the mass that can be stored in a source zone and in the resulting dissolved-phase plume, for hydrophobic, as well as hydrophilic, contaminants and reagents. We will introduce four types of source mass for consideration:

Neat chlorinated solvent — A neat chlorinated solvent liquid, such as perchloroethene or trichloroethene, forms a DNAPL that is hydraulically distinct from groundwater (density, viscosity, surface tension). Examples of neat solvent spills include tank leaks or dry cleaner releases, which have relatively little non-solvent mass content.

Chlorinated solvent/waste oil mixture — Many large chlorinated solvent plumes originate from the loss of solvent waste mixtures that contain the original solvent, along with oil and grease and other materials dissolved in the solvent. These materials may be near neutral density, with relatively high viscosities.

Soluble inorganic salt — The large storage capacity offered by immobile porosity generates persistent sources of soluble inorganic salts, such as ammonium perchlorate, at the location of a concentration solution spill.

Miscible organic compounds — Immobile porosity also provides large storage capacities for organic compounds that are fully miscible in groundwater, such as alcohols, ketones and 1,4-dioxane.

It was necessary to develop a common measurement for each of these sources, to reflect the mass in each compartment. We settled on the approach used by Suthersan and Payne (2005), which expresses distributions as mass per aquifer volume (e.g., kg/m^3). **Table 7.1** provides a summary of the mass storage capacities for several compounds of potential interest, including neat solvents, solvent-oil mixtures, miscible organics and a soluble salt. The residual NAPL was assumed to occupy 2 percent of the aquifer volume, in the same range as the emplaced NAPL source studied by Rivett, et al. (2001).

The data given in **Table 7.1** were used to generate mass distribution curves for a generalized contaminant plume structure (**Figure 7.1**), using a hydrophobic organic compound and a soluble salt as examples. Perchloroethene was used to represent a hydrophobic organic compound (**Figure 7.2**) and ammonium perchlorate was used as the representative soluble salt (**Figure 7.3**). The contaminant plume was assumed to have developed from a concentrated source mass embedded in the aquifer, with zones of decreasing dissolved-phase saturation extending along the groundwater flow axis.

TABLE 7.1

Estimates of maximum attainable mass (g/m³) in each of the five contaminant reservoirs in a cubic meter of aquifer matrix containing residual NAPL at 2 percent of the aquifer volume.

Compound	Aqueous Solubility (g/L) At 20 C	Aqueous-Phase Compartments		Sum of Aqueous-Phase	Aqueous-Phase Mass at MCL	Sorbed-Phase Compartments		Sum of Sorbed and Aqueous Compartments At Saturation	Non-Aqueous-Phase Compartment Residual NAPL @ 2 % of aquifer volume	Sum of All Compartments At Saturation
		Immobile	Mobile			Adsorbed	Absorbed			
Hydrophobic organics										
Perchloroethene	0.2	40	20	60	0.0015	0.25	48	108	32,400	32,508
Trichloroethene	1.1	220	110	330	0.0015	0.25	93	423	29,200	29,623
cis-Dichloroethene	3.5	700	350	1050	0.021	0.25	113	1,163	25,200	26,363
1,1,1-Trichloroethane	1.3	266	133	399	0.060	0.25	163	562	26,800	27,362
Naphthalene	0.031	6	3	9	NA	0.25	33	42	19,200	19,242
Vinyl chloride	2.76	552	276	828	0.0006	0.05	47	875	18,200	19,075
Solvent-Oil Mixture										
Weathered mix (1% CVOC as TCE)	0.01	2	1	3	0.0015					
Fresh mix (17% CVOC as TCE)	0.19	38	19	57	0.0015					
Miscible organics										
1,4-Dioxane	1,030	206,000	103,000	309,000	NA	0.05	370	309,370	0	309,370
Acetone	790	158,000	79,000	237,000	NA	0	0	237,000	0	237,000
Salts										
Ammonium perchlorate	209	41,800	20,900	62,700	NA	0	0	62,700	0	62,700

Notes: These calculations assume a total porosity of 300 L/m³, a migratory pore fraction of 33% (100 L/m³), and an aquifer matrix foc of 0.0005 (500 mg/kg TOC and 10% of the TOC is hard carbon). The k_{oc} and aqueous-phase solubility were taken from U.S. EPA (1996) for all compounds except 1,4-dioxane and ammonium perchlorate, which were drawn from Suthersan and Payne (2005). For NAPL-forming compounds (hydrophobic organics), the residual NAPL mass was assumed to be 2 percent of the aquifer pore volume (6 L/m³). All values are g/m³.

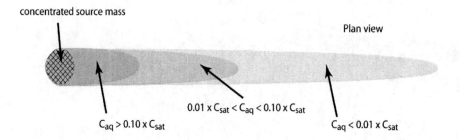

concentrated source mass

Plan view

$0.01 \times C_{sat} < C_{aq} < 0.10 \times C_{sat}$

$C_{aq} > 0.10 \times C_{sat}$

$C_{aq} < 0.01 \times C_{sat}$

FIGURE 7.1 Generalized groundwater contaminant plume development from a concentrated source mass.

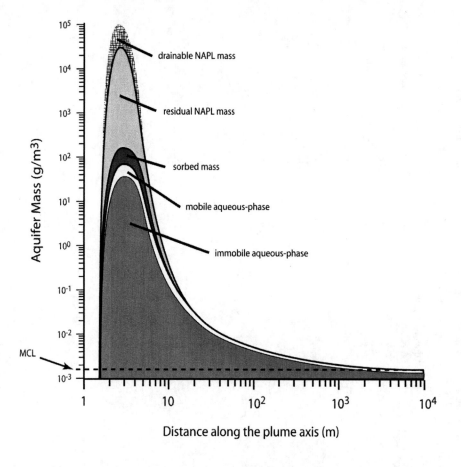

FIGURE 7.2 Mass storage capacities for hydrophobic organics, along a plume such as the example given in Figure 7.1.; constructed using the values for perchloroethene as an example (refer to Table 7.1 for details).

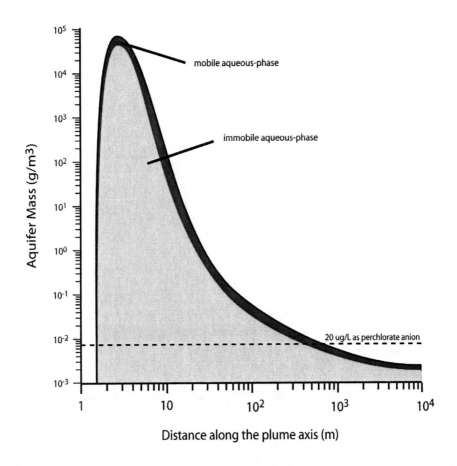

FIGURE 7.3 Mass storage capacities for a soluble salt, along a plume such as the example given in Figure 7.1.; constructed using the values for ammonium perchlorate as an example (refer to Table 7.1 for details).

The hydrophobic organic source depicted in **Figure 7.2** comprises drainable and residual perchloroethene over a 10-m dimension along the flow path[2]. The source mass is comparable in size to those at large-plume sites we have encountered, and a 10^7 to 10^8-fold concentration reduction would be required to return the source area to groundwater concentrations meeting current drinking water supply standards (noted by the "MCL" line on the figure). In an aerobic aquifer, a source mass of this dimension can generate a dissolved-phase contaminant plume extending well over a kilometer in length. A large fraction of the contaminant mass resides in the immobile pore spaces — two-thirds of the mass stored outside the source zone in this example.

The relatively low aqueous-phase solubilities for hydrophobic organic compounds constrain the mass that can be stored in the immobile porosity, although a source that persists for long periods can easily generate dissolved-phase concentrations in the immobile pore spaces that exceed relevant compliance criteria by 10,000-fold. In aerobic aquifers, there is very little destructive attenuation and the areal extent of highly contaminated immobile pore space can grow to large dimensions.

A large mass of soluble salts, such as ammonium perchlorate, can also be stored in an aquifer matrix. **Figure 7.3** shows a hypothetical source zone that is developed from a saturated salt solution

---

[2] For comparison, the residual DNAPL source studied by Rivett, et al. (2001) and Rivett and Feenstra (2005) totaled less than 1 m³ in volume and generated a contaminant plume that traveled more than 50 m in its first year.

that has displaced groundwater over a dimension of less than 10 m. The mass that can be stored in the immobile porosity is quite large for soluble salts, which explains how persistent groundwater contamination plumes can be formed by compounds that would otherwise wash quickly through an aquifer.

The dissolved-phase mass storage capacity for any compound is related to its organic carbon partition coefficient, K_{oc}, the organic carbon fraction in the aquifer matrix, f_{oc}, and the aquifer volumetric water content, θ_{tot}. We can calculate the f_{oc} that will generate an equal mass in the dissolved and sorbed phases for any compound, if we assume a value for the aquifer matrix bulk density and θ_{tot}. The mass in dissolved phase is given by the product of θ_{tot} and the dissolved-phase concentration. The mass in sorbed phase is given by Equation 6.4. Setting the two equal and simplifying,

$$C_{aq} \cdot K_{oc} \cdot f_{oc} \cdot \rho_{bulk} = C_{aq} \cdot \theta_{tot} \qquad \frac{mg}{L} \frac{L}{kg} \frac{kg}{kg} \frac{kg}{m^3} = \frac{mg}{L} \frac{L}{m^3} \qquad (7.1)$$

$$f_{oc\,breakeven} = \frac{\theta_{tot}}{K_{oc} \cdot \rho_{bulk}} \qquad \frac{L}{kg}$$

Figure 7.4 graphs the breakeven line for an aquifer with $\theta_{tot} = 300$ L/m³ and $\rho_{bulk} = 1,800$ kg/m³. For any aquifer with an organic carbon content below 0.001 (1,000 mg/kg TOC), mass stored in the dissolved-phase exceeds that in the sorbed phase, for any compounds with an organic carbon partition coefficient (K_{oc}) of less than 100 L/kg. This includes trichloroethene, cis-dichloroethene and vinyl chloride. For perchloroethene, aquifers with organic carbon fractions below 0.004 (400 mg/kg TOC) and for 1,1,1-trichloroethane, below 0.007 (700 mg/kg TOC) store more mass in dissolved-phase. The organic carbon content of many aquifers falls below the "breakeven" line and, consequently, it is likely that dissolved-phase mass storage is a very significant factor in many contaminant plumes.

Mixtures of chlorinated solvents, waste oils, grease and other hydrocarbons are often encountered at contamination sites. The solubilities of compounds in the mixture may be lower than for neat solutions, as suggested by Raoult's Law [Equation (7.2)]:

$$C_{eff\,i} = C_{sat\,i} \cdot X_i \qquad (7.2)$$

where $C_{eff\,i}$ is the effective solubility for compound i, $C_{sat\,i}$ is the aqueous-phase solubility for compound *i* and X_i is its mole fraction in the mixture. Equation (7.2) is a valid approximation for mixtures without significant co-solvent effects. However, as waste solvent mixtures weather in aquifers, it is likely that the mixture changes over time, particularly in cases where hydrocarbons can be degraded by aquifer bacteria populations. The lighter hydrocarbons in these mixtures are generally more degradable and fermentation processes that occur as part of the decomposition process generate hydrogen that, in turn, supports reductive dechlorination of the chlorinated alkenes and alkanes. Decomposition processes may lead to changes in the nature of source mixtures that influence the rate of dissolution for the solution components. **Figure 4.3** described a possible weathering sequence for a solvent waste mixture that initially contained a high percentage of chlorinated volatile organics and weathering decreased both the lighter hydrocarbon and chlorinated solvent components, leaving a small mole fraction of chlorinated mass, dissolved in a heavy hydrocarbon mixture. This weathering process would be accompanied by a significant

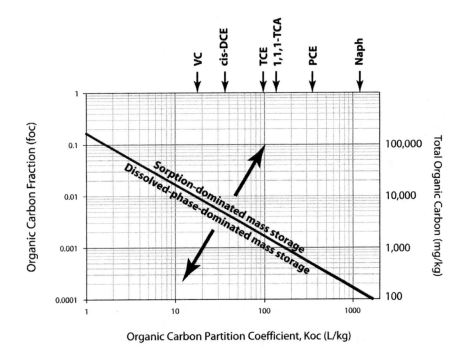

FIGURE 7.4 Analysis of the organic carbon content at which the dissolved-phase and sorbed-phase mass storage capacities are equal, in an aquifer with total porosity of 300 L/m^3 (conservatively low) and a bulk density of 1,800 kg/m^3. For Koc-foc combinations above the line, sorption will be the dominant mass storage compartment and for combinations below the line, dissolved-phase storage will be the dominant compartment.

increase in NAPL viscosity, as was observed at a large NAPL release in the Midwestern United States (refer to Chapters 1 and 4).

There are important deviations from the simple weathering process outlined above, caused by co-solvent effects of waste mixtures and by the action of surfactants produced by bacterial populations that decompose hydrocarbon mass. Each of these processes can increase the solubility of individual mixture components to levels higher than those predicted by Raoult's Law. Co-solvency is described in detail by Schwarzenbach, et al. (2003). The propagation and impact of biosurfactants has been studied in the context of petroleum hydrocarbon degradation (Barkay, et al., 1999, for example), where bacterial populations associated with fuel and crude oil weathering typically force an increase in the solubility (and bioavailability) of compounds such as poly-aromatic hydrocarbons.

7.2 SOLUTE TRANSFERS BETWEEN MOBILE AND IMMOBILE POROSITIES

The effective distances over which aqueous-phase diffusion can drive solute mass transfers are quite small, which makes it easy to overlook the role that can be played by diffusion in contaminant plume development. Solutes carried in strongly heterogeneous groundwater flows, especially when the mobile strata are less than 10 cm thick, expose an invading contaminant front to clean groundwater at distances of a few centimeters. In these cases, aqueous-phase diffusion can be a significant process, slowing the invasion rate for a contaminant or reagent and driving a notable solute mass into dissolved-phase storage.

Figure 7.5 shows a conceptual diagram of advective flow in a mobile stratum and the diffusive interaction that can occur with adjacent immobile porosity. We know from calculations in Chapter 5 (Equation 5.10, e.g.) that diffusion gradients flatten over periods of days, to levels that support only minimal mass flux. We can develop approximations of the mass flux densities to determine when diffusion can influence the progression of a contaminant front. Earlier, **Figure 5.9** showed the development of a diffusion gradient across a sharp concentration boundary, with plots of concentration versus distance at several elapsed times after the contact was initiated. **Figure 7.6** shows a similar plot, for a 200 ug/L trichloroethene solution, 1 day after contact is initiated with water containing no contaminant. The diffusive mass flux density can be calculated from concentration gradients, as shown in Equation (7.3). J is the flux density (mass per area per time), D_0 is the standard diffusivity (length squared per time) and dC/dz is the solute concentration gradient (concentration per length).

$$J = \theta_{total} \cdot D_0 \frac{dC}{dz} \quad \frac{\mu g}{cm^2 \cdot s} \tag{7.3}$$

The gradient strength, dC/dz, decreases over time at the interface and, as a result, the mass flux density decreases.

Standard diffusivity values must be adjusted to account for path tortuosity that restricts diffusion somewhat in porous media. D_{eff} is the effective diffusivity, adjusted to account for the effects of matrix tortuosity, as described by Parker et al. (2004):

$$J = \theta_{total} \cdot D_{eff} \frac{dC}{dz} \quad \frac{\mu g}{cm^2 \cdot s} \tag{7.4}$$

where

$$D_{eff} = D_0 \cdot \tau \tag{7.5}$$

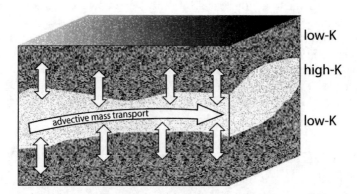

FIGURE 7.5 Interaction between advective and diffusive mass transport mechanisms at the stratigraphic scale. High-conductivity strata are the primary conduits for solute mass transport; lower-permeability strata become solute reservoirs, due to equilibration between high- and low-permeability strata through aqueous-phase diffusion. This diffusive interaction decreases the velocity of solute migration, relative to groundwater velocities in the higher-conductivity strata.

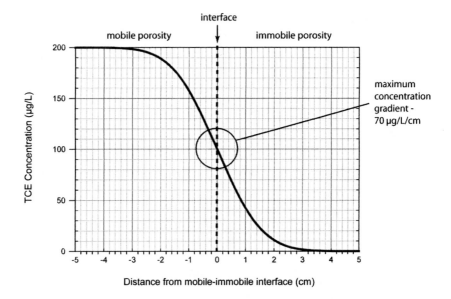

FIGURE 7.6 Concentration gradient 1 day after invasion of clean sector by groundwater containing 200 µg/L trichloroethene. The maximum concentration gradient at this time is indicated by the circled area.

The effective diffusion coefficient (D_{eff}) is reduced from the standard diffusion constant in free water (D_0) by a tortuosity coefficient, τ, reflecting constraints on diffusive movement of solutes in porous media. Values for τ are a function of the aquifer matrix structure and are likely to depend on the total porosity and the shapes and interconnections between pores. Parker et al. (1994) indicated that τ ranged from 0.27 to 0.63 in their studies of silty and clayey deposits, for example.

The mobile-immobile interface, at the center of the developing concentration gradient, is where the gradient is strongest, as shown in **Figure 7.6**. Flux densities were calculated from concentration profiles calculated for days 0.1 through 10,000 of gradient development and the results are plotted on **Figure 7.7**. The flux density at the first instant is undefined, because the concentration (theoretically) changes from 200 to 0 µg/L over "zero" distance. The very high flux densities of the earliest diffusion quickly slow to the levels approximated by **Figure 7.7**. The flux density at 1 day elapsed time, 17 µg/cm²/day, provides a reasonable maximum value for examining whether diffusive losses can influence the invasion rate for solutes in aquifers.

In earlier chapters, we showed that groundwater flows in natural aquifers tend to be heterogeneous and anisotropic and even the most uniform systems exhibit sharp contrasts in hydraulic conductivity at scales of a few centimeters. **Figure 5.2,** for example, showed an estimate of the varying rates of solute invasion for a highly uniform aquifer. Tracer studies and plume delineation data shown in later chapters all ratify the notion that the invasion of a solute in any aquifer system is likely to proceed along pathways that occupy a small portion of the total aquifer pore space and that solute-containing groundwater is exposed to solute-free groundwater along the invasion path. Because the scale of fast-flowing pathways may be small, on the order of centimeters in width or thickness or both, diffusive interchange between the solute transport pathways (the mobile porosity) and near-stagnant groundwater astride those pathways (the immobile porosity) may be intense.

There are at least two potential consequences of diffusive exchange between mobile and immobile pore spaces:

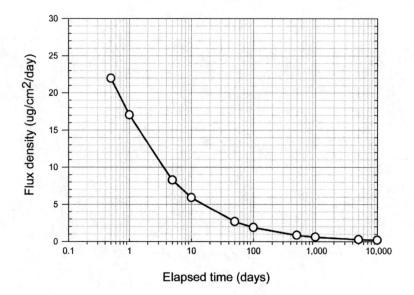

FIGURE 7.7   Diffusive flux density for a developing trichloroethene gradient with 200 ug/L initial concentration on the source side.

- Diffusive loss of solute from the mobile porosity slows the solute velocity, relative to the mobile-fraction groundwater velocity, and
- The immobile porosity serves as a storage bank for dissolved-phase solute that enters via aqueous-phase diffusion.

Calculations of chemical concentration gradients and diffusive mass flux densities that develop at the solute invasion boundary provide a basis for approximating the magnitude of aqueous-phase diffusion effects.

The concentration gradient, estimated from data prepared for **Figure 7.6**, was 70 ug/L/cm, which represents a high gradient strength, as would be expected when a solute front begins invading clean pore spaces with a total porosity of 0.35. For trichloroethene, for which we will assume D_{eff} = 9.8 x 10^{-6} cm/s, the resulting flux density calculation is as follows,

$$J = 0.35 * 9.8 \times 10^{-6} \cdot 70 \quad \frac{cm^3}{cm^3} \cdot \frac{cm^2}{s} \cdot \frac{ug}{cm^3} \cdot \frac{1}{cm} = 1.97 \times 10^{-4} \frac{ug}{cm^2 \cdot s} \tag{7.6}$$

Converting the result to daily flux, J = 17 ug/cm^2/day.

Figure 7.8 shows a conceptual set of interfaces for a zone of mobile porosity in contact with zones of immobile porosity above and below, and the diffusive flux is depicted for each of the interfaces. The thickness of the mobile porosity is represented by the variable, z, in the figure. Although the initial mass flux density would be high, the solute concentration gradient would decrease over time, as indicated by **Figure 7.6**. Also, the mass transfer rate estimated from the gradient in **Figure 7.6** is too high, because it assumed semi-infinite media on each side of the interface. A simple calculation of the mass in a mobile pore volume shows that the strong gradient

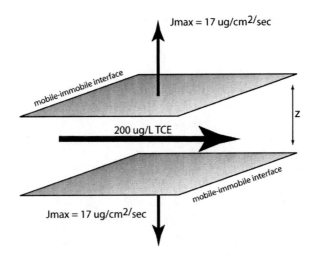

FIGURE 7.8 Conceptual diagram showing mass flux from a trichloroethene front, traveling through a mobile porosity of thickness, z. The maximum flux density was calculated for a linear concentration gradient of 70 ug/L/cm.

and high mass flux rate cannot be sustained. With an aqueous-phase concentration of 200 ug/L and a total porosity of 0.35:

$$Mass = \theta_{total} \cdot C_{solute} \cdot volume = 0.35 \frac{cm^3}{cm^3} \cdot 200 \frac{ug}{L} \cdot 0.001 \frac{L}{cm^3} = 7 \times 10^{-2} \frac{ug}{cm^3} \qquad (7.7)$$

The dissolved trichloroethene mass in a 1-cm cube of aquifer matrix is less than 1-400th of the mass that could diffuse off its upper and lower surfaces daily, if the early-time gradient could be sustained. Obviously, this cannot occur, so it is clear that the initial rate of mass transfer cannot be sustained, even for a day. However, it is also clear that diffusive flux can draw mass out of the leading edge of an invading solute plume, reducing the propagation rate for solute entering a clean zone.

We can obtain a more realistic approximation of diffusion effects by assuming that the porous matrix is bounded, above and below, by media that severely restrict diffusion or by parallel strata containing comparable solute concentrations, either of which act as reflecting boundaries. **Figure 7.9** shows the concept of reflective diffusion boundaries, formed midway between two high-permeability strata with invading solute fronts. As solute migrates from the source mass, it reaches solute diffusing from the parallel stratum and the net effect is to reflect the solute reaching the boundary back toward the source. This is exactly the same result we obtain if the diffusing solute reaches a solid barrier.

We can solve the bounded case using the infinite series approximation presented in Crank (1975). Recalling Equation (5.13),

$$C = \frac{1}{2} \cdot C_0 \cdot \sum_{n=-5}^{5} \left(erf \left(\frac{h + 2 \cdot n \cdot L - x}{2 \cdot \sqrt{D \cdot t}} \right) + erf \left(\frac{h - 2 \cdot n \cdot L + x}{2 \cdot \sqrt{D \cdot t}} \right) \right) \qquad (7.8)$$

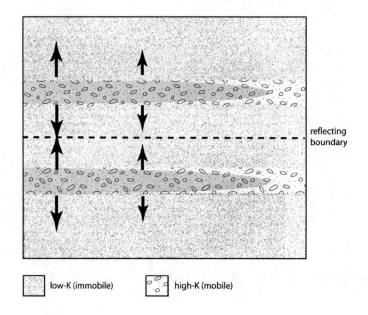

□ low-K (immobile)   □ high-K (mobile)

FIGURE 7.9 Conceptual model for dissolved-phase invasion of solutes into immobile porosities. Solutes advance rapidly in the mobile pore space and diffuse slowly into adjoining immobile pore spaces. Diffusive reflection is generated midway between mobile strata and solute infusion can be estimated with bounded-volume diffusion solutions [Equation (5.13), for example].

The concentration is given for a specified time, t, and distance, x, from the centerline of a diffusing stratum. The variable L is the distance from the centerline to the reflecting boundary and h is the distance from the centerline to the edge of the initial solute source.

We used Equation (7.8) to analyze diffusive intrusion of trichloroethene in two hypothetical aquifer matrix segments, each 1 m in thickness. The two matrix blocks are shown on **Figure 7.10**. The upper aquifer segment consists of 33 cm (average) of high-permeability material, bounded by 33-cm layers of lower-permeability material on each side. The lower aquifer segment has the same overall thickness and the same average hydraulic conductivity. The high-permeability material in the lower segment is divided into 3, 11-cm units, evenly distributed within the low-permeability material. The high-permeability strata in each of the aquifer units was flooded with water containing 200 ug/L trichloroethene, while the lower-permeability strata contained none. In the upper unit, it was assumed that diffusive reflection would occur at the upper and lower bounds of the volume. In the lower unit, diffusive reflection occurs at the mid-point between high-permeability strata, as well as at the upper and lower bounds of the system. For these calculations, we assumed that the effective diffusivity for trichloroethene is 5×10^{-6} cm²/s, less than 50 percent of its standard diffusivity. For each aquifer unit, trichloroethene concentration profiles were calculated at 1 day, 10 days, 100 and 1,000 days, to show the progression of the concentration profiles.

Diffusive mass transfer occurs more than 10 times faster in the lower unit. This illustrates the qualitative descriptor we began to define in Chapter 5, *mass-transfer geometry*. Both aquifer segments in **Figure 7.10** have identical average hydraulic conductivities and identical mobile and immobile pore fractions (hence, identical mobile-fraction groundwater velocities, as well). However, the lower unit exhibits a "high-mass-transfer" geometry, relative to the upper segment, and aqueous-phase diffusion affects solute distributions much more quickly in the high-mass-transfer geometry. These calculations were simplified by assuming a static water case. The

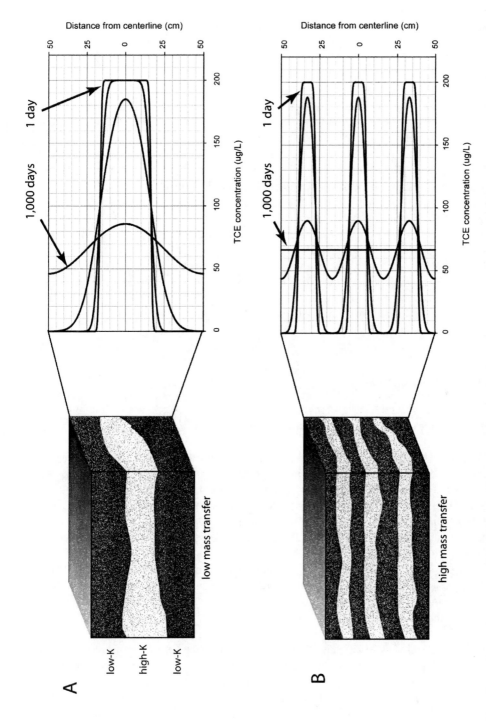

FIGURE 7.10 Two volume elements, A and B, with equal average hydraulic conductivities and mobile fraction groundwater velocities, but different mass transfer geometries. Equation (7.8) was used to calculate the progression of trichloroethene (TCE) concentrations at 1, 10, 100 and 1,000 days after a 200 ug/L source was placed in the high-K zones of each 100-cm-thick aquifer matrix segment. Equilibration proceeds approximately 10-fold faster in the porous matrix described in block B. The assumed D_{eff} for trichloroethene was 5×10^{-6} cm^2/s.

migrating-water case is much more complicated to calculate, but the diffusive interactions between solute-transmitting and static aquifer zones are clearly significant in that case, also. Readers may wish to consult Crank (1975) or Treybal (1980) for further reading on the mathematics of diffusion and mass transfer processes.

The mass distribution in both systems is an important point of interest. In both cases, when solutes in the high and low-permeability strata reach equilibrium levels, two-thirds of the solute will reside in the lower-permeability material. The high-mass-transfer geometry system reached equilibrium in less than 1,000 days and by 100 days, more than 50 percent of the mass had moved into the lower-permeability strata. For the low-mass-transfer geometry, roughly 50 percent of the mass migrated into the lower-permeability strata within 1,000 days. The dissolved-phase mass storage that results from diffusive invasion of lower-permeability strata in aquifers is a major factor in plume expansion and in the rate at which contaminant source removal efforts are reflected in improved groundwater quality.

The basic process of diffusive invasion of solutes into lower-permeability strata has been validated through meso-scale laboratory experiments at Colorado State University. **Plate 10** shows the experimental setup, a tank with alternating sand and clay strata, creating a heterogeneous aquifer matrix, as we might encounter in natural systems. Groundwater flow was established across the porous matrix and fluorescein dye was added to the water for 24 days. The upper panel of **Plate 10** shows the setup in visible and ultraviolet light, showing the sand and clay layers and that the primary flow of fluorescein occurs in the sand layer. However, fluorescein diffused into the clay zone, which became evident when clean water flow was restored. The lower panel of Plate 10 shows diffusion from the clay zone, back into groundwater flowing through the sand strata. In only 24 days of diffusive interaction, the clay became a significant reservoir of solutes, capable of maintaining a residual solute concentration in the sand strata, after clean water flushing has been established.

7.3 STEADY-STATE CONCEPTS

The chemical mass associated with each instance of groundwater contamination or reagent deployment can be divided into three main compartments:

Source mass — The source comprises a persistent, concentrated mass of contaminant or reagent, for which the equilibrium dissolved-phase concentrations exceed the levels in groundwater flowing through the zone. The concentrated mass serves as a persistent source and that aquifer volume is a net exporter of solutes to areas downgradient.

Primary transport conduit — Each aquifer contains mobile and immobile porosities. The primary solute transport occurs in the mobile portion of the porosity, at velocities exceeding the average groundwater velocity.

Dispersed storage — Diffusive infusion of solutes into immobile porosity and into the sorbing organic carbon fraction that may be present, especially in the finer-grain aquifer matrix materials, serve as a dispersed storage mechanism for solutes, along the entire length of a contamination plume.

The source mass is always in an "out-of-steady-state" condition, in the sense that it contains reagent or contaminant mass that does not occur naturally in aquifers, so the long-term steady-state groundwater concentration is zero. For large, slowly dissolving source masses, a quasi-steady-state condition can be achieved, with a steady shedding of mass to the passing groundwater, generating a relatively constant source to the aquifer system. We can use the conceptual aquifer matrix blocks, introduced above, to introduce concepts of steady-state in plume and reagent transport.

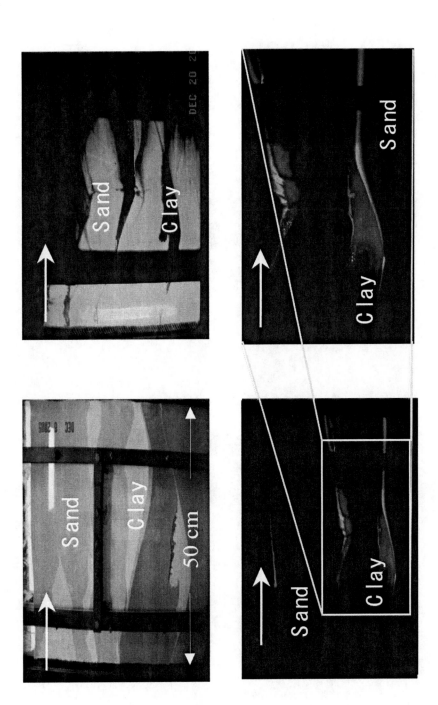

PLATE 10 Results of back-diffusion experiments at Colorado State University. A heterogeneous porous medium was built with sand and clay strata (upper left, in visible light). Groundwater flow was established from left to right and the system was perfused with a fluorescein solution for 24 days (upper right, in ultraviolet lighting). Clean water was then flushed through the system and fluorescein that had penetrated the clay through aqueous diffusion formed a persistent source mass that re-contaminated the clean water flush (lower left and expanded view in the lower right). Photos courtesy Ms. Lee Ann Doner and Dr. Thomas Sale, Colorado State University. *(See Plate section for color version of Plate 10.)*

Expanding Solute Plume

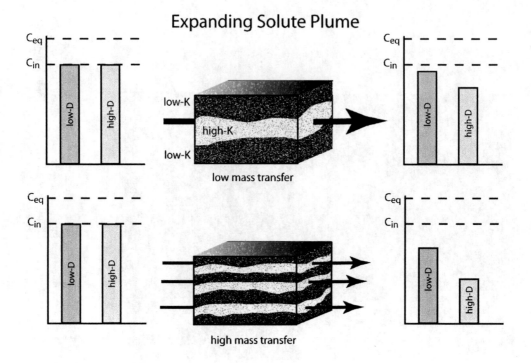

Contracting Solute Plume

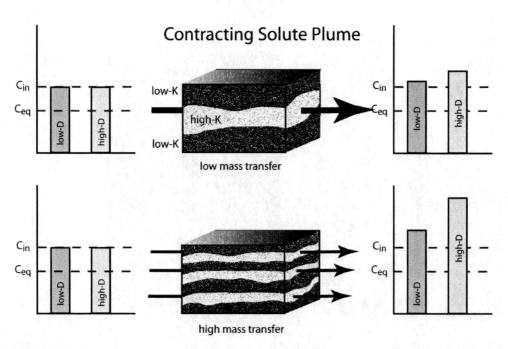

FIGURE 7.11 Comparative effects of aquifer matrix geometry and diffusivity contrasts on expanding-plume and contracting-plume mass transfers, in multi-porosity matrices. In areas of plume expansion, the solute input to any volume element is below plume equilibrium and above local equilibrium; in areas of plume contraction, the solute input to any volume element is above plume equilibrium and below local equilibrium. Both solute diffusivity and matrix mass transfer geometry influence the rate of equilibration.

Figure 7.11 provides comparisons showing the inflows and outflows from aquifer matrix blocks in several scenarios. For each of these, groundwater flow occurs in the high-permeability (mobile) fraction, which is the primary transport conduit for solutes.

Expanding solute plume — The upper two blocks show the inflows and outflows during plume invasion. During plume invasion, the dispersed storage is not fully loaded, so the influent concentration to any block (C_{in}) is less than the steady-state concentration (C_{eq}). We show high and low-diffusivity solutes, arriving at the same concentration, C_{in}, into each block. The effluent concentrations are reduced for both solutes in high and low-mass-transfer matrix structures, although diffusivities and mass-transfer geometries control the rate of mass loss that occurs on passage through each matrix block: the higher-diffusivity solute transfers more mass into each block and both solutes transfer more mass into the high-mass-transfer block.

Contracting solute plume — The lower two blocks show patterns of inflow and outflow that would occur during plume retreat. We assume for this panel that the source mass has been reduced, which generates a lower long-term equilibrium concentration. We also assume that the system had reached quasi-steady-state with the original, higher mass flux rate, so the mass in dispersed storage exceeds the new equilibrium value. The dispersed mass stored in the aquifer system is now unloading into the primary transport conduit, so the influent concentration exceeds C_{eq} and the outflowing concentration exceeds C_{in}. As above, the aqueous-phase diffusivities and mass-transfer geometries influence the mass transfer rates between the immobile and mobile porosities.

These steady-state concepts can be applied to develop a sense of what we are likely to observe in detailed cross-sectional aquifer transects, downgradient from a persistent contaminant source mass. **Figure 7.12** shows a cross-section at three stages of groundwater plume invasion and retreat.

Invasion phase — In the top panel of **Figure 7.12**, high solute concentrations are observed in the primary flow conduits, which are structured by the depositional environment. The system is heterogeneous and anisotropic, typical of natural aquifers. The high solute concentrations drive diffusive invasion of the immobile pore spaces that lie along the primary flow conduits.

Steady-state — The middle panel depicts a steady-state condition. If the primary conduit carries solute steadily over an extended period, diffusive transport reaches a near-steady-state condition, and solute concentrations in the immobile porosity are a close match to concentrations in the mobile porosity. At this point, a very large solute mass resides in dispersed storage in the formation.

Retreat phase — In the lower panel, the source mass has been eliminated and solute concentrations in the primary flow conduits are less than those in the adjacent immobile porosity. The dispersed storage has become a source of solute to the primary transport conduits that may persist for an extended period.

The mass that resides in dispersed storage increases over extended periods and the exposure time determines: 1) whether steady-state is reached, and 2) the time required to draw down the dispersed storage reserves to acceptable levels. Longer exposure times generate more "mature" plumes and the level of effort expended to treat matured plumes will be much greater than for early-stage plume. **Figure 7.13** suggests that the return on invested cleanup efforts will not be the same for early-stage and mature plumes. For an early-stage plume, there is likely to be a significant progress associated with first efforts (it should be noted that most laboratory systems have limited maturation and are likely to behave like early-stage plumes). For a mature plume, contaminant

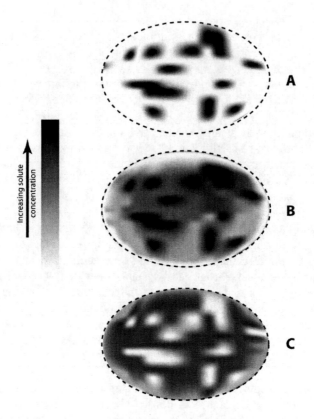

FIGURE 7.12 Conceptual diagram showing the initial spreading (A), maturation (B) and depletion (C) of a groundwater contamination plume that develops from a concentrated source mass such as that shown in Figure 4.1. In the early stage of plume development, contaminant migration occurs along flow tubes that develop in zones of higher hydraulic conductivity, shown in A. Over time, aqueous-phase diffusion spreads solute into the less permeable zones, as shown in B. After the contaminant source is eliminated, cleaner water travels through the flow tubes and solute from the low-permeability zones re-contaminates the higher-flow zones, through aqueous-phase diffusion, as shown in C.

rebound from solutes returning from dispersed storage is likely to frustrate early restoration efforts.

This conceptual model explains the common observation that source mass removal typically does not generate rapid declines in solute concentration in downgradient areas. The rate of progress of the clean water front is much slower than the maximum or average groundwater velocities, its velocity slowed by the same mechanisms that slow plume invasion rates. We can now carry the steady-state concepts into a conceptual analysis of contaminant plume development and retreat.

7.4 PLUME DEVELOPMENT AND RETREAT

The shape of the bounding envelope and the concentration patterns that form a contaminant plume is determined by the rate at which mass is shed from the source (source mass flux) and the rate of contaminant loss to destructive processes and dilution along the flow path. When the rate of loss is in balance with the source mass flux, the plume could be characterized as "at steady state." For any well-defined source mass, we can speculate as to the dimension and concentration distribution

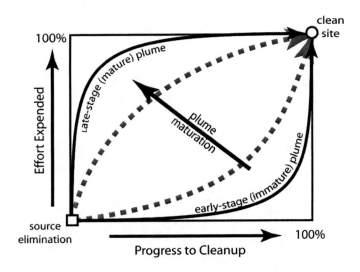

FIGURE 7.13 Impact of plume maturation on the rate at which cleanup efforts produce observable progress toward site restoration.

of groundwater contamination that would develop, over an extended time span, *if the source remained constant.* However, because sources typically aren't stable over time, it is unlikely that a true steady state can be achieved.

Figure 7.14 sets up the first of several plume development scenarios that are built on the concept of a persistent source mass, feeding a fast-flowing primary transport conduit, with losses from the transport conduit into dispersed storage in static groundwater along the transport corridor. In the top panel of the figure, we see a persistent source mass, from which an estimate has been made for the maximum bounding volume that might be reached by the plume. The maximum bounding volume is constrained by diffusive losses to the immobile porosity and solute destruction that may occur by physical, chemical or biological mechanisms. The plume is shown in a pre-steady-state condition, reflecting the observation that few contaminant plumes are likely to reach steady state for any extended period in their life cycle.

As diffusive invasion of solutes into the immobile porosity flattens diffusion gradients, mass losses to storage decrease near the source and the near-steady-state front moves progressively in the downgradient direction. If we remove a part of the source mass, the mass flux decreases and a new steady-state plume can be determined. It will occupy a smaller bounding volume, as indicated in the middle panel of **Figure 7.14**. Over an extended time span with a constant source mass flux, the new steady state could be reached, with the cleaner water front moving from upgradient to downgradient, as shown in the bottom panel of the figure.

If a small source mass enters an aquifer (a reagent pulse is a good example), the initial mass flux is likely to be large, which would generate a large bounding envelope for the steady-state project, as shown on **Figure 7.15**. In this case, the source mass erodes relatively quickly and the steady-state projection is largest at the outset and shrinks continually over time. The source may have been totally eliminated when the contaminant plume is still expanding its bounding envelope. The clean water front progresses from upgradient to downgradient.

If a source mass has been in place for an extended period and has reached a steady-state bounding envelope, shrinkage of the plume may occur very slowly, as mass trickles out of the

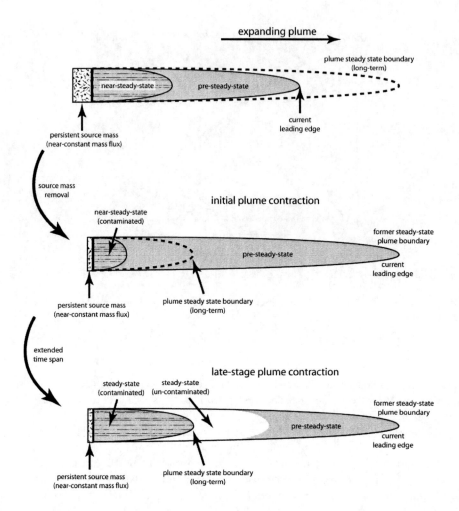

FIGURE 7.14 Plume maturation concepts. Upper panel — a solute plume is developing from a persistent source mass. The long-term (plume) steady-state is indicated by the dashed line. A zone near the source has locally reached near-steady-state mass loading. Middle panel — after the upper plume reached steady-state, the source mass was reduced, leading to a smaller long-term steady-state plume, indicated by the dashed line. Lower panel — the clean water front progresses from the new steady-state plume boundary in the downgradient direction. In the expanding plume, pre-steady-state areas are locally below the long-term mass equilibrium and in a contracting plume, the pre-steady-state areas are locally above the long-term mass equilibrium.

dispersed storage along the flow path, as shown in **Figure 7.16**. We expect the clean water front to be observed first in the upgradient areas, then near the downgradient edge, following source elimination.

Finally, a pattern that is often observed is a period of net source mass accretion, during which the plume and the steady-state bounding envelope are constantly expanding, as in **Figure 7.17**. Many of the sources we examine have undergone long histories of accretion — waste disposal pits, leaky tanks or piping, poorly-tended waste storage – and when groundwater contamination is

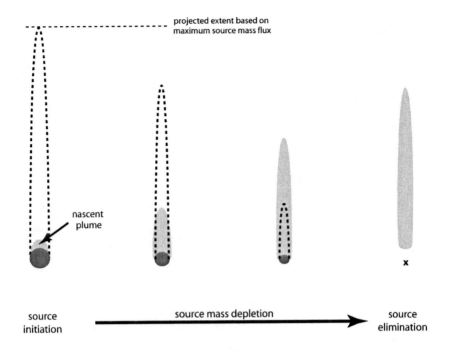

FIGURE 7.15 Life span of a small, pulse-loaded contaminant source and its associated groundwater contaminant plume. If the source is well-exposed to groundwater flow, mass flux is high and rapidly depletes the source. The source is fully depleted before the plume footprint reaches its maximum extent and a steady-state balance between source mass flux and plume dimension is not reached. Dashed lines represent the steady-state plume dimension that would be estimated from the source mass flux rate and light shading represents the plume extent.

discovered, the accretion is halted and some amount of source mass removal is attempted. If the discharge that formed the source is eliminated, a period of net source mass depletion would follow, along with a slow contraction of the occupied aquifer volume. These cycles could easily be many decades long.

Another useful way to view contaminant source and plume life cycles is a plume maturation and attenuation arc, as shown on **Figure 7.18**. A contaminant plume reaches maturity when it has fully occupied the steady-state bounding volume indicated on the previous figures, and mass stored in the immobile porosity is at equilibrium with the mobile porosity. The time scale for maturation and attenuation is shown in log scale, to make it clear that the time spans for maturation and natural attenuation of most plumes we work with appear to be many decades long. Also, the natural attenuation portion of the arc is dramatically longer than the maturation portion of the arc, due to the lesser chemical gradients at work unloading the dispersed storage, relative to the gradients that formed the storage, initially.

If the source is eliminated before the plume reaches its steady-state bounding envelope, clearly we can shorten the time span for attenuation, as indicated on **Figure 7.19**. Attacking the source mass at earlier times in the maturation process is expected to yield faster responses in passive cleanup. Finally, if source removal is accompanied by aggressive plume intervention, as suggested in **Figure 7.20**, it may be possible (although generally quite costly) to push the attenuation process more quickly.

For all the cases we examine in the field, it is near impossible to quantify any of the key elements of a contaminant plume — the total source mass is usually unknown and its distribution is uncertain. The primary transport conduits are difficult to locate[3], although more effort in this area is likely to provide an excellent return on invested effort at most sites. The breakdown between mobile and immobile porosities is generally not clear, either. All of these investigative limitations make it unlikely that we can develop an accurate, high-resolution quantitative model of any site, but as we probe the site and test the conceptual site model, these general behavior patterns are helpful in understanding and classifying the system responses we observe.

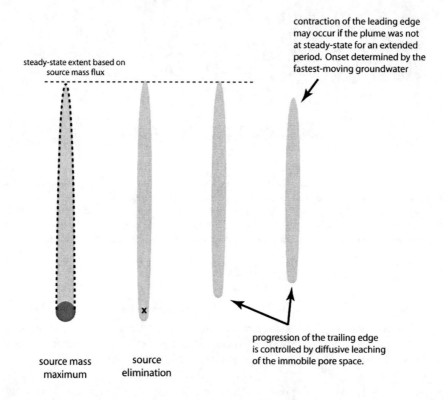

FIGURE 7.16 Contraction of a contaminant plume from steady-state, after source elimination. The clean water front progresses from the former source area, at a rate controlled by diffusion of contaminants from immobile pore space. If the plume was not at steady-state for an extended period, fast-flowing groundwater begins to decrease concentrations at the leading edge, although the pace of contraction at the leading edge may be slower than the progression of the clean water front from the former source zone. Dashed lines represent the steady-state plume dimension that would be estimated from the source mass flux rate and light shading represents the plume extent.

---

[3] Hydrostratigraphic analysis, described in Chapter 9, provides an effective method for understanding aquifer structure, which is essential for efficiently locating flow conduits.

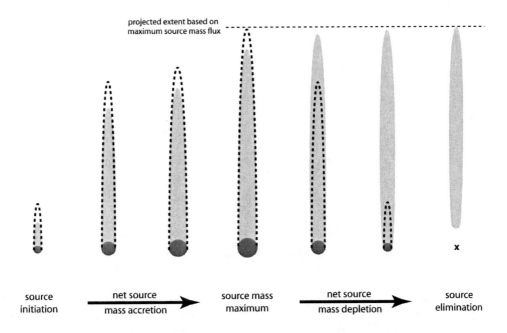

FIGURE 7.17 Life span of a large, accretion-loaded contaminant source and its associated groundwater contaminant plume. If a large source mass is poorly-exposed to groundwater flow, mass flux is low, relative to the mass, and slowly depletes the source. Dashed lines represent the steady-state plume dimension that would be estimated from the source mass flux rate and light shading represents the plume extent.

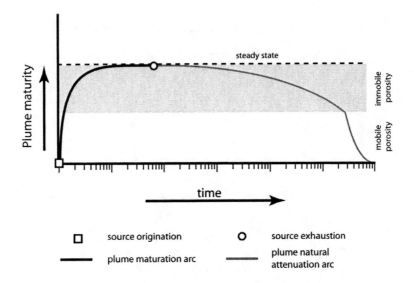

FIGURE 7.18 Plume maturation and natural attenuation arc. When a contaminant source reaches groundwater, contaminant migration proceeds along mobile porosity pathways. Over extended periods, immobile porosity becomes contaminated through diffusive mass transfer from the mobile to immobile pore space. After the source is exhausted, clean water begins to perfuse the mobile porosity and is re-contaminated by mass stored in the immobile pore space, extending the recovery arc, relative to the maturation arc.

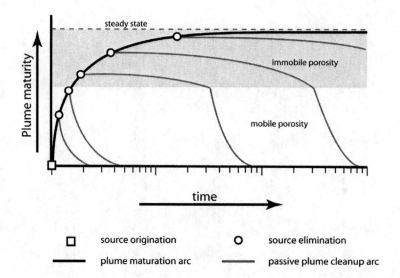

FIGURE 7.19 Plume maturation and passive cleanup arcs. Contaminant migration proceeds along mobile porosity pathways. Over extended periods, immobile porosity becomes contaminated through diffusive mass transfer from the mobile to immobile pore space. After source elimination, clean water flow proceeds along mobile pathways. If cleanup is initiated before the immobile pore space is highly contaminated, the cleanup can be completed quickly. If cleanup starts after the immobile pore space is highly contaminated, the passive cleanup arc, which relies on clean water flushing of the mobile pore space, may extend over very long time spans.

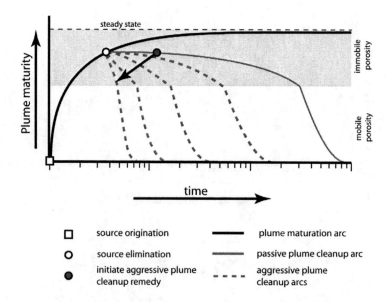

FIGURE 7.20 Plume maturation and aggressive cleanup arcs. After source elimination, aggressive cleanup processes that eliminate contaminants stored in the immobile pore space result in dramatic decreases in the cleanup time spans, as indicated by the arrow.

7.5 MIGRATION OF A CLEAN WATER FRONT

One of the most challenging tasks in remediation engineering is estimating the rate at which contaminant concentrations will decrease, downgradient from a treatment process. This estimate is an essential part of any remedial strategy development, because it supports long-term exposure assessments, guides the spacing of treatment barriers and monitoring locations, and provides the basis for estimating time-to-completion and project life-cycle costs. The contaminant concentration reductions that follow the invasion of clean water usually propagate at a pace much slower than the mobile fraction groundwater velocity. A direct calculation of the clean water propagation velocity is impractical, because we can't collect sufficient information on aquifer matrix structure, groundwater flow and contaminant distributions from natural aquifers. We can, however, predict the patterns of clean water propagation in a diffuse contaminant plume, based on knowledge gained from natural-gradient tracer studies and comparisons of contaminant concentrations in mobile and immobile aquifer strata.

From the time a source mass removal or treatment barrier becomes fully effective, clean water flow is re-established into the head of the contaminated aquifer segment, immediately downgradient from the treatment. Typical aquifers comprise mobile and immobile porosities and the invasion of a clean water front is countered by back-diffusion of contaminants that entered the immobile pore spaces when contaminants were present in the mobile pathways. These processes were depicted in **Figures 7.10**, and **7.11** and **Plate 10** and they can significantly slow the propagation of clean water if the immobile pore space is exposed to contamination for extended periods (allowing large mass accumulations to build).

The clean water invasion front travels in the mobile pore space at a velocity given by Equation (3.36) (the mobile pore space velocity). This is the fastest possible rate of clean water invasion and the time of travel to any distance, x, along the flow path can be calculated as shown in Equation (7.9).

$$t_{invasion\ front} = \frac{x}{V_{avg} \cdot \dfrac{\theta_{total}}{\theta_{mobile}}} \tag{7.9}$$

Equation (7.9), therefore, provides the earliest possible arrival time for an indication of clean water invasion. It is important to note that the porosity driven retardation processes discussed in Chapter 6 (Equation (6.12), in particular), describe the migration rate for the *center of mass* of a solute pulse, but a *solute peak* is slowed to that extent only when mass transfer is perfectly efficient (mass transfer is so fast that mobile and immobile strata are always at equal concentrations). Natural-gradient tracer studies (see Chapter 12) can be used to quantify the mobile pore space velocity and to determine the nature of diffusive interchange between mobile and immobile porosities.

Figure 7.21 examines a hypothetical contaminant plume in which a treatment barrier has been installed to generate a clean water flush. The figure shows solute concentration patterns that might be observed at a point located at a distance, x, from the treatment barrier. The upper graph shows the range of breakthrough curve behavior that can be observed in natural-gradient tracer studies. The log-normal tracer breakthrough curve, indicated by the dark line, occurs with limited mass transfer between mobile and immobile strata, which is typical behavior for natural aquifers. The peak arrival time reflects the groundwater velocity in the mobile porosity.

If the diffusive mass exchange between the mobile and immobile porosities is atypically high, the two compartments would be at equilibrium along the flow path and the solute peak arrival would coincide with the center of mass arrival. This case is indicated by the gray line in the tracer graph.

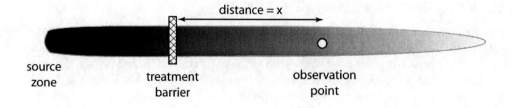

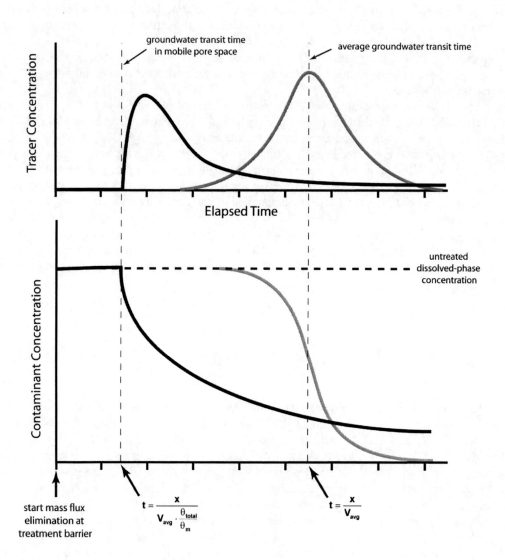

FIGURE 7.21 Migration of a clean water front following contaminant mass flux elimination. The upper panel shows a hypothetical contaminant plume, with a treatment barrier and a downgradient observation point. The upper graph shows tracer pulse breakthrough curves that would be associated with low-mass-transfer (dark line) and high-mass-transfer (gray line) conditions. The lower graph shows the contaminant reductions that might be expected for the high and low-mass-transfer conditions. In the low-mass-transfer case, contaminants slowly leach from the immobile pore space, back into groundwater flowing through the mobile pore space. In the high-mass-transfer case, equilibration between mobile and immobile pore spaces is rapid. Most aquifers behave as low-mass-transfer systems.

For both mass transfer cases, the center of mass for a tracer pulse travels at the average groundwater velocity. The arrival time for the center of mass is provided by Equation (7.10).

$$t_{center\ of\ mass} = \frac{x}{V_{avg}}$$ (7.10)

The shape and timing of contaminant concentration reduction curves in **Figure 7.21** are also determined by the ratio of mobile to total porosity and the rate of diffusive mass transfer between mobile and immobile segments of the aquifer. In the atypical case of extremely high mass transfer rates, the clean water front arrival would be centered at the average groundwater transit time, as shown by the gray curve on the lower panel of **Figure 7.21**. For the more typical case, limited mass transfer, concentration reductions would begin more quickly, as indicated by the dark line on the figure. Earliest indications may occur at the mobile porosity transit time, calculated by Equation (7.9), above.

Although the first indications of clean water invasion occur more quickly in the case of limited mass transfer, the tailing portion of the concentration curve may extend for long periods after the first reductions are observed. The shape of the tailing curve is determined by the mass transfer characteristics of the aquifer and the extent of plume maturation at the observation point, with longer maturation times causing longer-duration concentration tailing.

Figure 7.22 compares the general shape of contaminant concentration reduction curves that would be observed, as a function of the plume maturity at the observation point. If the clean water invasion begins shortly after contaminants first arrived at the observation point, there will be very little contaminant mass stored in the immobile porosity and the clean water invasion is completed quickly. This scenario is represented by the lower curves on **Figure 7.22**. If the aquifer matrix has been exposed to contaminants for an extended period, the recovery process will be quite slow, as represented by the upper curves in the figure. Near the leading edge of a plume, we would expect to achieve rapid cleanup; near the source of an extensive plume, clean water invasion would proceed more slowly, as the longer exposure time would generate larger mass storage in the immobile pore spaces.

The shape and timing of contaminant reduction curves are expected to be very site-specific and we aren't able to provide a calculation that can be broadly applied. This is an area of active research in the academic and practitioner communities. Natural gradient tracer studies are very useful in determining the mass transfer behavior of the matrix, an estimate of the mobile pore fraction, mobile groundwater velocity and average groundwater velocity. Direct measurements of contaminant concentrations can be made in the mobile and immobile strata in a contaminated aquifer matrix, to determine whether a plume is expanding or contracting, as indicated in **Figures 7.11** and **7.12**. At this time, numerical simulation models (ideally calibrated with field observations after clean water invasion starts) provide the best means to estimate clean water invasion rates. The curves are expected to follow the general shapes shown in conceptual **Figures 7.21** and **7.22** and by modeling results given in Chapter 11 (**Figures 11.13** through **11.15**, for example). It is important to note that acceptable groundwater concentrations may be observed in the mobile strata (which dominate monitoring well results with some sampling techniques), even though higher concentrations remain in the immobile strata.

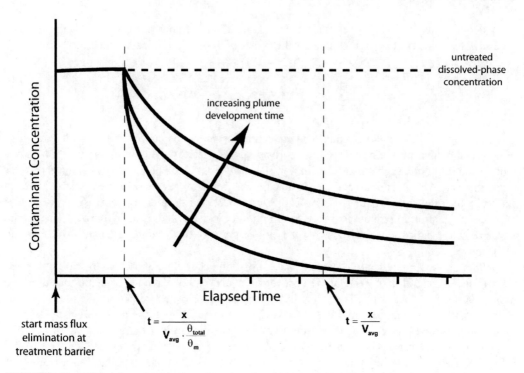

FIGURE 7.22 Effect of contact time on the migration of a clean water front following contaminant mass flux elimination. The tailing concentration in a low-mass-transfer case is determined by the duration of contaminant exposure at the observation point. As the exposure time increases, the mass stored in immobile pore spaces increases as well, increasing the duration and concentration of leaching back into groundwater flowing through the mobile pore spaces.

8 Conceptual Site Models

"Those who assume hypotheses as first principles of their speculation, although they afterwards proceed with the greatest of accuracy, may indeed form an ingenious romance, but a romance it will still be."

Roger Cotes, Preface to Principia Mathematica, Second Edition, 1713[1]

The conceptual site model must, ultimately, answer one fundamental question: *How will the site contamination respond to prospective remedial treatment strategies?* Developing and rigorously testing a conceptual site model is a prerequisite for remedial system design and implementation and jumping into remedial action without an adequate conceptual site model may be the most common source of failure for aquifer restoration projects. Developing *and maintaining* a solid conceptual site model is worth the effort.

We recently undertook a performance review for remediation projects across our company, to determine why some projects were technically outperforming others. Projects were judged on how well they met technical, regulatory and client objectives. We hoped to identify key characteristics of successful projects that could be extended to all projects, to more reliably select the most appropriate remedy for a given set of circumstances. One characteristic of successful projects stood out: all of the high-performance projects followed a pattern of interpretation, analysis, review and revision, *at all stages of project development and operation.* The most successful project teams were highly adaptive and were consistently able to develop and apply technically sound solutions to groundwater restoration problems. Underlying each of the best projects was a good site model that evolved from highly conceptual at the outset, into a more quantitative, predictive form as the project matured. Most importantly, the site models on the most successful projects continued to evolve during remedial system operation.

But what is a conceptual site model? In many ways it can be conceived as a framework for site investigation, evolving from the establishment of project objectives toward the development of a remedial solution for a fully-defined contamination problem. **Figure 8.1** summarizes our view of the development and application of conceptual site models in site investigations and remedial system design and operation. The following sections of this chapter present the details of the process represented in the figure. The focal point of the process is the construction of an integrated understanding of site conditions, inclusive of, and consistent with, all available information. The model must recognize the limitations of our investigative methods and the superficiality of our observations, in order to develop a remedial strategy that can be adapted in response to system behaviors which emerge during implementation.

8.1 ELEMENTS OF THE SITE CONCEPT

The site concept begins to form with answers to a series of fairly standard questions that can be separated into four main categories: the facility and its operations, the site and regional geology,

---

[1] Translated from the original Latin by Andrew Motte in 1729. Quoted from the 2003 edition, Kessinger Publishing.

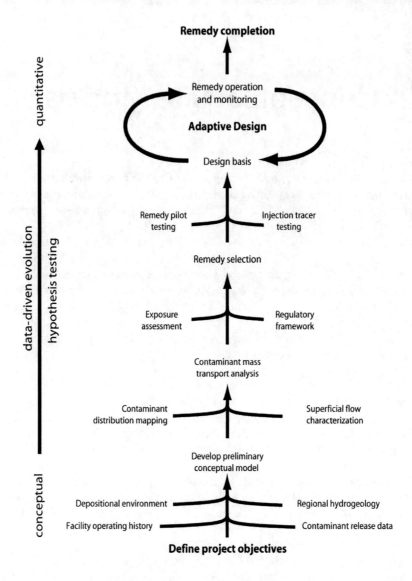

FIGURE 8.1 The conceptual site model development process, highlighting the data-driven evolution of site understanding that develops as the model adapts to information obtained through site investigation and remedy design testing. The evolution continues during system operation, as operational results guide adaptations to the remedy, leading to the most efficient arrival at remedy completion.

the hydrogeology and environmental impacts. The first answers will form a basis for setting project objectives and will guide further investigation. The conceptual model builds through cycles of queries, answers, model revisions and further queries. Each cycle strengthens the model and it evolves to become more quantitative and predictive. Remedial system design and operation provide additional knowledge of the site that should be incorporated into the conceptual site model. These activities continue through the entire site remediation process. The following is a synopsis of the conceptual site model elements:

Site History and Facility Operations — The backbone of a conceptual model is in understanding a facility's history. It establishes the basis for understanding a site, and places reasonable bounds on contaminant distributions. The period of facility operations sets the time frame within which impacts could have occurred. Understanding the use of the facility allows focused assessment of potential impacts related to manufacturing, storage or shipping. Facility management is important as it relates to maintenance of operations (e.g., Was infrastructure replaced on a schedule or only after failure?). This leads to the identification of potential sources related to constituent storage, transfer and use. The use of chemicals always generates a waste stream. Were wastes recycled? How were un-recyclable wastes managed? Were they stored on-site (landfilled or containerized) or were they shipped for off-site disposal? All facilities generate wastewater. Is there an onsite wastewater treatment and was effluent discharged to sewers, surface water, leach fields, infiltration lagoons or disposal wells? Have there been events, either documented, undocumented, or general practice related, where by contaminants could have been released to the environment?

Site Geology and Classification of Site Soils — Site geology should be assessed from two perspectives. The first is at the regional level — understand the geological setting in which the site is located. A summary of regional observations places bounds on likely soil conditions, depositional environment and properties that could be encountered during site investigation. The second is at the site level — Consolidated formations should be identified as sedimentary, igneous or metamorphic and their condition assessed regarding weathering, fractures and competency. Unconsolidated materials should be characterized as to their origin (alluvial, fluvial, eolian or glacial), and all fill placed on site and beneath structures should be identified and characterized, to the extent possible. A summary of the geologic conditions needs to be placed on a geologic cross-section to show the relationships between geologic materials as well as their approximate thicknesses and extents. Geological information-gathering should be quickly narrowed to a site-relevant focus, unless information is developed to indicate large-scale contamination or neighboring sources.

Surface Mapping — There is much valuable data that can be mined from surficial mapping. Geologic contacts are often apparent on surficial topography as changes in strike and slope. Historical quad sheets can be used to identify former locations of streams as well as changes in land surface that could identify potential flow pathways through recent fill material. Digital elevation data and aerial photography are now widely accessible and can provide valuable interpretive support, as well as material for presentations. Aerial photography can provide historical information and helps place the site into context.

Hydrogeology — The hydrogeology characterizes interactions of the groundwater (and surface waters, in many cases) with the geology. The characterization should summarize the saturated and unsaturated formations, identifying aquifers and aquitards, noting whether aquifers are confined, unconfined or perched. The hydrostratigraphy of the site should be developed, as it is one of the most valuable assessments that can be provided by the hydrogeologist. Water elevations (surface and groundwater) should be summarized in plan view and in cross-sections, repeated seasonally to develop an understand of cyclical and long-term trends in site hydrology. Groundwater elevations should be hand contoured to understand the magnitude and direction of hydraulic gradients. Our experience has been that computer generated contours are too often accepted uncritically. We recommend using computer contouring as a support tool in water level interpretation, but the site hydrogeologist cannot abdicate responsibility to analyze the information brought from the field and laboratory.

The physical properties should be quantified for all porous media. The most important properties include a sieve analysis, the hydraulic conductivity, porosity, bulk density and the fraction of organic matter in the soil. The groundwater flux and the groundwater velocity should be estimated. The sources and sinks of water, natural and anthropogenic, also need to be identified to understand the water balance for the site. The available information regarding wells, drains, buried utilities, streams, lakes, precipitation, and percolation (recharge) should be summarized. The background geochemistry and native mineralogy needs to be summarized as it can affect remedy design.

Contaminants — The extent of contamination needs to be established and contoured for extent and magnitude. The contouring, similar to water levels, must be hand drawn by a hydrogeologist to incorporate the effects of geology, groundwater and transport behavior to estimate the extent of contamination. Documenting the extent and magnitude of groundwater plumes and how concentrations change over time is essential to assessing natural attenuation processes and remedy performance. The conceptual site model should document all known source and impacts and the correlation between plumes and source areas. A lack of correlation between ongoing sources and groundwater plumes is an indication of undefined pathways, and poorly defined sources. The plume should be considered to be a long term tracer test and used to estimate minimum contaminant velocities — a 1000-foot plume originating from a facility which operated for less than 10 years is moving at a velocity in excess of 100 feet per year. A summary of the documented fate and behavior for identified constituents is similarly important as it bounds reasonable expectations for residual mass and rates of degradation.

8.2 ACCOUNTING FOR REALISTIC AQUIFER BEHAVIORS

A major reason for developing this book and the focus of the preceding chapters has been to re-cast the way we view groundwater flow and contaminant migration patterns, recognizing that superficial averaging does not provide sufficient detail to support contaminant source and plume mapping and remedial system design. We strongly believe there is a need to restore the role of hydrogeologists as members of multi-disciplinary science and engineering teams, in the examination of aquifers and contaminant transport. Hydrogeologist input is critical to obtaining the realistic and detailed understanding required to build successful groundwater restorations. Here are a few key points in support of this contention:

Anisotropic groundwater flows — The data we see in high-resolution aquifer characterization suggest that a very large fraction of the groundwater flows through a very small fraction of the aquifer volume in many, perhaps most, aquifers. **Figure 8.2** shows a cumulative flow distribution for detailed core data from the Borden aquifer, given in Rivett, et al. (2001). This is the most homogeneous distribution of permeabilities we're likely to see, with 20 percent of the aquifer matrix carrying 45 percent of the flow. Most aquifers we examine are likely to have much greater concentrations of flow (although we're unlikely to have such detailed data). Environmental professionals are somewhat accustomed to anisotropic flows in the vertical dimension, but comparable anisotropies exist on the horizontal axes, as well. The chance of capturing the most permeable fraction of a formation with a small number of randomly placed borings is essentially nil. We get the best chance of correctly assessing contaminant migration patterns when we incorporate knowledge of aquifer matrix depositional processes at the site and use that as a guide to map the aquifer structure, placing a priority on locating the high-flow pathways.

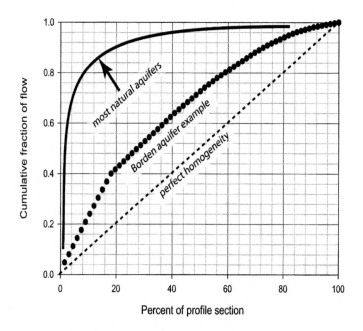

FIGURE 8.2 Distribution of flow in vertical profiles. The dashed line shows the flow distribution for a perfectly homogeneous aquifer. The circles show the flow distribution for a core from the Borden aquifer, calculated from the Core 1 data in Figure 3.11. In this very homogeneous natural aquifer, 45 percent of the flow would occur in 20 percent of the vertical profile (data extracted from Rivett, et al., 2001). For most natural aquifers, flow will be concentrated in a smaller fraction of the aquifer pore space, as indicated by the solid line.

Groundwater velocities — Real velocities are typically much higher than aquifer average values — this impacts plume prospecting as well as remedial technology planning. Contaminants have often traveled farther and faster than project teams are prepared to observe.

Near-zero transverse dispersivity — A wide spread on the contamination plume may indicate multiple source points. Historically, assumptions of high transverse dispersivity led project teams to assume that a single source could explain a wide-swath of contamination. We have seen multiple sites where the conceptual site model understated the scope of the contaminant source area, due to an unrealistic assumption for transverse dispersivity.

No outliers, no censoring — Avoid declaring any data points to be outliers. All data should be accommodated in the conceptual site model and if site data doesn't fit the conceptual model, the problem is more likely with the model than with the data. Above all, never use a site conceptual model to determine whether to accept or reject any data. The conceptual site model is not a filter or censoring device – if the data doesn't fit, adapt the model. We have observed many cases in which data critical to understanding a site were set aside because they conflicted with the conceptual site model.

Don't trust mysterious disappearance — Some of the most costly mistakes we have observed were generated by project teams that were willing to accept observations that didn't make sense technically, but the results were too good to question. Make sure that all observations are explainable within the conceptual site model.

8.3 TESTING CONCEPTUAL SITE MODELS

Successful conceptual site models are not static — they must be continually refined and tested, by data that arrives in the course of investigation and remedial activities and by specific data collections indicated by inconsistencies in the model. All stakeholders need to be aware of the evolving nature of high-quality conceptual models and revisions need to be welcomed. The following three case studies are offered as examples of some of the critical determinations and testing that accompany site model development. In the first, we compare the predictions of dispersion modeling to actual plume conditions, reinforcing our expectation that transverse dispersivity should be assumed to be near-zero. This has a significant impact on where we look for contaminants and how fast they may be moving (narrow plumes, fast movement – that is the norm). The second case study demonstrates a model that needed to be tested, due to logical inconsistencies and the risk associated with an incorrect interpretation of plume migration patterns. The third case study shows the importance of testing critical design assumptions, groundwater velocity in this case, prior to finalizing a remedial design. This case study utilized some of the advanced tracer study approaches outlined later, in Chapter 12.

Case Study 1

The site for the first conceptual site model case study was a military training base in the Northeastern United States, active from 1968 to 1996. This site provides a good example of a large source mass that has reached a quasi-equilibrium with the groundwater system. The area is largely undisturbed by pumping and is not confounded by other contaminant sources. It provides a good opportunity to see some of the plume behaviors we've been describing in earlier chapters.

Environmental surveys that began following site closure in 1996 noted that perchloroethene had been used at the site and that handling practices at the site discharged perchloroethene to the subsurface, through a dry well and leach field. Shipping manifests indicated solvent was used at the site between 1969 and 1992. Solvent handling practices would have first introduced perchloroethene to the subsurface in 1969.

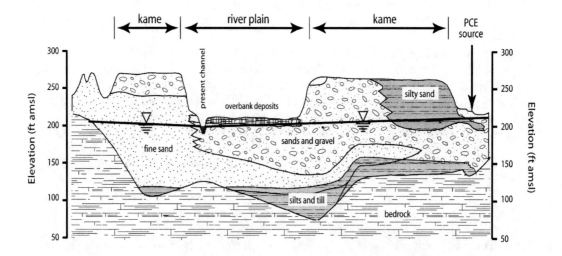

FIGURE 8.3 Site geology cross-section for the conceptual model in Case 1. Groundwater flows from the perchloroethene (PCE) source area to the river floodplain in permeable sands and gravels of the kame deposit.

The base was set on a glacial kame deposit, a large mound of stratified ice contact deposits. **Figure 8.3** gives a cross-section of the site geology, indicating the perchloroethene source area. The groundwater slopes at a 0.004 m/m gradient, to a discharge interface in a river plain, approximately 900 m from the source area. A perchloroethene source mass was discovered in the complex post-glacial aquifer system, on the upgradient side of the kame deposit, as shown in **Figure 8.3**.

Figure 8.4 shows the zone where we might prospect for the presence of perchloroethene, based on advection-dispersion approach, expressed in Equation (5.4). This approach, which we do not recommend, suggests that transverse dispersivity spreads a contaminant plume laterally. As can been seen in the figure, the plume would be expected to spread significantly as it moved toward a discharge boundary at the river. The concentration would also be expected to decline broadly, due to the dilution incurred in the dispersion process. If the plume reaches the river, Equation (5.4) (assuming transverse dispersivity is 1 percent of the plume length, a standard assumption) suggests that the 5 ug/L concentration would spread laterally more than 900 feet to each side of the plume axis. The calculation also suggests that the peak concentration would be reduced to less than 100 ug/L within the first 1,700 feet of travel.

Figure 8.5 shows the actual plume, which has reached the river flood plain and clearly exhibits a near-zero transverse dispersivity. In Chapter 5, we argued that transverse dispersivity is likely to be near-zero and this contaminant plume is an example of what we should expect, instead. The 1,000 ug/L concentrations reach most of the 900 m length and the plume retains the same width along its entire length.

The source mass began to accumulate in 1968 and the plume had reached its current extents by 2000. We can deduce from this that the contaminant front traveled at a velocity greater than 28 m/year (91 ft/year). Considering the retardation of solute migration that occurs due to sorption and porosity effects, the groundwater velocity required to drive the observed contaminant migration must be significantly greater than 91 ft/year.

> ***Testing the model for consistency*** — Aquifer permeability testing and elevation gradient data indicate the average groundwater velocity is 170 ft/year in the downgradient portion of the plume, and less than 20 ft/year near the source. These observations are consistent with the observed plume extent. Subsequent reagent injections and passive in-well flux meters showed groundwater velocities ranging from 1 to 2 ft per day which is also consistent with the earlier velocity estimates, understanding that the peak groundwater velocities are always expected to be significantly higher than the average values.

The selected remedy for the plume is enhanced reductive dechlorination, covering the source zone and deployed in reactive barriers that intersect the plume at two locations along its path. The time to achieve treatment is determined by the rate of propagation of the clean water front departing each reactive barrier. The fast-flowing groundwater allows the reactive barriers to be spaced at larger intervals than would be possible in slow-flowing systems. However, bleed-back of dissolved-phase mass stored in the immobile aquifer pore space and sorbed in the aquifer organic carbon still will control the time required to achieve complete restoration.

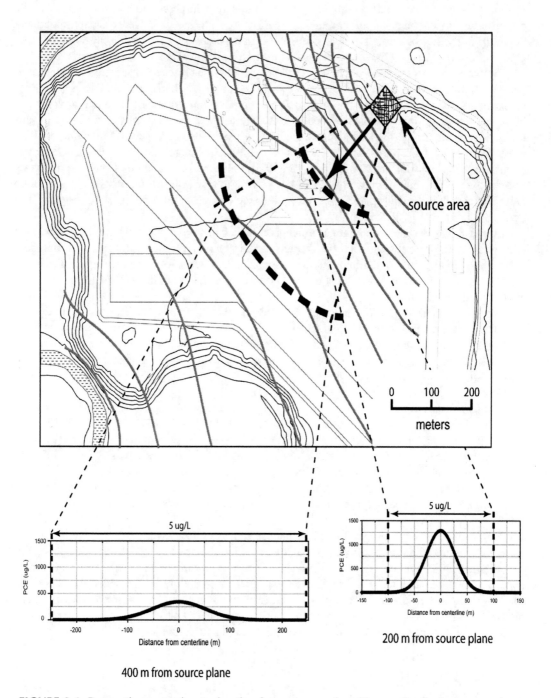

FIGURE 8.4 Prospective contaminant migration from a source plane 30 m perpendicular to groundwater flow. Graphs built from Equation (5.4), with transverse dispersivity assumed as 0.01 x distance along the plume axis. Comparing to Figure 8.5 (the actual plume), there was effectively no transverse dispersion and the dispersion-guided search pattern could miss the plume.

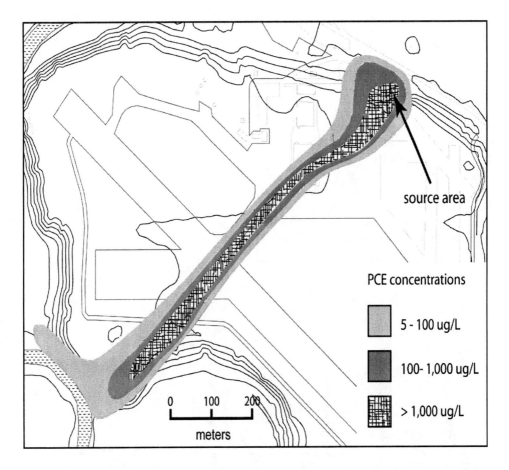

source area

PCE concentrations

[ ] 5 - 100 ug/L

[ ] 100 - 1,000 ug/L

[ ] > 1,000 ug/L

0   100   200

meters

FIGURE 8.5 Plume map for perchloroethene release in Case 1. The plume is approximately 900 meters long and less than 100 meters wide along its path, until it reaches the discharge boundary in the river plain.

CASE STUDY 2

The second case study calls attention to the importance of evaluating groundwater flow as a vector, rather than a scalar. The study site is in the Midwestern United States, at a location where 1,1-dichloroethene was released to groundwater. The plume reached an area bordering a wetland, at concentrations exceeding 1,000 ug/L. Concentrations just above the detection limit (1 to 5 ug/L) were observed in seepage entering the wetland. The initial conceptual site model considered the wetland as a discharge boundary for groundwater in that area. However, if the plume continued its migration beyond the wetland, sensitive receptors could be at-risk. This critical assumption of the conceptual site model required testing, due to the potential risk associated with an incorrect assessment.

The site is a relict dune of the early post-glacial era and although average hydraulic conductivities are generally high, there is significant cross-bedding in many areas. Earlier reagent injection testing at another area of the site showed that flow pathways are complex and it is not safe to assume homogeneous permeability distributions.

The area near the groundwater-wetland interface was logged with a cone penetrometer and sufficient monitoring wells and piezometers have been placed in the area to support preparation of a vertical groundwater elevation and soil permeability profile near the area of interest. **Figure 8.6** summarizes the information available to test the key conceptual site model assumptions. **Figure 3.13** showed the analysis of groundwater flow vectors for this location. The vertical component of the groundwater elevation gradient is 20 times the magnitude of the horizontal elevation gradient, so the driving force clearly favors discharge to the wetland and provided the basis for the original concept of the wetland as a discharge boundary for this portion of the aquifer. However, when the detailed permeability distribution became available with the cone penetrometer study, the flow vector could be calculated and it showed that the horizontal flow path would be dominant. Only a small portion of the flow discharges to the wetland interface, which helps explain why the contaminant concentrations have never approached the levels observed at depth (an inconsistency in the conceptual site model that should be taken as an indication the model may need revision).

The project team extended the subsurface exploration beyond the wetland, an area that had been considered safe from contamination, due to the presumed surface water discharge of the plume. Groundwater sampling indicated that the plume had traveled more than 1,000 feet farther than expected and was nearing a sensitive receptor. This experience reinforces the notion that groundwater flows are determined by permeability patterns in the aquifer matrix, often to a greater extent than by elevation gradients.

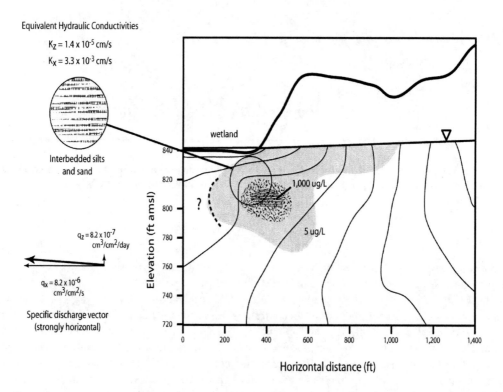

FIGURE 8.6 Revision of a conceptual model to account for specific discharge vector. The aquifer matrix structure in this location is the dominant factor controlling groundwater flow direction. Although small amounts of contaminant had been discovered in the wetland, the revised understanding of groundwater flow allowed re-direction of sampling efforts. Contamination was discovered 1,000 feet farther, along the horizontal flow component. Refer to Figure 3.13 for more information on the specific discharge vector analysis.

Case Study 3

The final case study is from a prospective site for an *in situ* reactive zone technology in the Southwestern United States. A geochemical reactive zone strategy (chemical precipitation) was under consideration after preliminary aquifer testing had been conducted at the site. The remedy requires that precipitation-inducing chemical conditions be sustained for a period of weeks to months, so the cost of the selected remedy is sensitive to groundwater flow through the treated volume (reagent mass and injection frequency are both dependent on groundwater flow rates). Early testing indicated groundwater velocities near 18 cm/day, but the aquifer segment where the technology application is planned lies in a different area of the site. Consequently, it was necessary to conduct detailed tracer studies to characterize groundwater flows in the targeted area and to obtain final remedy design parameters.

A three-fold tracer package was used to quantify groundwater flow:

Negative conductivity — Site groundwater is high in dissolved solids, so the background electrical conductivity exceeds 3,000 uS. A salt-based positive tracer would generate a very high density solution that may have flow characteristics distinct from local groundwater. De-ionized water was used as a negative tracer (read more on negative tracers in Chapter 12), with a conductivity of 18 uS.

Fluorescein — The organic dye fluorescein was added to the de-ionized water to generate a UV-spectrometer signal (1,000 ug/L). Fluorescein concentrations were quantified in off-site laboratory analyses.

Thermal — The injected tracer solution was warmer than the groundwater by approximately 10 °C, which generated a temperature signal near the injection zone.

Approximately 38.5 m^3 of tracer was injected over an 18-hour period and monitoring was conducted from 5 hours prior, to approximately 500 hours after, injection startup. The tracer injection well and all the observation wells were screened over 4.6-meter intervals. The injected chemical tracers reached 50 percent of their peak responses after 7.7 m^3 of injection, at the response well located 4.6 m from the injection well. From these data, we can determine the mobile porosity is approximately 3 percent of the aquifer volume [based on Equation (12.2)]. This is at the low end of the range of mobile porosities we have observed for unconsolidated aquifer matrices and leads us to expect high groundwater velocities. The tracer results for observation wells located 10 and 15 meters from the injection well did, in fact, show very high transport velocities for all tracers.

Figure 8.7 shows the tracer study layout. The injection well and two observation wells (4.6 and 10 m from the injection point) were aligned on the estimated groundwater flow path. Two additional observation wells were placed off the flow path, at approximately 15 m from the injection point. Tracer was only observed in observation wells A, B and C, indicating a slight bias in the flow direction. **Figures 8.8** and **8.9** show breakthrough curves for the observation wells located at 10 and 15 m, respectively, from the injection point. We calculated the center of mass for each breakthrough curve, which is indicated by the vertical dashed line on each graph. Centers of mass for the two chemical tracers agreed closely at each location. The center of mass (equals average groundwater) velocity for well B was 3 m/day and for well C was 6 m/day. The center of mass for the well B breakthrough curves was influenced by slight increases in concentration, arriving late in the observation period. This may represent a thin layer of slower-moving tracer that did not arrive at well C during the study period. Peak velocities were approximately 10 m/day at both of the study locations.

The tracer study made it clear that groundwater flows are highly concentrated in the target area, as indicated by the low mobile porosity, and the remedy design must be re-evaluated to determine whether the selected strategy can be accomplished cost-effectively. The site conceptual model process, shown in **Figure 8.1,** requires testing of critical hypotheses — in this case, that the groundwater velocity was 18 cm/day. That element of the conceptual site model process saved the project from a significant remedy design error.

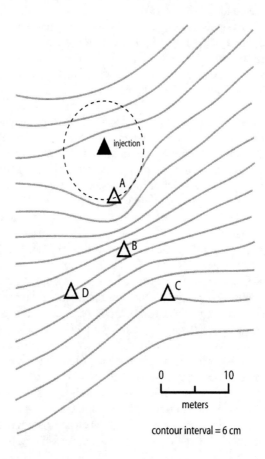

FIGURE 8.7 Layout for Case 3 tracer injection and observation wells. Approximate groundwater elevation contours are shown at 6-cm intervals. Tracer was injected until 60 percent of the injected reagent strength was observed at well A. The dashed circle indicates the hypothetical radius of full-strength reagent injection.

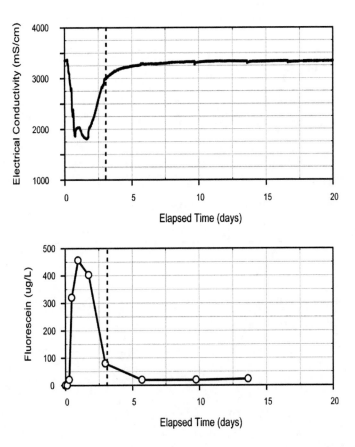

FIGURE 8.8 Tracer breakthrough curves for observation well B in Case Study 3. De-ionized water was used to generate a negative conductivity tracer and fluorescein was used as a confirming chemical tracer. The dashed lines indicate the center of mass for each curve, which arrived at approximately 3 days for each tracer. This well is located 10 m from the injection well. The average groundwater velocity was approximately 3 m/day to this location and the peak velocity was almost 10 m/day.

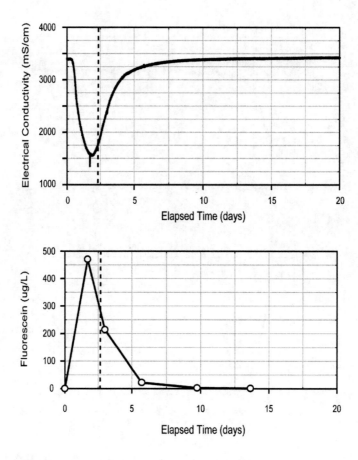

FIGURE 8.9 Tracer breakthrough curves for observation well C in Case Study 3. De-ionized water was used to generate a negative conductivity tracer and fluorescein was used as a confirming chemical tracer. The dashed lines indicate the center of mass for each curve, which arrived at approximately 2.5 days for each tracer. This well is located 15 m from the injection well. The average groundwater velocity was approximately 6 m/day to this location and the peak velocity was approximately 10 m/day.

9 Hydrostratigraphic Characterization

> "Hydrogeology has been too much inclined toward *hydraulics* and solving of the flow equations, and not enough toward *geology* and understanding/describing the rock structure, facies and properties in a geologically realistic manner, thus proposing "exact" solutions, but to poorly posed problems"
>
> *Ghislain de Marsily[1]*

Hydrostratigraphy is a term that has increasingly gained traction by practioners of *in situ* remediation methods, primarily because of the need to develop quantitative means of describing behavior of contaminants or reagents in aquifer systems. The genesis of **hydrostratigraphy** stems from "hydro" meaning water and "stratigraphy" meaning study of sedimentary sequences or even more simply, sedimentary layers. As introduced in Chapter 2, the sedimentary sequences comprising aquifer systems constitute the megascopic building blocks that control three-dimensional aquifer architecture. While sedimentary deposits are typically laid down in more or less horizontal layers, the individual facies comprising sequences and depositional elements demonstrate a self-organizing three-dimensional structure that translates into a complex, but characteristic relationship that is determined by the sequence of depositional and erosional processes from the macro- to the micro-scale. The result is that the "layer-cake" approach to stratigraphy is oversimplified, lacking the detail in mapping permeability structures that control reagent distribution and contaminant transport.

Depositional models provide insight into how the depositional processes associated with particular geologic environments impart texture and structure on sedimentary basins, petroleum reservoirs and aquifers. The primary sedimentary depositional environments of interest to remediation hydrogeologists include: alluvial systems (rivers and alluvial fans), deltaic and coastal systems, aeolian systems (wind) and glacial systems. Entire books and even series of books have been devoted to the depositional processes and stratigraphy related to subsets of these environments. Among the depositional environments, alluvial systems are considered most important to stratigraphy on the continental mainland because of their widespread influence, but also because alluvial processes contribute to erosion and sedimentation in deltaic, coastal, and glacial systems. In this book, we survey the methods of hydrostratigraphic analysis using alluvial/ fluvial systems to illustrate the concepts. The interested reader is encouraged to seek out literature for other depositional environments and to probe deeper in sedimentary processes through the many textbooks and monographs available in the literature.

We advocate a stratigraphy approach that has been successfully used in the petroleum industry for many years. Reservoir engineers use depositional models to establish a framework for interpreting reservoir architecture and developing strategies to locate and produce petroleum and natural gas reserves. Their approach emphasizes the distinction between "reservoir quality

---

[1] de Marsily, G. et al., 2005. Dealing with spatial heterogeneity. *Hydrogeology Journal*, 13: 161-183.

formations," which are high-energy deposits with well-developed, coarse grain-size distributions resulting in sufficient permeability for cost-effective petroleum recovery, and "stratigraphic traps", which are low-energy deposits with fine grain-size distributions that are characterized by low permeability and serve as boundaries for the reservoirs.

Reservoir engineers and petroleum geologists develop their interpretations using vertical profile facies relationships that are demonstrated in down-hole geophysical logs and facies relationships between boreholes to map the structure and geometry of the reservoir. In the petroleum vernacular, a well-sorted, coarse-grained facies is termed **well-developed**, referring to its capacity to function as an efficient reservoir because the coarse-grain fraction is well-developed in the formation. By similar convention, fine-grained facies are termed **poorly-developed**. Seismic data are used to add three-dimensional controls in interpreting the structure of the reservoirs, including stratigraphic tops and bottoms and geometry of faults. Key among the factors in their interpretation is the co-location of petroleum in the reservoir quality formations.

In remediation hydraulics, we advocate an approach that utilizes depositional models and facies relationships to map the aquifer characteristics between boreholes and provide a framework for evaluating how the three-dimensional aquifer architecture controls the flow of groundwater, reagents and contaminants within the system. To meet this objective, our approach to hydrostratigraphy integrates the mapping of the well-developed, coarse-grained facies within aquifers and poorly-developed, fine-grained facies within the aquifers and aquitards, with the three-dimensional distribution of contaminants in the system. We also recommend focused hydraulic testing and injection tracer testing methods to correlate permeability and facies trends comprising the aquifer, so that quantitative interpretations regarding flow and transport can be made.

In order to develop successful *in situ* remedy solutions, we have to refine the scale of our interpretations to understand how reagents and contaminants move within the depositional forms that make up the highest permeability segments in the aquifers. This can only be accomplished using high-resolution soil and groundwater sampling and hydraulic testing techniques. Soil sampling should be completed using either continuous sample collection and characterization, or cone penetrometer testing methods for stratigraphic interpretations. It is also necessary to go beyond typical soil classification methods and develop stratigraphic interpretations that focus on vertical-profile facies relationships and facies trends between borings, using knowledge of the depositional environment to aid in the interpretation process. We recommend the use of stratigraphic logging techniques to facilitate this analysis when soil sampling and logging methods are employed. To correlate the plume distribution with the hydrostratigraphy, it is necessary to employ vertical aquifer profile groundwater sampling at multiple depth intervals in the aquifer system.

High-resolution hydraulic- and tracer-testing methods are required to develop quantitative understanding of distribution of flow and transport within the higher-permeability facies and to characterize the interaction with the lower permeability facies. We provide a discussion of stratigraphic contrasts at the end of this chapter to aid practitioners in understanding the relevance of descriptive hydrostratigraphic logging and how it supports quantitative mass transport assessments. The concept is based on the idea that three-dimensional aquifer architecture leads to anisotropy in permeability in plan view, much as it does in the vertical profile. **Chapter 11** provides an overview of hydraulic testing methods that are well-suited to mapping out the aquifer architecture and developing quantitative conceptual site models. **Chapter 12** discusses the methods and techniques of tracer testing, which we find indispensable for understanding contaminant transport and reagent distribution at the remediation hydrogeology scale.

Ultimately, these methods enable practitioners to map out the architecture of the well-developed facies comprising the mobile porosity (advective migration pathways) and the poorly developed facies that comprise the immobile porosity (less mobile, diffusive pathways and storage reservoirs). To effectively apply these techniques, it is essential that interpretations are made by trained geologists and hydrogeologists who not only understand the concepts through their academic training and experience, but also approach it with enthusiasm when given the opportunity to practice quantitative geology. We feel strongly that our industry needs to recognize the importance of providing career opportunities for field geologists and hydrogeologists, if we expect to be successful in the endeavor of *in situ* remediation.

9.1 SOIL CLASSIFICATION SYSTEMS

Several different classification systems have been developed to provide quantitative descriptions of the nature of soils and sediments. Two of the most relevant to practitioners in the environmental industry are the United States Department of Agriculture's (USDA's) *Soil Taxonomy* (NRCS, 1999) and the American Society for Testing of Materials (ASTM's) *Unified Soil Classification System* (USCS) (ASTM, 2000).

The USDA's *Soil Taxonomy* (NRCS, 1999) was developed for agricultural drainage and agronomy purposes. The method provides a detailed classification approach to soils that emphasizes characteristics that are important to plant growth and agricultural management. It is often used in ecological assessments, but has limited utility in remediation hydraulics. One of the key criteria for soil classification under the USDA's taxonomy approach is grain size and relative proportions of sand-silt-clay, as illustrated in **Figure 9.1**, after Shirazi and Boersma (Shirazi and Boersma, 1984). The ternary sand-silt-clay soil classification diagram uses the term **loam** in its nomenclature to denote soil mixtures that are predominantly sand and silt, with lesser fractions of clay. Loam is broadly used to denote soils that are conducive to agriculture, in part because of the balance between

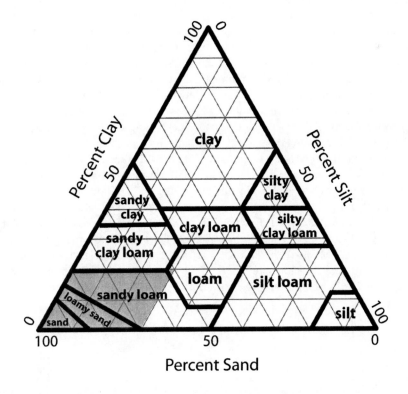

FIGURE 9.1 Soil classification as a function of sand, silt and clay composition. Redrawn from Shirazi and Boersma, 1984. The shaded area shows the range of soil composition most supportive of groundwater flow.

moisture retention and drainage. From the prespective of remediation hydraulics, it is important to note that only the sand, loamy sand and sandy loam classifications have the permeability necessary to enable effective *in situ* remediation. The permeability associated with the rest of the soils in the ternary diagram is marginal, or too low to allow *in situ* remediation by conventional means.

The most widely used classification system in engineering disciplines is the Unified Soil Classification System (USCS), which was developed by Arthur Casagrande in 1942 for construction of airfields during World War II (Das, 1986). The ASTM *Standard Practice for Description and Identification of Soils (Visual-Manual Procedure) — ASTM D-2488* (ASTM, 2000) updated and extended the initial USCS system to the current form that is used today, primarily for engineering and construction applications. **Table 9.1** provides a summary of the unified soil classification system as described in the ASTM standard.

The fundamental basis for classification in the USCS and ASTM system is grain size: soils are classified as coarse-grained if greater than 50 percent of the material is retained on the number 200 sieve (0.074 mm diameter or silt using the USCS basis for classification); soils are classified as fine-grained if more than 50 percent of the grains pass through the number 200 sieve.[2] Considering the USDA ternary diagram in **Figure 9.1**, it is apparent that the generic coarse-grained classification provides little insight regarding the permeability of the soils, as the "coarse grained" soil could contain nearly 50 percent silt or clay.

Subdivision of the coarse-grained soil classification is based on relative fractions of gravel, sand and fines. Subdivision of the fine-grained soils is based on whether the primary fraction is silt or clay, and the relative fraction of coarse-grained minor fractions (sand and gravel). The method relies on visual examination of grain-size fractions and morphology, and simple manual tests to standardize field-classification of fine-grained soils. However, the method requires geotechnical testing of the soil to determine liquid- and plastic-limits, which are required to definitively distinguish between organic and inorganic silts and clays.

The ASTM method prescribes a systematic review of morphological characteristics including angularity, shape, color, odor, moisture content, consistency, cementation, structure and grain-size distribution. Fine-grained soils are also field-tested using manual procedures to evaluate dilatancy, plasticity, dry strength and toughness — characteristics that aid in field-classification of silts and clays. In general, the ASTM approach promotes a uniform engineering classification of soils through consistent use of terminology. The primary deficiencies of the method from a remediation hydrogeology perspective are that the taxonomy is not geared toward evaluating permeability and engineering terms are favored over geologic terms in describing the morphological characteristics, and there is a particular lack of emphasis on stratigraphy and facies trends.

Two fundamental differences between the ASTM USCS standard and geologic terminology are grain-size classification and characterization of the uniformity of grain-size distributions. **Table 9.2** compares the grain-size classification among the ASTM USCS and Modified Burmeister (Burmeister, 1979)(engineering classifications), USDA (agronomy classification) and Udden-Wentworth (geologic classification) systems. The ASTM USCS approach leads to the fewest grain-size classifications, as the primary objective is to distinguish between gravel, sand and silts and clays. The Udden-Wentworth classification method leads to finer divisions among the sands and silts because the class divisions are based on the phi-unit ($\log_2$ of grain diameter). There are also subtle differences in minor grain-size class divisions between gravel and sand and very-fine grained sand and silt.

---

[2]   Under the ASTM method, the number 200 mesh size grain diameter of 0.074 mm corresponds to silt. Referring back to Table 2.1, the Udden-Wentworth classification based on fractional diameters characterizes this grain size as very-fine sand.

TABLE 9.1

Summary of major-division soil classifications using the Unified Soil Classification System (USCS) from ASTM D 2487 (ASTM, 2000). Classification is based on whether the primary fraction (greater than 50%) of the soil is coarser or finer than silt, using the USCS grain-size designation. Coarse-grained subdivisions are based on fraction of gravel and sand; fine-grained subdivisions are based on fraction of silt and clay. The reader should consult the ASTM standard for a detailed field methodology and taxonomy for descriptive soil classification.

Major Divisions			Group Symbol	Typical Names
Coarse-Grained Soils More than 50% retained on the 0.075 mm (No. 200) sieve	Gravels 50% or more of coarse fraction retained on 4.75 mm (No. 4) sieve	Clean Gravels	GW	Well-graded gravels, gravel and sand mixtures — little or no fines
			GP	Poorly-graded gravels, gravel and sand mixtures — little or no fines
		Gravel with Fines	GM	Silty-gravels; gravel, sand, and silt mixtures
			GC	Clayey-gravels, gravel-sand-clay mixtures
	Sands 50% or more of coarse fraction passes the 4.75 mm (No. 4) sieve	Clean Sand	SW	Well-graded sands and gravelly-sands - little or no fines
			SP	Poorly-graded sands and gravelly sands — little or no fines
		Sand with Fines	SM	Silty-sands, sand and silt mixtures
			SC	Clayey-sands, sand and clay mixtures
Fine-Grained Soils More than 50% passes the 0.075 mm (No. 200) sieve	Silts and Clays Liquid Limit 50% or less		ML	Inorganic silts, very-fine sands, rock flour, silty- or clayey-fine-sands
			CL	Inorganic clays of low- to medium-plasticity; gravelly/ sandy/ silty/lean clays
			OL	Organic silts and organic silty-clays of low-plasticity
	Silts and Clays Liquid Limit greater than 50%		MH	Inorganic silts, micaceous or diatomaceous fine-sands or silts, elastic-silts
			CH	Inorganic clays or high-plasticity, fat clays
			OH	Organic clays of medium- to high-plasticity
Highly Organic Soils			PT	Peat, muck, and other highly-organic soils

Prefix: G = gravel, S = sand, M = silt, C = clay, O = organic

Suffix: W = well-graded, P = poorly-graded, M=silty, L=clay with liquid limit less than 50%, H = clay with liquid limit greater than 50%

ASTM, 2000. Standard Practice for Description and Identification of Soils (Visual-Manual Procedure), West Conshocken, PA.

TABLE 9.2

Comparison of grain-size classification systems for sediments (references footnoted).

US Standard sieve mesh	Millimeter scale Decimal	Fraction	ASTM USCS [1]	Burmeister [2]	USDA [3]	Udden-Wentworth [4]
	4096		boulders		cobbles	boulders
	1024					
	256	256				
	64	64	cobbles			cobbles
	16		coarse gravel	coarse gravel	medium gravel	pebbles
5	4	4		medium gravel		
6	3.36		fine gravel		fine gravel	
7	2.83			fine gravel		granules
8	2.38		coarse sand			
10	2	2				
12	1.68		medium sand	coarse sand	very coarse sand	very coarse sand
14	1.41					
16	1.19					
18	1	1				
20	0.84				coarse sand	coarse sand
25	0.71					
30	0.59					
35	0.5	1/2				
40	0.42		fine sand	medium sand	medium sand	medium sand
45	0.35					
50	0.3					
60	0.25	1/4				
70	0.21			fine sand	fine sand	fine sand
80	0.177					
100	0.149					
120	0.125	1/8		fine sand		
140	0.105				very fine sand	very fine sand
170	0.088					
200	0.074					
230	0.0625	1/16	silts and clays	coarse silt		coarse silt
270	0.053				silt	
325	0.044					
	0.037			fine silt		
	0.031	1/32				medium silt
	0.0156	1/64				
	0.0078	1/128				fine silt / very fine silt
	0.0039	1/256				
	0.002			clay	clay	clay
	0.00098					
	0.00049					
	0.00024					
	0.00012					
	0.00006					

(US Standard sieve mesh lower rows: *use pipette or hydrometer*)

[1]ASTM, 2000. *Standard Practice for Description and Identification of Soils (Visual-Manual Procedure)*, West Conshocken, Pennsylvania.

[2]Burmeister, D.M., 1979. *Suggested method of test for identification of soils*, Symposium on identification and classification of soils.

[3]NRCS, 1999. *Soil Taxonomy - a Basic System of Soil Classification for Making and Interpreting Soil Surveys. Agriculture Handbook.* U.S. Government Printing Office, Washington, DC 20402.

[4]Leeder, M.R., 1982. Sedimentology: Process and Product. George Allen & Unwin Publishers, Ltd., London, England, 344 pp.

Depositional Forms and Processes

As indicated in **Chapter 2**, we prefer the Udden-Wentworth classification because the geometric progression of grain-size diameter also reflects relationships that are important when considering the erosion and transport of sediments during the depositional process. The correlation between increasing grain size and degree of sorting and permeability is most important, as permeability structure is responsible for the mobile and immobile porosity within aquifer systems. Therefore, it is critical to develop soil descriptions that elucidate the link between depositional process and hydrostratigraphy: the energy and variability of the sediment transport process impart the permeability structure that creates the aquifer architecture.

$$\text{Energy} \Rightarrow \text{grain size} \Rightarrow \text{permeability}$$

High-energy depositional processes form well-developed facies that comprise mobile porosity; low-energy depositional processes form poorly-developed facies that comprise immobile porosity.

$$\text{Variability in energy} \Rightarrow \text{sorting or fine-scale structure}$$

Gradual fluctuations in energy lead to fine-scale variability, such as interbedding, or gradational grain size changes in the profile (grading); rapid fluctuations in energy lead to poorly-sorted, complex bedforms. Fine-scale and micro-scale depositional elements control the degree of communication between the advective pathways (mobile porosity) and diffusive transport and storage reservoirs (immobile porosity).

As discussed in **Chapter 2**, incipient motion of grains occurs when the gravity force is balanced with the drag force. From a macroscopic perspective, the relationship between sediment transport and erosion is dictated by the stream power and shear force acting on the grains. For a river, the stream power, P_s, is related to the volumetric discharge rate, Q_s, fluid density including suspended sediment, ρ, and channel width, w, as shown in Equation (9.1) after Raudkivi (Raudkivi, 1990):

$$P_s = \frac{\rho \cdot g \cdot Q_s}{w} \qquad (\frac{W}{m^2}) \qquad\qquad (9.1)$$

Note that the volumetric flow rate is the product of the mean discharge velocity, V_m, and the river stage depth, d, and width of the channel.

The shear stress, τ, acting on the river bed depends on the stage depth, density of the fluid including entrained sediment, and the slope of the channel, α, as shown in Equation (9.2) after Knighton (Knighton, 1987):

$$\tau = \rho \cdot g \cdot d \cdot \tan(\alpha) \qquad \frac{dynes}{cm^2} \qquad (9.2)$$

The common variables in Equations (9.1) and (9.2) imply that the mean velocity and shear stress control the process of sediment transport and erosion. Equation (9.3) after Simons and Richardson (Simons and Richardson, 1966), shows that the product of shear stress and velocity equals the product of stream power and the slope of the channel, S.

$$\tau \cdot v_r = \frac{\rho \cdot g \cdot Q_s}{w} \cdot S \qquad (9.3)$$

This equality, termed unit stream power, implies that the stream power increases with velocity, stage depth and slope of the channel.

Figure 9.2 illustrates the relationship between stream velocity and sediment transport processes after Hjustrom (Hjulstrom, 1935) and Raudkivi (Raudkivi, 1990). The empirical relationship

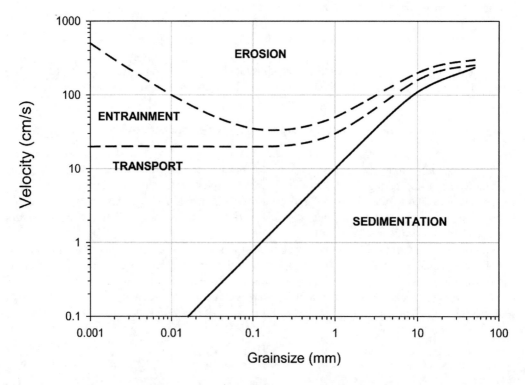

FIGURE 9.2 Erosion and deposition criteria for sediment, as a function of stream flow velocity and mean grain size. After Hjulstrom (1935) and Raudkivi (1990).

illustrates the transition from sedimentation to grain transport, in which grain movement is initiated by rolling or saltating (bouncing) along the bed, through **entrainment**, which consists of bulk movement of grains in suspension without net losses of sediments, and **erosion**, where the grains are fully suspended and there is a net loss of sediment. It is notable that the diagram uses a log-scale for both the velocity and grain size. The u-shaped curve demarcating the transition between transport and erosion reflects the resistance of clays to erosion (on the left) and the geometric progression of grain mass with increasing diameter (on the right). The reverse transition from erosion, entrainment and transport to sedimentation with decreasing velocity is relatively abrupt for coarse-grains because of the mass relationship. Finer-grained sediments, such as silt and clay, remain entrained for a much larger range of velocity conditions.

The implication of this empirical relationship is that steady or gradually varying velocity conditions result in hydraulic sorting of the sediment grains, which produces well-sorted deposits or gradational trends in grain-size distribution that accompany bedding forms. Abrupt changes in the flow velocity can result in sedimentation spanning a broad range of grain sizes because interparticle interference enhances sedimentation rates. Poorly sorted deposits reflect dynamic changes in velocity and stream power. We advocate that sorting be used to describe the uniformity of grain-size distribution because it is a reflection of the variability of the energy associated with the depositional process. We also recommend that the term "grading" should be used to describe grain-size distribution trends in the vertical profile, as these morphological characteristics provide insight into the facies trends that are so important in making stratigraphic interpretations.

Figure 9.3 illustrates the evolution of bedforms that accompanies increasing unit stream power for grain-size classes ranging between coarse-grained sand and very-fine-grained sand. This diagram reflects sediment transport processes that occur in the transport and entrainment regions shown on **Figure 9.2**. At low-levels of unit stream power, grains begin motion by rolling and saltating (**bedload transport**), forming plane beds under conditions of limited transport. With increasing stream power in the low-flow regime, ripples form and give way to larger amplitude dunes as increasing grain sizes are transported in the bedload. As stream power increases to the upper-flow regime, dunes transition to upper plane beds and ultimately antidunes, which migrate upstream under the turbulent flow conditions characterized by standing waves in the river. The height of ripples, dunes and antidunes typically correlate with depth of flow in the channel (Knighton, 1987).

FACIES AND DEPOSITIONAL MODELS

The bedforms shown in **Figure 9.3** represent the most general relationship between stream power and fluvial sedimentation processes. The nature of the river or fluvial style determines the characteristics of the sedimentary deposits that are formed by rivers over long periods of time. The fluvial style or river form is influenced by the size and volume of sediment, slope of the channel, volume and variability of discharge and characteristics of the basin through which the river is evolving (Collinson, 1996). The basic patterns of fluvial style include braided, anastomosing, meandering and straight rivers, as illustrated on **Figure 9.4**. **Table 9.3** compares relationships among the fluvial styles.

Braided rivers form in the steepest reaches where coarse-grained sediment is abundant and discharge varies widely. As a result of the dynamic sediment load and flow, braided river channels separate and recombine frequently, often forming bands of roughly parallel braids of channels. Alluvial fans form at the margins of mountains and canyons, eroding the coarsest sediments during ephemeral flood events. Channels formed on alluvial fans are typically braided, but often distributary (diverging) in nature. In middle river reaches where slopes are modest and finer-grained sediment is abundant, rivers develop a single winding channel in the alluvial plain — hence the name meandering river. Anastomosing rivers exhibit the characteristics of braided and meandering rivers because of the dynamic range of flows and finer grained sediment load. Straight

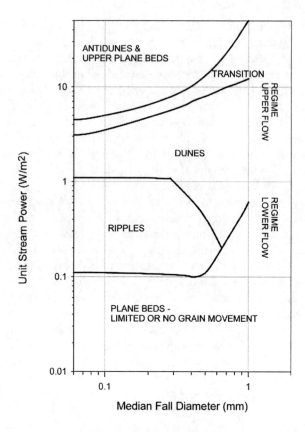

FIGURE 9.3 Bedform evolution and stability, plotted as function of the unit stream power and median fall diameter for bed materials [after Simons and Richardson (1966)]. As transport capacity increases with stream power in the lower flow regime, bedforms evolve, increasing the height above the bed to form plane beds, ripples and dunes. With continued increases in stream power in the upper flow regime, dunes transition to plane beds and antidunes.

rivers commonly occur in upper reaches within canyons, or in the lower reaches of rivers where slopes are very flat.

Figure 9.5 illustrates a conceptual depositional model for a meandering river system. As shown in the three-dimensional model, the sediments that are deposited by the system range from gravel and coarse-grained sands (shown as gravel and sand patterns) to fine-grained silts and clays (shown as dark shaded areas). In a meandering river system, the locus of coarse-grained deposition (accretion) is at the point-bar, which is the concave side of each meander. As the meander migrates laterally within the basin, erosion occurs at the convex side of the river bend and the point bar is extended in the direction of the meander migration. The process of horizontal accumulation of sediments during channel migration is termed **lateral accretion**. Channel migration results in lateral continuity of the point bar in the direction of the meander migration, or transverse to the general river flow direction. In general, the thickness of the channel facies and characteristic grain size decreases with distance up- and downgradient from the point bar. This is due to erosional scouring at the convex side of the meander, which cuts the channel deeper into the underlying sediments, and decreased velocity and depth of flow at the point bar.

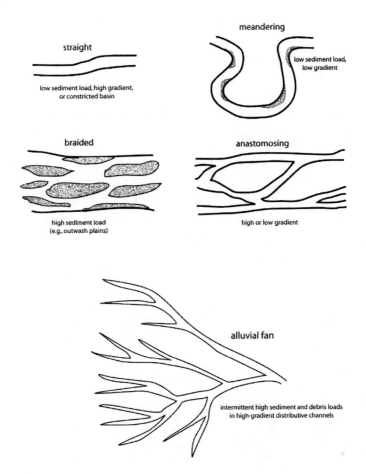

FIGURE 9.4 Basic fluvial styles.

The degree of point bar continuity in the direction parallel to flow (longitudinal direction) depends on the spacing and sinuosity of the meanders (analogous to phase and amplitude of sine wave): if meanders are tightly spaced, accretion over time at adjacent point bars can merge and even overlap, resulting in longitudinal point bar facies continuity; however, if meanders are separated, there will be discontinuity of the point bar facies in the longitudinal direction. Temporal variability in flow and sediment supply and spatial variability in the underlying sediment's resistance to erosion result in variability in the meander shape and migration path within the meander belt in the basin. The fluvial meander and point bar depositional process typically imparts an architecture that results in higher-degrees of lateral continuity in well-developed, coarse-grained facies in the direction of the point bar migration compared to the longitudinal direction. As a consequence, an aquifer created by this process would tend to exhibit higher permeability in the direction of the point bar migration, potentially leading to channels that are transverse to the overall flow direction of the river.

As the meander channel migrates away from aggrading point bars, overbank flood events deposit fine-grained sands, silts and clays in the flats comprising the flood plain. Because the stream power dissipates rapidly away from the channel during these events, hydraulic sorting results in deposits that grade from sand, nearest the channel, to silt and clay at distal points. This

TABLE 9.3
Fluvial style relationships after Collinson (Collinson, 1996) and Raudkivi (Raudkivi, 1990).
River forms can change along the length of the river from its headwaters, where slopes are highest and basins are smaller, through its middle reach, where basin size is increasing and slopes are decreasing, to its mouth, where the basin is largest and slope is least. Grain size typically decreases from headwater to mouth. Fluvial form at river mouth depends on tidal influence.

Fluvial Style	Environment	Slope	Nature of Flows	Sediment types	Examples
Alluvial Fan	headwaters; canyons and mountains; small basins	> 0.02%	Flash floods, ephemeral	bed-load; boulders, cobbles, gravel, and sand	Margins of Rocky Mountains, western US.
Braided River	headwaters; glacial and mountain; constrained basins	>0.015%	Dynamic	bed-load to mixed-load; boulders, cobbles, gravel, sand	Bramaphutra, India Platte, Nebraska
Anastomosing River	middle to mouth; tropical, vegetated; very large basins	0.015% to 0.002 %	Very high, dynamic	mixed-load to suspended-load; Sand, silt, and clay	Amazon
Meandering River	middle to mouth; continental; very large basins	>0.002%	varies	Mixed-load to suspended load; sand, silt, and clay	Mississippi
Straight River	middle; constrained basins	<0.002%	varies	suspended-load; sand, silt, and clay	Columbia, Oregon

Collinson, J.D., 1996. Alluvial sediments. In: H.G. Reading (Editor), *Sedimentary environments: processes, facies, and stratigraphy.* Blackwell Science, Ltd., Malden, Massachusetts, pp. 37-82.
Raudkivi, A.J., 1990. *Loose Boundary Hydraulics.* Pergamon Press, Inc., Elmsford, New York, 538 pp.

lateral grading process results in sediment accumulation in the vertical profile that transitions from gravel and sand at the point bar, to finer sand during proximal overbank deposition, to silt and clay during distal deposition. Accumulation of overbank deposits leads to **vertical accretion**, or basin fill away from the active channel. When the sedimentary sequence of deposits is viewed in profile, one sees gravel and sand at the bottom transition to fine-sand, silt, and then clay near the top. This vertical facies profile is the idealized **fining-upward sequence** that typifies fluvial depositional processes, as shown on **Figure 9.6.**

In real fluvial systems, the energy levels can vary gradually or fluctuate dynamically. Gradual variations in energy result in gradational changes in grain-size distribution (grading) within the fining-upward sequence; episodically varying energy levels result in stratified grading; dynamic,

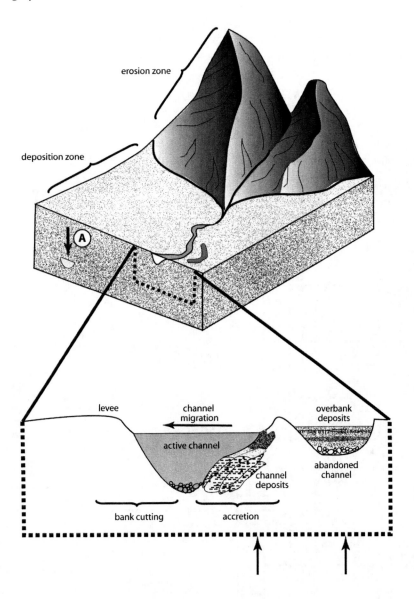

FIGURE 9.5 Diagram showing some of the erosion and deposition processes that generate hydraulic conductivity patterns in aquifers. A. Over long time intervals, deposition causes subsidence and older channels can be encountered at depth, relative to the active channel. Arrows indicate location of conceptual stratigraphic logs shown in Figures 9.6 and 9.7.

cyclic variations result in oscillations between sedimentation of sand and silt, creating interbedding at the top of the fining upward sequence. **Figure 9.7** shows two stratigraphic logs that might be obtained from an aquifer formed by the meandering river depositional environment. The log to the left illustrates a well-developed channel that demonstrates the idealized channel fill succession and fining-upward sequence, which one would expect if the boring were placed at the point bar. The log to the right illustrates a fining upward sequence, but notice that the thickness of the channel facies is less and the overbank facies thickness is greater than the log to the right.

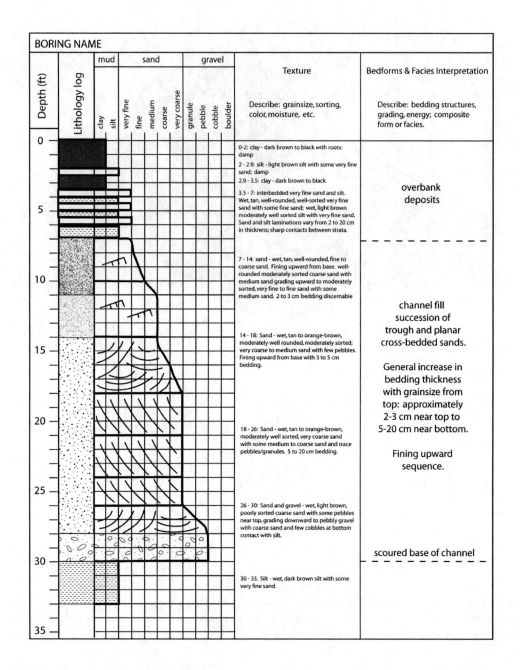

FIGURE 9.6 Stratigraphic log for idealized fining upward sequence, resulting from channel succession in a meandering river depositional environment.

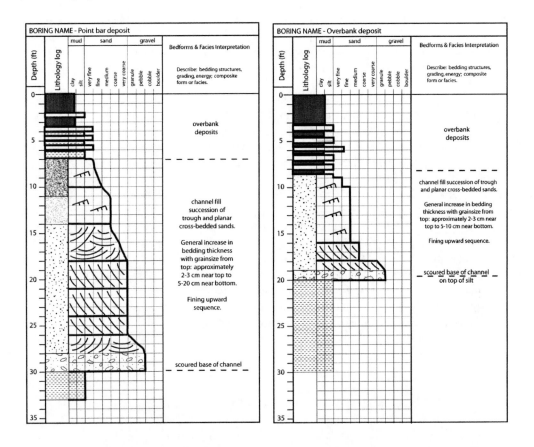

FIGURE 9.7 Stratigraphic logs showing lateral facies trends in a meandering stream depositional environment. The log on the left corresponds to the point bar area and the log on the right corresponds to the area of overbank deposits, highlighted in Figure 9.5.

The facies trend between these boreholes indicates that the coarser-grained facies are less well developed at the boring log on the right, which suggests that the channel was located toward the boring to the left. When developing cross-sections and considering flow in such an aquifer, this facies trend would suggest that the permeability of the point bar facies is decreasing to the right, as a result of decreasing grain size that accompanies the "pinch out" of the point bar and replacement by the less-permeable overbank facies away from the point bar. The implication is that gravel and sand bodies formed through the fluvial sedimentation process are expected to be lens- and pod-shaped, resulting in permeability structure that reflects the architecture of the well-developed depositional elements comprising the point bar, rather than a horizontal layer-cake grading from gravel and sand to silt and clay.

Periodically during storm events, higher flows overtop the natural levees on the river banks, creating a temporary breach in the levee and depositing sands as a **crevasse splay** deposit. Crevasse splay processes dissipate energy rapidly as the flow area expands away from the channel into the flats, resulting in poorly sorted, fan-shaped deposits. During periods of exceptional flow, the channel jumps its bank creating a new channel and abandoning the old, leaving behind an oxbow lake. The oxbow lakes and distal-point bars then accumulate silt and clay during overbank episodes because the energy dissipates rapidly away from the active channel. Oxbow lakes and slack-water depressions in the floodplain also accumulate vegetation, forming organic-rich deposits and peats. **Plate 12** shows the current and past meandering river channel for the Holitna River, in Western Alaska. The historical trace of the Holitna's channel meanders are shown clearly by the arcing striations inside each river bend and the oxbow lakes, which resulted from the past channel abandonment.

The frequency of **avulsion** (channel abandonment) and vertical accretion/basin subsidence rates control the macroscopic architecture of the aquifers formed through the fluvial process. **Figure 9.8** (after Leeder, 1999) illustrates the lateral and vertical connectivity between channel facies as a function of avulsion and vertical accretion/basin subsidence. The cross-section view is taken from a vantage point that looks parallel to the river flow. The avulsion rate affects the width of the channel facies and vertical accretion/subsidence rates affect the thickness of overbank facies that separate the individual channels. Systems that have high frequency of avulsion and vertical accretion rates (upper left) result in relatively narrow channels that are separated in the lateral and vertical directions. Under similar vertical accretion rates, less frequent avulsions would result in wider channels that overlap (upper right), potentially showing continuous channel facies in the horizontal direction. At the lower left, where avulsion and subsidence rates are low, the individual channel facies overlap and abut each other, resulting in a high degree of vertical and lateral connectivity between channels. When viewed in the vertical profile, the individual channel facies would show the fining-upward sequence, but the thickness and characteristics of overbank facies would vary based on the location of the profile relative to the channel and balance between avulsion and subsidence rates.

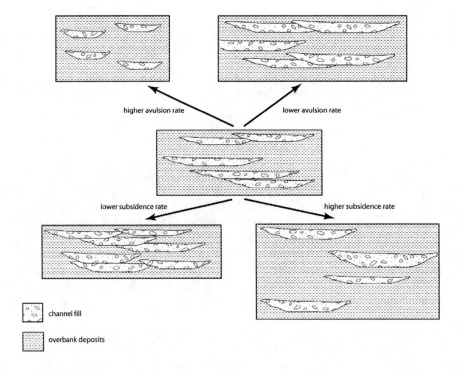

FIGURE 9.8 Conceptual relationship between channel avulsion (abandonment) rate and the vertical accretion (subsidence) rate in fluvial formed aquifer architecture. The channel avulsion rate affects the width of the channel facies; the vertical accretion rate affects the lateral and vertical connectivity of channels. Redrawn from (Leeder, 1999).

Plate 13 shows the Delta River near Fort Greely in Alaska. The Delta River is situated in an active outwash plain, downgradient of the many glaciers associated with the mountains of the Alaska Range. The active channels within the braided river system are shown in the darker gray-scales, while the inactive channels are shown in the lighter gray-scales. The dramatic range of flow and sediment load associated with this river system results in dynamic river channel formation and abandonment that occurs daily and seasonally with the melting of the headwater glaciers. Under high-flow conditions, braided rivers such as the Delta are capable of moving boulders, cobbles and gravel, as well as significant amounts of sand and silt. When channel flow ebbs, the largest

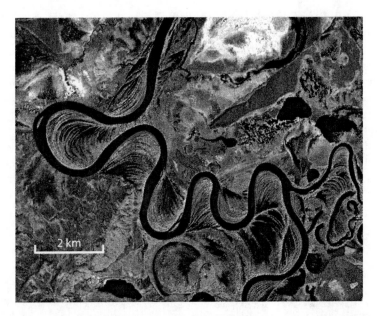

PLATE 12 Depositional patterns associated with river meanders; Holitna River, Alaska. *(See Plate section for color version of Plate 12.)*

PLATE 13 Braided stream channels on the Delta River near Fort Greely, Alaska. *(See Plate section for color version of Plate 13.)*

particles drop-out forming gravel bars within the channels, later being overtopped by successively finer sands and silts. During the next episode of high-flow conditions, the presence of the recently formed gravel bars can divert flow, creating new channels with successive gravel bar deposits forming during the cyclic waxing and waning of flow. This cyclic process repeats itself across a spectrum of spatial and temporal scales, resulting in complex tributary (converging confluences of channels), distributary (diverging confluences of channels) and anastomosing (reconnecting channels) patterns.

9.2 STRATIGRAPHIC LOGGING TECHNIQUE

Stratigraphic logging techniques were developed by geologists to map facies relationships in the vertical profile at individual borings or bedrock outcrops. The utility of the method lies in the fact that vertical trends in grain size and bedforms provide an indication of the depositional process that formed the sedimentary deposit. Taken in the context of an appropriate depositional model based on a site's geologic setting, the facies relationships between adjacent borings provide a rational basis for interpolating the hydrostratigraphy at our remediation sites. Stratigraphic logging techniques provide a framework for evaluating the three-dimensional structure and continuity of the coarse-grained sediments comprising the aquifer matrix, which advectively transport the majority of contaminant mass and reagents, and distinguishing it from the finer-grained aquitards that act as reservoirs for contaminant mass and exert a controlling influence on transport pathways.

CREATING STRATIGRAPHIC LOGS

Figure 9.7, presented earlier, showed the fining upward sequence and lateral facies trends that accompany channel abandonment in a meandering fluvial system. The graphical logging techniques that were used to illustrate the grain-size trends facilitate interpretation of the vertical facies trends at each boring and provide a rationale for interpreting the hydrostratigraphy between borings.

Obviously, the first step in preparing stratigraphic logs is to develop geologically based interpretations of soils. High-quality, continuous soil sampling methods, such as rotosonic, geoprobe or CPT, are required to provide a continuous record of the soil in profile. The geologist's focus in characterizing the soil should be to develop accurate descriptions of facies trends that are manifested through grain-size distribution, sorting, grading, bedforms and interbed contacts. From a practical perspective, the Udden-Wentworth system is required to resolve the classes of sand that are important when equating grain size to permeability and distinguishing the mobile and immobile portions of the aquifer. Sorting and grading enable an assessment of the variability of the permeability structure within individual facies, particularly when bedforms are not discernable in soil samples. Bedforms provide additional clues about the permeability structure of the aquifer, as they comprise the depositional elements that make up the coarse-grained and fine-grained bodies that are the mobile and immobile porosity at the remediation hydrogeology scale. The nature of the contacts between beds indicates whether an abrupt change in conditions created a formation of interest, or if there was a gradual change. **Erosional contacts** are sharp, often crossing boundaries of underlying sedimentary deposits, indicating the high-level of energy associated with the sedimentation process. **Conformable contacts**, where successive deposits are draped upon one another, indicate lower energy or resistance of the underlying bed.

The second step is to use a stratigraphic log to graphically depict the grain-size distribution, sorting, grading and bedforms to facilitate the interpretation of the vertical facies profile. The format for the stratigraphic log presented in **Figure 9.7** was adapted from those that are commonly used by stratigraphers and presented in most stratigraphy and sedimentology books. It includes a conventional lithology log to illustrate dominant grain size graphically, a grain-size scale that graphically depicts the vertical grain-size trends and two columns for soil descriptions. The texture column can be used for the descriptors that are conventionally used in the ASTM USCS approach, recognizing that geological terminology is most important. The second column is used to interpret facies relationships and infer boundaries of composite bedforms. It is common to include facies code descriptions in the lithology log, such as those proposed by Miall (1985) and summarized on **Table 9.4**.

Grain-size plots illustrate facies relationships, in the same way that downhole gamma logs show clay content, and electrical resistivity logs show permeability (using Archie's Law). **Figure 9.9** illustrates common examples of gamma ray response curves for different type depositional environments. Here, we use the grain size as a surrogate for permeability and we use the shapes of the curves to infer grading and facies trends, making interpretations regarding the aquifer quality and architecture in a manner similar to geophysicists interpreting logs to evaluate reservoir quality and structure.

When developing stratigraphic logs, bear in mind that many of the features are manifested at the centimeter and decimeter scale, making explicit logging of clay varves or interbedded silt and sand practically impossible for a 10 meter (33 ft) or longer boring. A useful convention is to represent these fine-scale and micro-scale features using serration patterns (zig-zags) on the interval of interest, to illustrate the cyclic nature of the process. For those trained in stratigraphy, the zig-zag shape conjures mental images of cyclic facies trends, while blocky shapes indicate channels or uniform facies.

Gamma Response

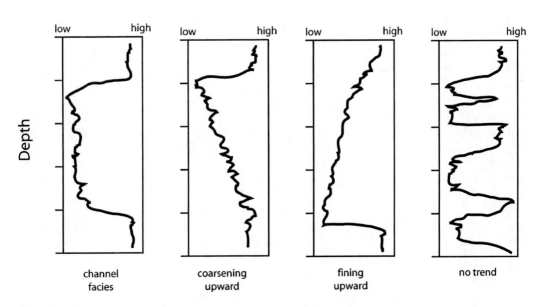

FIGURE 9.9 Relationship between gamma ray response and clay content, to illustrate vertical facies profiles. The shape of the gamma ray response curve shows the vertical trends in clay content, with clay content positively correlated to the gamma response. Sand content is inferred to be negatively correlated with gamma response. The leftmost frame illustrates a block shape, which implies channel facies; the second frame shows an upward-decreasing gamma response, indicating a coarsening-upward sequence; the third frame shows an upward-increasing gamma response, which implies a fining-upward sequence. The fourth panel shows no trend in gamma response, consistent with flood plain or coastal margins.

TABLE 9.4

Facies classification system after Miall (Miall, 1985). The facies classification extends lithologic classification of sediments by using morphological characteristics and trends, including grain-size distribution, bedding forms and interbed contacts to infer depositional processes. The examples provided below were developed to aid in the interpretation of fluvial systems.

Facies code	Facies	Sedimentary structures	Interpretation
Gmm	Matrix-supported, massive gravel	Weak grading	Plastic debris flow (high-strength, viscous)
Gmg	Matrix-supported gravel	Inverse to normal grading	Pseudoplastic debris flow (low strength, viscous)
Gci	Clast-supported gravel	Inverse grading	Clast-rich debris flow (high strength), or pseudoplastic debris flow (low strength)
Gcm	Clast-supported massive gravel		Pseudoplastic debris flow (inertial bedload, turbulent flow)
Gh	Clast-supported, crudely bedded gravel	Horizontal bedding, imbrication	Longitudinal bedforms, lag deposits, sieve deposits
Gt	Gravel, stratified	Trough cross-beds	Minor channel fills
Gp	Gravel, stratified	Planar cross-beds	Transverse bedforms, deltaic growths from older bar remnants
St	Sand, fine to very coarse, may be pebbly	Solitary or grouped trough cross-beds	Sinuous-crested and linguoid (3-D) dunes
Sp	Sand, fine to very coarse, may be pebbly	Solitary or grouped planar cross-beds	Transverse and linguoid bedforms (2-D dunes)
Sr	Sand, very fine to coarse	Ripple cross-lamination	Ripples (lower flow regime)
Sh	Sand, very fine to coarse, may be pebbly	Horizontal lamination parting or streaming lineation	Plane-bed flow (critical flow)
Sl	Sand, very fine to coarse, may be pebbly	Low-angle (< 15°) cross-beds	Scour fills, humpback or washed-out dunes, antidunes
Ss	Sand, very fine to coarse, may be pebbly	Broad, shallow scours	Scour fill
Sm	Sand, fine to coarse	Massive, or faint lamination	Sediment-gravity flow deposits
Fl	Sand, silt, mud	Fine lamination, very small ripples	Overbank, abandoned channel, or waning flood deposits
Fsm	Silt, mud	Massive	Backswamp or abandoned channel deposits
Fm	Mud, silt	Massive, desiccation cracks	Overbank, abandoned channel, or drape deposits
Fr	Mud, silt	Massive, roots, bioturbation	Root bed, incipient soil
C	Coal, carbonaceous mud	Plant, mud films	Vegetated swamp deposits
P	Paleosol carbonate (calcite, siderite)	Pedogenic features: nodules, filaments	Soil with chemical precipitation

Miall, A.D., 1985. Architectural-element analysis: a new method of facies analysis applied to fluvial deposits. Earth Science Review, 22: 261-308.

9.3 SAMPLING METHODS

Hydrostratigraphic mapping requires the use of high resolution soil and plume characterization methods. Our goal is to provide a detailed mapping of the aquifer architecture that differentiates between the coarse-grained facies, which constitute the mobile porosity in the aquifer, from the fine-grained facies, which constitute the immobile porosity, at the fine and stratigraphic-scales. We superimpose the distribution of the plumes on the architecture to interpret how contaminant mass moves and is stored in the system. This approach is essential in the interpretation of plume geometry and migration rates when hydraulically-based groundwater-flow directions do not match, but also to develop quantitative conceptual site models that are necessary in framing reasonable expectations regarding reagent distribution for *in situ* remediation and time to achieve clean-up goals.

SOIL SAMPLING AND LOGGING METHODS

Hydrostratigraphic logging and facies interpretations require continuous observations over the entire interval of interest. For physical examination, continuous sample collection is required and for geophysical observation, continuous logging is required. The best drilling and sampling methods cause minimal disturbance of the soil and afford an opportunity to recover the maximum volume from each sample. For soil sampling, we recommend direct-push or sonic methods and for geophysical logging, we recommend cone-penetrometer testing (CPT). These methods are better-adapted to continuous, minimum-disturbance sampling than conventional hollow-stem auger or rotary drilling techniques. Each of the recommended methods has limitations and advantages that need to be evaluated when planning subsurface characterization.

DIRECT-PUSH TECHNIQUES

In shallow conditions, direct-push methods provide for continuous soil sampling and high-quality soil samples in sandy and finer-grained environments. Recovery is limited in coarse-grained deposits, as the inside-diameter of the sampling tool is normally 5 cm (2 inches) or less. Direct-push rigs can be mounted on skids, tracks and all-terrain vehicles enabling access where larger sonic and CPT rigs cannot go. Direct-push rigs are typically capable of drilling to depths of 25 meters (80 feet) in normal situations, and do so without use of water or drilling fluids. Exploration depths can be limited in glacial environments with abundant gravel or very stiff tills. Direct-push rigs can also be outfitted with probes that aid in lithologic characterization (e.g., conductivity) or plume delineation (i.e., membrane interface probes, etc.) and can be used in the installation of small diameter wells. Several different types of screens and samplers can also be used to collect groundwater samples for vertical aquifer profiling.

SONIC TECHNIQUES

At significant depth, or in cases of heaving sands beneath the water table, sonic methods are preferred, as the method is capable of advancing an outer casing to minimize this interference when the core barrel is advanced and retrieved. The outer casing is a particular advantage when advancing borings across multiple aquifer units, as it serves as a temporary casing to isolate upper formations. Sonic rigs typically advance a 10-centimeter (4-inch) core barrel and are capable of retrieving gravel-sized grains when a basket is used. In dense, cobbly-gravels it is not uncommon to retrieve rock plugs resulting from coring of the cobble-sized fractions. Large sonic rigs are capable of drilling more than 100 m (300+ ft) with very high production rates, as the larger rigs typically advance 3- or 6-m (10- or 20-ft) rods.

Water is typically used to advance the casings, but it can be avoided at the cost of decreased production rates. Sonic rigs are capable of installing large-diameter wells; however, rotary methods might be preferred for well installation in certain difficult drilling conditions. Although the unit rates are higher for sonic rigs, their high production rates typically lead to least cost exploration when considering duration of the project. The mini-sonic rig overcomes the primary obstacle of the customary large drill rig, which can limit access to important locations in many cases. Because of the competitive cost and high-production rates, mini-sonic rigs can compete with direct-push rigs in difficult or deeper drilling conditions.

Cone penetrometer testing

Cone penetrometer testing (CPT) methods use sensors and indirect sampling methods to characterize soils by their behavior under probe-induced stress, without actually retrieving samples. CPT methods were originally developed for geotechnical applications, but have become increasingly popular for environmental characterization methods since the Site Characterization and Analysis Penetrometer System (SCAPS) program was demonstrated for the Department of Defense (Myers et al., 2002). CPT has the particular advantage that many different probes and sensors have been developed to aid in the classification of soils (gamma, resisitivity, conductivity, seismic) and provide screening-level assessments of water quality (laser induced fluorescence for aromatic hydrocarbons, membrane interface probe for a variety of volatile organic compounds, Raman spectral analysis — calibrated to particular compounds).

Conventional application of CPT has been used for characterizing soils in the vertical profile in geotechnical engineering applications since the early 1960s. Since the early 1980s, practitioners have utilized the **soil behavior type** methodology after Robertson and Campanella (1983) and Robertson (1986) to classify soils using CPT. The soil behavior type is derived from measurements obtained by three primary sensors as the tool is advanced in soils: cone tip pressure – measures resistance to penetration; cone sleeve friction – measures friction along the wall of the cylinder of the probe; and pore pressure – measures the hydrostatic pressure in the saturated soils. Measurements are obtained from the three sensors at 3 to 4 centimeter intervals (8 to 10 samples per foot). The method is ideally suited to differentiating between aquifer matrix (sands and gravels) from aquitards (silts and clays) because of the way the sensors respond to soil characteristics:

Tip stress (resistance) is proportional to grain size, when soils exhibit comparable levels of compaction or in-place density. Higher tip stress implies increased grain size, but can also indicate cementation, very dense soil conditions, and over-consolidated conditions in clays.

Sleeve stress (friction) is proportional to clay and silt content because of their cohesive nature. Generally, higher sleeve friction implies greater silt and clay content. The ratio of sleeve friction to tip stress, termed the friction ratio, is commonly used with tip resistance in the soil behavior type algorithm.

Pore pressure – hydrostatic pressure increases linearly with depth below the water table. As the probe is advanced in the saturated zone, it must displace groundwater. High permeability sands and gravels readily dissipate the water, typically yielding measurements close to the hydrostatic pressure. Silts and clays do not readily dissipate the water resulting in negative and positive excess pore pressure readings at the sensor. A key feature of silts is that they are dilatant — that is they are cohesionless to the extent that they expand when disturbed, but are characterized by relatively low permeability. As a result, silts commonly show "negative" pore-pressure deviations from hydrostatic pressure during probe advancement. In contrast, clays do not dilate and are characterized by low permeability. When the CPT probe is advanced in clays, excess pore pressure cannot dissipate and positive pressure spikes are measured in excess of hydrostatic pressure. The pore-pressure response is a key factor in distinguishing permeable sand and gravel aquifer matrix from less permeable silts and clays comprising aquitards. When quantitative permeability estimates are required, the CPT probe can also be used to measure the rate of pore pressure dissipation in moderately permeable soils[3] and estimate hydraulic conductivity.

CPT probe responses must be calibrated against continuously-sampled soil borings to ensure accurate interpretation. While the Robertson and Campanella soil behavior type classification algorithm is useful as a frame of reference, the divisions are generally more applicable to engineering classification than hydrostratigraphy. We favor making manual correlations between tip stress, sleeve friction, friction ratio and pore pressure responses to develop site-specific interpretations. CPT's utility for distinguishing between aquifer matrix (sands and gravels) and aquitards (silts and clays) is improved by adding electrical resisitivity or conductivity probes to the instrumentation. Sands and gravels exhibit higher electrical resisitivity (lower conductivity) than silts and clays because of the cations in the aquitard materials, typically resulting in a contrasting electrical signature between the soil types under normal conditions.

The utility of the CPT logging technique is that, once calibrated to site conditions, it is possible to use the tip stress, sleeve friction and pore-pressure responses in the vertical profile much like a stratigraphic log. **Figure 9.10** shows an example CPT sounding from a site in Florida that is approximately 1 kilometer (0.6 mile) from the Gulf of Mexico, where the present day depositional model is barrier islands and lagoons. The format of the CPT output varies by vendor, but this one is simple to use and interpret. Note that the units are tons per ft[2] (TSF). The first column in the chart shows the sleeve stress, a measure of friction, which is used as one indicator of silts and clays. Silt and clay content is also evaluated using the third column. Ratio COR (a vendor-specific term) is the ratio of the sleeve stress to the tip stress (more commonly known as the friction ratio). Ratio COR is a more sensitive indicator of silt and clay than sleeve stress. The fourth, Pore Pressure, is the most important sensor for discriminating between silts and clays (due to dilatancy differences). The second column shows the tip stress or resistance profile, which is an indication of grain size. The fifth and sixth columns graphically depict the soil behavior type, after Robertson and Campanella (1986). We typically use the soil behavior type only as a preliminary guide to interpreting the soils, then develop detailed interpretations using the responses of the CPT tools after calibrating the responses to site conditions.

Figure 9.11 shows an expanded view of the CPT vendor report for the 15 to 45-ft interval of the log, using a slightly different combination of plots to facilitate stratigraphic interpretations. We prefer to combine the friction ratio and sleeve stress on one graph, the tip stress on one graph, and the pore pressure and hydrostatic pressure plots on a third panel, to emphasize the positive and negative deflections that indicate clays and silts. The simplest way to interpret CPT profiles is to start with the pore pressure and friction ratio (Ratio COR) plots to identify "marker beds" for stratigraphic correlation and facies interpretations.

---

[3] Pore-pressure dissipation tests can be applied to soils with hydraulic conductivity values ranging from 10^{-3} cm/s to 10^{-6}, using a premature truncation procedure for analysis that is similar to slug testing. The pore pressure probe is used to record the recovery of pore pressure to hydrostatic by holding the probe at a fixed location. Higher permeability soils do not build excess pore pressure, so lack of excess pore pressure implies $K > 10^{-3}$ cm/s. Lower permeability soils require too long a time to recover, making quantification impractical; however, the test can be used to show qualititatively that $K < 10^{-6}$ cm/s.

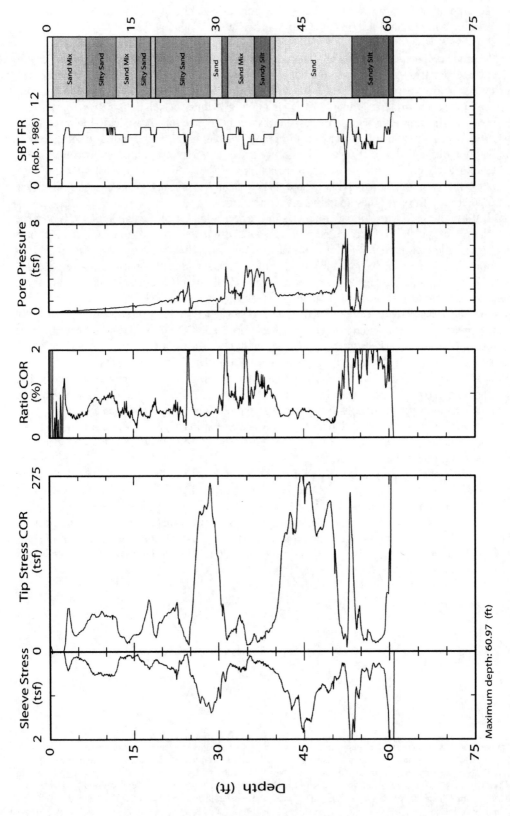

Maximum depth: 60.97 (ft)

FIGURE 9.10 Example of standardized commercial reporting for a cone penetrometer log.

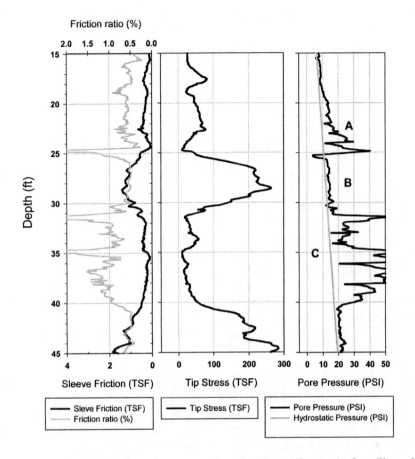

FIGURE 9.11 Expanded view and alternative presentation of CPT sounding results from Figure 9.10. Sleeve friction and tip stress are reported in tons per ft^2 (TSF); pore pressure is reported in pounds per in^2 (PSI). Sleeve friction and friction ratio are plotted on the same chart to emphasize facies within intervals. Pore pressure is plotted with hydrostatic pressure to emphasize negative and positive deflections. Points A, B and C show pore pressure responses that call attention to facies changes that are not resolved in typical soil type charts or vendor generated plots.

1) Scan the pore pressure plot for significant deviations from the hydrostatic baseline — this response is the signature of silts and clay. Points A and C on **Figure 9.11** show significant positive pore-pressure spikes, but also sharp negative spikes within the overall positive pore pressure response. The pore pressure tool responds to fine-scale bedding features and is useful for identifying interbedded zones, such as those highlighted near A and C. Bear in mind that the degree of deflection is a combined response, due to the thickness and degree of contrast in the permeability (clays) or dilatancy (silts). As a result, there is ambiguity in the response – thin beds with very high contrasts will show minor deflections, whereas thicker beds with the same degree of contrast will show more significant deflections. The increasing pore pressure trend at point A suggests increasing clay content, while the negative deflections at approximately 22, 23, 24 and 25 ft suggest potential silt interbeds. Note that the soil behavior type from the vendor log simply lumps this interval as silty-sand. Similar negative pore pressure spikes at point C near 33, 36 and 38 ft suggest potential silt in the interbeds.

2) Correlate the pore pressure spikes and trends with friction ratio spikes and trends to interpret the interbeds. The scale of the friction ratio plot is often selected to amplify the signature of silts and clays, as the changes in the tip stress are more difficult to see, because the scale is typically quite large. Identification of silts or clays is confirmed at intervals where the increasing friction ratio curve deflection coincides with the pore pressure deflection. When the friction ratio and pore pressure both increase, this indicates the presence of clay, as in the 25 ft interval beneath point A. Silts are indicated when the friction ratio increases and the pore pressure sharply decreases, as in the 36 and 38 ft intervals beneath point C. The presence of sand is indicated when the friction ratio decreases sharply coincident with a pore pressure decrease, as in the 32 to 34 ft interval near point C.

3) Sand facies interpretations can be made using the shape of the tip stress curve and pore pressure response. Point B indicates well developed sand that is indicated by a bell shaped tip stress response between 26 and 30 ft and a pore pressure response that is near hydrostatic pressure. The modest negative pore pressure responses at the top and bottom of the sand unit coincide with increasing friction ratios, suggesting potential for silt beds in transition to the overlying and underlying clay-rich deposits. Note that the vendor's soil behavior type classification for this interval is silty sand from 19.5 to 29 ft and sand at the bottom foot. The best way to pick the top and bottom of the sand unit is to correlate the friction ratio and pore pressure responses — between 26 and 30 ft the friction ratio is at a minimum and the pore pressure is moderately-well dissipated. The slight degree of pore pressure above the hydrostatic reference line indicates potential for clay or/and very-fine-grained sand content.

Cone penetrometers provide a very consistent, objective response to aquifer soils and can usually distinguish soil characteristics with much greater resolution than can be achieved by hydrogeologists examining samples of the same material. However, cone penetrometers sometimes generate soil behavior types that misrepresent the aquifer matrix. CPT output must always be calibrated by a hydrogeologist examining site soil samples.

9.4 COORDINATED HYDROSTRATIGRAPHIC AND GROUNDWATER SAMPLING

The dominant flow pathways supporting contaminant mass transport typically align more closely with facies trends than with groundwater elevation gradients, which confounds contaminant plume mapping efforts based solely on elevation gradients. Hydrostratigraphic studies give us the tool needed to develop facies maps and the hydrostratigraphy becomes particularly relevant when it is used to go beyond a mapping of local stratigraphy, to develop an accurate model of the site depositional environment. Combining hydrostratigraphic and groundwater contaminant surveys makes it possible to identify mobile porosity trends and to resolve the location of plume transport corridors with fewer samples than would be required in an uncoordinated sampling program.

Recent advances in field screening and down-hole sampling tools have improved the reliability and cost-effectiveness of vertical aquifer profiling. It is now possible to superimpose groundwater sampling on hydrostratigraphic surveys, to provide high-resolution contaminant mapping and transport corridor identification. We recommend that exploratory investigations use these high-resolution groundwater sampling methods, especially near source zones, to establish the relationship between the plume distribution and facies trends within the aquifer.

VERTICAL AQUIFER PROFILING METHODS

Vertical aquifer profile groundwater sampling, or **vertical profiling**, is a technique where multiple groundwater samples are taken sequentially with depth in an aquifer and used to map the vertical distribution of contaminants in the aquifer system. Vertical profiling can be conducted during initial exploration or in permanent well sets, after an initial delineation has been established. To combine hydrostratigraphy and groundwater sampling during the exploratory phase, direct-push or CPT rigs can be equipped with discrete-interval sampling devices or specialized sensors that can be used to superimpose stratigraphic and contaminant mappings.

Discrete-interval sampling – Groundwater samples can be collected over short vertical intervals, with either conventional CPT or direct-push rigs used for hydrostratigraphic logging. For both systems, screen sections up to 1 m in length can be pushed behind expendable drive tips. Small-volume groundwater samples can be analyzed in on-site laboratories, guiding exploratory sampling quite effectively. These samples are typically somewhat turbid, because it isn't possible to fully develop the formation with this equipment. Groundwater sampling equipment is not compatible with the complete CPT tool string, so discrete interval samples are normally collected in secondary probe holes, adjacent to fully-instrumented log holes.

Membrane interface probes — Membrane interface probes draw vapor samples from a gas-permeable membrane exposed to groundwater in a heated probe tip. The vapors are analyzed with a detector that matches the target compounds — electron capture detectors (ECD) are used for chlorinated alkene delineation, for example. Membrane interface probes are most effective at concentrations typically encountered near contaminant source zones.

Chemical sensors — Several real-time *in situ* water chemistry sensors are available for CPT tool assemblies. These include laser induced fluorescence for aromatic compounds and Raman spectrometers featuring tunable-frequency lasers that can be used to delineate chlorinated alkenes. *In situ* chemical sensors are most effective at relatively high aqueous-phase contaminant concentrations and are therefore most effective for detailed mapping near source zones.

Video logging — Small aperture side-view video cameras are available for CPT tool strings. These devices provide an extreme close-up of the aquifer soil structure and allow visualization of dilatancy in silty strata. Residual NAPL is also visible with these cameras.

COORDINATED STUDY EXAMPLES

Example 1 – CPT with discrete-interval sampling

Groundwater sampling for chlorinated solvents was coordinated with cone penetrometry hydrostratigraphic logging to support design of a permeable zero-valent iron reactive wall in the Eastern United States (described further in **Chapter 14**). The site is located in a river flood plain and the reactive wall was installed to capture solvent contamination reaching the river. **Figure 9.12** shows a hydrostratigraphic interpretation that was developed using CPT to map the facies trends and vertical aquifer profiling to map the plume distribution. The site geology consisted of fine- to medium grained sand with interbedded zones of fine- to very-fine grained sands, silts and clays. The pore pressure and friction ratio responses were correlated to distinguish the interbedded zones from the more permeable sands comprising the aquifer. The plume distribution shown on **Figure 9.13** was interpreted using existing monitoring wells (not shown) and vertical aquifer profiling data collected from the CPT soundings at 1.5 to 3 meter (5 to 10 feet) intervals. As shown, the downward vertical distribution of the plume appears to be limited by the presence of the interbedded zone, which occurs between 25 and 35 feet in depth. The occurrence of shallower interbeds also appears to correlate well with the lateral separation of the plume cores.

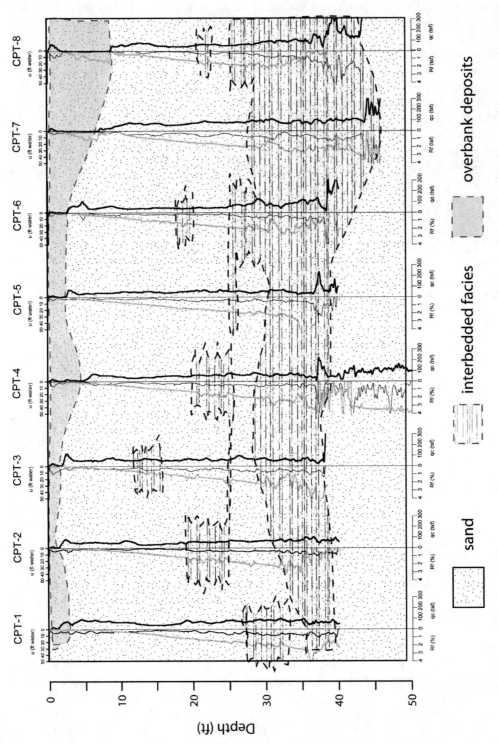

FIGURE 9.12 Superposition of facies interpretation on a CPT log set for a groundwater contamination site in the Eastern United States. Pore pressure and friction ratio response curves were used to interpret interbedded facies.

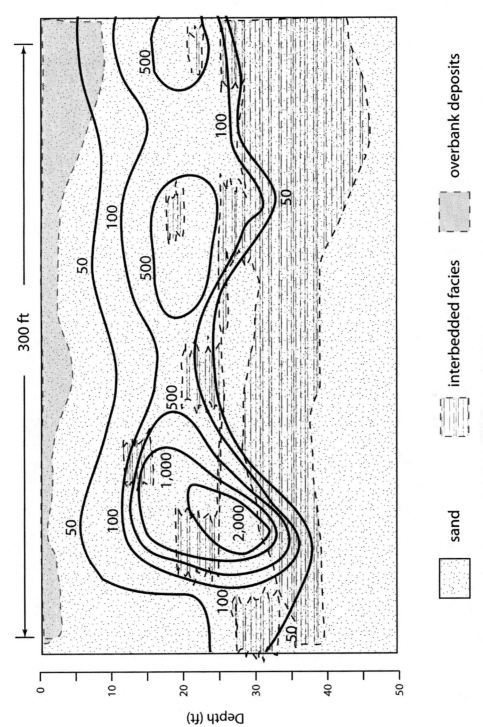

FIGURE 9.13 Superposition of trichloroethene concentration contours (µg/L) and facies interpretation on a CPT log set, for a groundwater contamination site in the Eastern United States. Contour lines represent trichloroethene concentrations, which were determined from discrete-interval groundwater samples collected from the cone penetrometer, combined with data from conventional monitoring wells.

Example 2 – CPT with membrane interface probe

A membrane interface probe was used to coordinate groundwater chemistry mapping and hydrostratigraphy at the early-stage plume mapping project described in **Figures 9.10** and **9.11**, above. The apparent (pre-CPT) contaminant plume migration did not match groundwater elevation gradients, indicating that aquifer matrix structure is the primary determinant in contaminant migration at this site. The coordinated CPT and membrane interface studies were designed to more accurately map the plume structure and to update the conceptual site model with a correct understanding of contaminant distribution mechanisms.

Figure 9.14 shows the results of membrane interface probe testing for the CPT sounding given in **Figure 9.11**. This figure shows the results for a calibration sounding that was installed near an existing conventional monitoring well pair that showed 60 mg/L of volatile chlorinated organic compounds (trichloroethene, 1,1,1-trichloroethane and associated degradation products) in the 30 to 35 ft zone and 3 mg/L total VOCs in the 55 to 60 foot zone. The CPT calibration sounding and the well pair were located approximately 330 ft (100 m) downgradient from the suspected source and the dissolved-phase concentrations suggest the presence of a NAPL-bearing source mass.

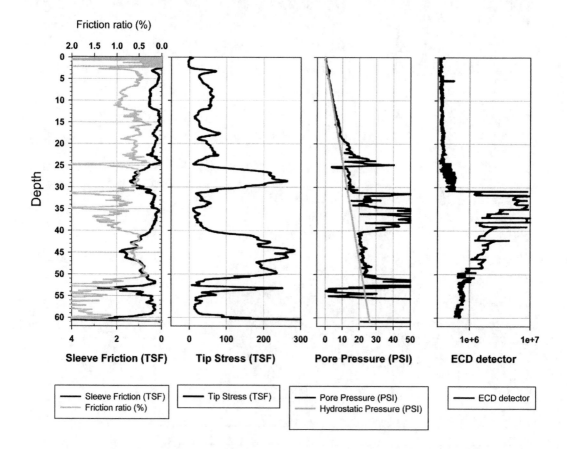

FIGURE 9.14 Membrane interface probe response for the sounding shown on Figure 9.11. The electron capture device (ECD) was calibrated to a 50 part-per-million (PPM) standard, because the well screened between 30 and 35 ft showed a total VOC concentration of 60 ppm. The ECD curve shows the highest readings in the interbedded zone, between 30 and 40 ft, and reduced responses in the sands above and below, due to higher groundwater flux in the more permeable facies.

The CPT log in **Figure 9.14** shows sand zones from 25 to 32 and 40 to 50 ft below ground surface, and sand interbedded with clay from 32 to 40 feet. From 50 to 60 ft, the CPT log shows interbedded silts and clays. Chemical logging was performed with a membrane interface probe, coupled to an electron capture detector for maximum sensitivity to chlorinated alkenes. The electron capture detector (ECD) was calibrated for maximum response at 60 mg/L, corresponding to the maximum groundwater concentration observed in the vicinity of the CPT log.

The ECD log shows that the highest chlorinated alkene concentrations occur in the interbedded sands and silts and that somewhat lower concentrations are likely in the lower sand zone, from 40 to 50 ft. The ECD showed a moderate response through the interbedded silt and clay zone, from 50 to 60 ft. These data provide a starting point for developing a site conceptual model that incorporates a deep source mass, parallel transport at multiple depths and groundwater flow paths determined by stratigraphic structure, rather than gradient control. A particular challenge for the project team will be to determine the mass flux rates in each of the strata and to map the full extent of the contaminant plume. As the project moves to remedy design, the large fraction of contaminant mass that resides in the upper interbedded sand and clay zone and in the underlying interbedded silts and clay will generate significant technical challenges.

Additional confirmation data are being collected via vertical aquifer profiling at 5-ft intervals between 30 and 50 ft. Although the field-screening techniques have advanced, we strongly recommend site-specific calibration and verification against vertical profile sampling to ensure that the methods are capable of meeting project data quality objectives.

SAMPLING BIAS ASSOCIATED WITH VERTICAL PROFILING

There are two important sources of bias in discrete-interval sampling associated with hydrostratigraphic logging — significant concentration variance for any interval and silt entrainment.

> *Concentration variance* — Samples collected from conventional monitoring wells that span 1.5 to 3-m aquifer intervals are usually a composite of many distinct groundwater strata that may carry significantly different contaminant concentrations. The discrete-interval sampling methods can capture these differences and the resulting sample results may be significantly higher or lower than would be reported from a conventional monitoring well that included the sampled interval. If we guide the discrete-interval sample collection with a chemical sensor log, we can intentionally bias the sample collection to the highest concentration strata, where transport may be occurring or stored mass treatment may be required to achieve remedy objectives.

> *Silt entrainment* — It is not possible to conduct well development during discrete-interval sample collection and the resulting samples often carry high turbidity. This may directly interfere with sample analysis, especially for dissolved metals. Care must be taken during sampling to purge the sediments from the sampling device to minimize turbidity prior to collecting samples for analysis. We recommend purging the sediments from the water column in the drill rods by slowly moving a pump down to the sampling device. During the purge cycle, the pumping rate is typically higher than during sampling to maximize the sediment recovery. The extra step involved in purging adds a limited amount of time to collect each sample and is likely to significantly reduce turbidity.

9.5 STRATIGRAPHIC CONTRAST

The importance of heterogeneity in aquifer architecture is obvious when considering vertical variation of permeability associated with facies changes. The common practice of using "layer cake" models to describe hydrostratigraphy likely reflects broad acceptance of this vertical anisotropy as a typical aquifer attribute. Throughout this chapter, we have introduced data that show how anisotropies occur in the horizontal plane, as well — a consequence of the depositional processes that form aquifer matrices. It should be our default assumption that aquifers are structured, at all scales, along all axes, by their formative processes and those structures embed significant permeability contrasts into the matrices. Conversely, the assumption that natural aquifers may be isotropic and homogeneous should be discarded.

Most natural aquifers exhibit 100- to 1,000-fold or greater ranges in permeability, over spans of a few meters or less, when high-resolution measurements are made. This is a result of grain-size variations and sorting, with variation in fine particulate content being a significant contributor to permeability variation (hence our desire to abandon the ASTM USCS soil classification scheme, in favor of the Udden-Wentworth system). Aquifer characterization tools, almost exclusively associated with drilling, are well suited to observe heterogeneities on the vertical axis, but we rarely have an opportunity to sample along the horizontal axes, so it has been somewhat difficult to dispense with the "layer cake" analogy for site conceptual models. The preceding discussion of lateral facies trends associated with various depositional environments makes it clear that aquifers will demonstrate a high-degree of variability and structural anisotropy in plan view, also. The key question is: "How much contrast in conductivity between the mobile and immobile porosity segments of an aquifer is required to create anisotropic behavior under flow and transport?"

Recall from **Chapter 3**, where we compared a uniform aquifer matrix and a three-layer system — consisting of a high-conductivity layer embedded in a lower-conductivity matrix — for which the effective hydraulic conductivity was calculated from the harmonic mean of individual conductivities. Both systems, shown on **Figure 3.10,** would yield the same volumetric flow when subjected to pumping, but the layered system would transmit water in the high-conductivity layer at a velocity that would be 3.3-fold higher than the average, and two orders of magnitude higher than in the lower-conductivity portion of the matrix. When the hydraulic conductivity of the mobile porosity segment is two orders-of-magnitude greater than the immobile, the contrast in transport velocity is so significant that the immobile segment advection rate is effectively zero, but diffusive transport and storage becomes significant. If the same 100-fold conductivity contrast occurred in plan view, hydraulic calculations show that the immobile porosity segment would behave as a horizontal barrier to groundwater flowing in the mobile porosity segment.

From a transport perspective, investigators at the MADE site[4] showed that lesser contrasts in hydraulic conductivity between simulated channels and bulk matrix (10- to 30-fold) led to asymmetric breakthrough curves and significantly increased longitudinal dispersion due to diffusive mass transfer between the advective-dominated channels (mobile porosity) and diffusive solute interchange with the bulk matrix (immobile porosity)(Liu et al., 2004). Results presented in **Chapters 5** and **7** indicated that connectedness of the mobile and immobile domain affects the mass transfer rate between the advective-dominated transport and diffusion-dominated transport segments of the aquifer. Highly connected systems result in rapid mass transfer, which leads to significant storage of contaminant mass in the immobile domain and solute migration velocity retardation in the mobile aquifer segment [as given by Equation (5.14)].

$$R_{f\theta} = 1 + \frac{\theta_i}{\theta_m}$$

---

[4] MAcro-Dispersion Experiment site, at the Columbus Air Force Base, northeastern Mississippi.

We have adopted the three-dimensional hydrostratigraphic aquifer architecture terminology to emphasize that real aquifer systems are complex, but structured. When significant contrasts in lithology or facies are viewed in a single boring log, it is obvious that the anisotropy due to stratification will control flow. It is more difficult to evaluate the anisotropy in the system arising from facies trends between borings, because the changes are more often gradational, or occur at such fine scales that it is impractical to map explicitly. The key to understanding transport behavior then is not simply to know the contrasts in hydraulic conductivity between the mobile and immobile segments of the aquifer, but to know the three-dimensional structure and how well connected the segments are.

Hydraulic testing methods can be applied in ways that measure the permeability characteristics of facies and expose contrasts that are otherwise undetectable. However, the conventional approach of using long-term, steady-state pumping tests is not the answer, as volumetric averaging progressively filters out the influence of the fine-scale variability on the conductivity estimate. From a water supply perspective, it is acceptable to perform large-scale pumping tests to estimate regional-scale average hydraulic conductivity. The superficial averaging is representative and appropriate for the purpose of evaluating hydraulic capture or water supply problems (de Marsily et al., 2005). These limitations were discussed in more detail in **Chapters 3** and **11**.

For remediation hydraulics, the localized heterogeneities that control transport and reagent distribution must be mapped three-dimensionally, if we aim to be successful at *in situ* remediation. We advocate using short-term, low-volume hydraulic tests that enable measurement permeability characterizing the fine-scale heterogeneities – both in the mobile and immobile matrix, so that we can understand the potential contrast in flow and transport potential in these segments of the aquifer. Rather than aim to collect enough samples to calculate an average value for each facies, the goal would be to characterize the variability of the individual facies that make up the mobile and immobile segments of the aquifer. Knowing the range of permeability values associated with a channel facies or overbank facies allow us to bracket our expectations regarding flow and transport. This is an acceptable approach in understanding sites during the investigation phase and in refining our conceptual site models as we pre-screen remedy options. However, based on the foregoing discussions of transport summarized above and detailed in **Chapters 5** and **7**, it is clear that transport testing is required to evaluate the combined influence of the contrast in conductivity and connectivity of the mobile and immobile porosity segments. **Chapter 11** discusses appropriate hydraulic testing methods that assist in developing conceptual site models. **Chapter 12** discusses the approach to tracer testing that is required to evaluate transport behavior and successfully apply *in situ* remedies.

10 Principles of Well Design

There are several useful guides to aid practitioners in the principles of well design for general water supply and environmental applications. Among the best, we recommend Driscoll's *Groundwater and Wells* (Driscoll, 1986), Neilsen's *Practical Handbook of Ground-Water Monitoring* (Nielson and Schalla, 1991), and the National Ground Water Association's *Handbook of Suggested Practices for the Design and Installation of Ground Water Monitoring Wells* (Aller et al., 1989). Our experience suggests that these and other available references provide a sound basis for the design of monitoring wells and groundwater recovery wells. However, the advent of *in situ* remediation methods and the need to characterize multiple intervals in aquifers places a premium on certain aspects of well design that are not normally covered in these references.

The recommendations provided in these key references are sometimes overlooked and wells are installed without regard to the characteristics of the formation, purpose and design objectives of the wells, long-term operability or data quality. Our experience indicates that common well construction practices have drifted somewhat from the detailed procedures of Driscoll and the others. In this chapter, we highlight some of the most important principles in well design, with an emphasis on matching each well's design and construction to its intended purpose. These issues were developed and emphasized in the references cited above and our aim is to reinforce their importance and provide guidance in the context of remediation hydraulics. We encourage the reader to revisit these topics in the recommended texts, to gain additional background that will further improve the design process.

10.1 ELEMENTS OF WELL DESIGN AND CONSTRUCTION

The primary elements of a well, whether designed for monitoring, injection or extraction applications, include: the well string components — casing, screen and end-cap; and the bore-hole components — annular seal and filter pack. Each element of a particular well construction requires proper design consideration to ensure that the well will perform according to expectations and provide cost-effective, long-term operability. **Figure 10.1** illustrates the basic configuration of the elements comprising a well designed for monitoring purposes.

From the outset, it is important to consider the nature of the geologic environment and depth of the well installation so that proper drilling methods can be selected to minimize disturbances to the formation and allow cost-effective installation. Often, drilling methods that enable the best characterization of the hydrostratigraphy do not provide for the most efficient and cost-effective installations, particularly when maximum well-yield or injection capacity are the prime objective. Driscoll's *Groundwater and Wells* (Driscoll, 1986) provides a comprehensive overview of conventional drilling techniques that have successfully applied to water supply problems and our experience indicates that these methods are most appropriate. When site hydrogeology requires drilling through multiple aquifer units without cross-contamination, other advanced methods such as sonic drilling techniques can prove more cost-effective, as the outer-casing can be used to temporarily isolate upper aquifers during the drilling process. Careful consideration of the site conditions and project constraints is warranted in selecting drilling methods to ensure that the best technical approach is selected. Our experience indicates that upfront costs are not the most

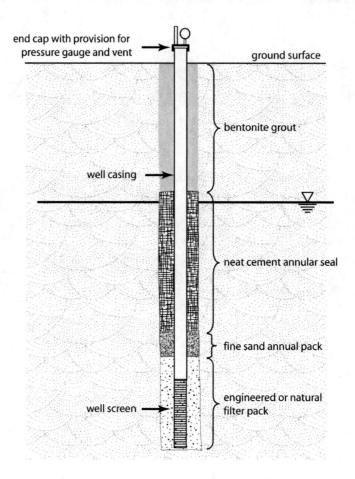

FIGURE 10.1 Schematic of a standard well construction. The well screen and filter pack are designed based on formation sieve analysis to balance hydraulic efficiency with filteration of fines. A coarse sand filter should extend 0.6 m minimum above the screened interval; a fine filter pack should extend 0.6 m above the coarse sand to retain the annular seal material. The annual seal should be neat cement in the saturated zone, or at least twice the screened interval length. Bentonite grout can be used from the top of the neat cement to the ground surface.

important factor, as the price for a failed well or extensive well rehabilitation and maintenance can off-set the short-term savings. It is also important to note that the conventional drilling methods, such as water or mud rotary, can produce efficient, cost-effective wells when proper well development techniques are employed.

The intended use of a well is the primary determinant in its design and construction. Other important factors include the depth of the target interval, the geochemistry of the formation, and chemical characteristics of contaminants and reagents. The well-string components and bore-hole must be sized appropriately to accommodate down-hole equipment for testing, maintenance and flow capacity. Well-string materials should be selected to provide adequate strength to support the installation of the well assembly as well as resist failure by formation collapse and compression. Material compatibility with the formation geochemistry and contaminants is also essential, as corrosion and chemical degradation of the well-string and annual seal can compromise the well integrity. It is also notable that combining dissimilar metals in the well-string can lead to accelerated corrosion due to galvanic potential. Nielson's *Practical Handbook of Ground-Water*

Monitoring (Nielson and Schalla, 1991) treats these issues in detail. Once these factors have been considered, one can begin the process of designing the well to match its intended purpose and optimize performance.

The starting point in designing a well is to understand the purpose of each element of the well construction. The bullet points below summarize the important design considerations for each element:

- *Well Casing* — The purpose of the well casing is to support the well during construction and convey piping, sensor lines and other equipment from the ground surface to the screened interval(s). In the case of wells that will be used directly for injection or extraction, without a separate riser or drop line, the diameter of the casing should be sized to ensure that the velocity of water does not exceed 1.5 meters (5-feet) per second to minimize frictional losses (Driscoll, 1986).

- *Well Screen* — The purpose of the well screen is to efficiently transmit water from the formation, while minimizing conveyance of fine particulates (fines). Optimal well-screen designs maximize the flow and communication with the formation in the target interval by balancing screen slot-size to minimize conveyance of fines while maximizing flow area per linear foot of screen. Continuous wire-wrapped screen construction provides the highest flow area per linear foot and minimizes frictional losses better than mill-slotted screens. Stainless-steel wire-wrapped screens are superior to PVC, both from an open area and long-term maintenance perspective. Well-screen areas should be designed to ensure that entrance velocities do not exceed 3.0 cm/s (0.1 ft/s) in extraction wells, or 1.5 cm/s (0.05 ft/s) in injection well applications (Driscoll, 1986). The optimal well screen design is obtained by considering slot-size, diameter and length. Well-screen design is discussed in the following section.

- *Annular Seal* — The purpose of the annular seal is to provide a high-integrity, low-permeability seal between the well casing and formation. For injection or extraction wells, the strength of the annular seal is critical, to avoid failure and short circuiting. While high-quality bentonite pellets might be suitable for monitoring wells in certain applications, injection and extraction wells require the use of neat-cement to maintain the structural integrity. In certain applications, it is advisable to apply additives to the cement, to accelerate setting and promote modest expansion. Neilson recommends adding calcium chloride (1 to 3% by weight) to shorten setting time in cold environments or gypsum (3 to 6% by weight) to promote expansion and enhance the integrity of the annular seal (Nielson and Schalla, 1991). The most important consideration in using cement is to avoid using too much water, as it can result in void formation, reduces strength, and excess losses to the formation. Bentonite-cement grouts are not advised in any case, as the ion exchange between the bentonite and the cement results in low strength and permeability characteristics that are poorly suited to any environmental application (Aller et al., 1989; Nielson and Schalla, 1991).

- *Annular seals* — high-quality bentonite grout or neat cement - should always be placed using tremie techniques from the top of the filter pack moving upward through the boring to avoid bridging and ensure seal integrity. Grout and cement seals typically require at least 48 to 72 hours to cure before well development can be completed. For deep well completions, the annular seal should be built in two lifts: the first lift should use neat cement throughout the entire saturated section, or at least 2-times the length of the screen in shallower completions; the second lift can be completed using high-quality bentonite grout to the ground surface. One advantage of the dual-lift annular seal is that it simplifies connections of distribution piping.

- *Filter Pack* — The purpose of the filter pack is to retain fine particulates from the formation, preventing their conveyance through the well screen, while providing a permeable hydraulic connection to the formation. Natural filter packs can be created through formation collapse and proper well development in some aquifers. However, when the formation consists of abundant fines, an engineered filter pack is required. We typically build the filter pack in two stages: the first stage, or **primary filter pack**, is constructed along the length of the screen, extending two to five feet above the top of screen, and designed to have comparable porosity and permeability as the formation (Driscoll, 1986). The upper stage, or **secondary filter pack**, is designed using fine sand to retain the annular seal well-above the primary filter pack. The thickness of the secondary filter pack depends on the stratigraphy and depth to the screen. It is advisable to build at least a 2- to 3-foot secondary pack to ensure performance when using neat cement or bentonite grouts. If cross-connection with another formation would be a problem, the design should be modified to minimize potential for cross-contamination. The design of the filter pack is discussed in more detail in the following section.

Some regulatory programs prescribe materials for annular seals (e.g., cement-bentonite grout) or set limits on filter pack construction (no more than 2 feet above well screens, for example). Our experience indicates that most regulatory programs can accommodate variances to these prescriptions when a defensible argument is made for construction alternatives that provide superior well performance and protect the long-term integrity of the well. We also acknowledge that using neat-cement (with or without additives), which we recommend in injection wells, is more difficult and costly than other grouting techniques. However, construction-phase costs are only one element of the total life cycle cost for a monitoring, extraction or injection well. When designers consider the full life cycle cost for a well installation, investing in higher-quality construction practices typically provides the best return.

10.2 SCREEN AND FILTER PACK DESIGN AND CONSTRUCTION

The primary function for a well screen is to provide an interface with the aquifer matrix, holding out fine particulates while allowing groundwater and injected reagents passage across the interface with minimal resistance. There are two main types of well screen used for environmental applications — mill-slotted PVC and wire-wound stainless steel (wound PVC construction is found in a limited number of applications). We recommend stainless-steel, v-wire-wrapped[1] screen as the default selection for well screen material and construction, rather than mill-slotted PVC, which gained popularity in the past 20 years. Although the material cost for stainless-steel is higher than for PVC, its cost is offset by significant savings in other elements of the well life-cycle, due to better hydraulic performance and long-term maintainability.

Flow Limitations and Screen Hydraulics

Well screen designs must strike a balance between screen open area, which directly controls fluid flow rates across the interface, and the requirement to hold back fine particulates from the formation. The fluid velocity across the screen provides a basis for striking that balance. We follow Driscoll's recommended maxima of 3.0 cm/s (0.1 ft/s) for extraction wells and 1.5 cm/s (0.05 ft/s) for injection wells (Driscoll, 1986). The total open area for a particular length of screen is determined by the slot-opening, slot-type, and diameter of the well and it is relatively simple to compute well-screen open-areas. However, most manufacturers provide well-screen specifications in their catalogues and websites. **Table 10.1** summarizes open areas and allowable flows for stainless-steel and PVC

---

[1] V-wire refers to the shape of the wire strands, which reduce blockage and enhance hydraulics.

TABLE 10.1

Summary of design parameters for screen selection in extraction and injection well applications. Screen transmitting capacity for extraction wells based on maximum entrance velocity of 0.1 ft/s (3.0 cm/s). Injection wells should reduce exit velocity by factor of 2, to yield maximum exit velocity of 0.05 ft/s (1.5 cm/s) for screen transmitting capacity. Apply formation blockage factor of 0.5 to evaluate actual injection/extraction well capacities.

2-inch (50-mm) Well Screen Specifications

Slot Size	Open Area (percent)			Screen Transmitting Capacity (gallons per minute)			Collapse Strength (pounds/inch^2)		
	SS V-wire	PVC V-wire	PVC Slotted	SS V-wire	PVC V-wire	PVC Slotted	SS V-wire	PVC V-wire	PVC Slotted
6	9.1	4.7	2.0	10.7	5.5	2.3	2,094	72.5	190
10	14.1	7.7	2.8	16.6	9.1	3.3	2,004	69.6	190
12	16.6	NA	NA	19.5	NA	NA	1,948	NA	190
15	19.8	NA	4.5	23.3	NA	5.3	NA	NA	190
20	24.8	14.3	5.8	29.1	16.8	6.8	1,754	65.3	190
30	33.0	19.9	8.3	38.8	23.4	9.8	1,559	60.9	190

Tensile strength: stainless-steel v-wire – 3,400 pounds; PVC v-wire 1,700 pounds; PVC slotted 2,100 pounds.

4-inch (100-mm) Well Screen Specifications

Slot Size	Open Area (percent)			Screen Transmitting Capacity (gallons per minute)			Collapse Strength (pounds/inch^2)		
	SS V-wire	PVC V-wire	PVC Slotted	SS V-wire	PVC V-wire	PVC Slotted	SS V-wire	PVC V-wire	PVC Slotted
7	8.2	4.1	2.9	19.3	9.6	6.8	340	75.4	97
10	11.3	6.7	4.1	26.6	15.7	9.6	326	72.5	97
12	13.4	12.5	4.5	31.5	29.4	10.6	317	NA	97
15	NA	17.7	NA	NA	41.6	NA	NA	NA	97
20	20.6	22.2	5.8	48.5	52.2	13.6	285	68.2	97
30	28.4	26.3	7.7	66.7	61.8	18.1	253	63.8	97
40	34.9	NA	8.0	81.9	NA	18.8	228	59.5	97
50	41.9	NA	8.5	98.4	NA	19.8	207	56.6	97

Tensile strength: stainless-steel v-wire – 4,800 pounds; PVC v-wire 2,100 pounds; PVC slotted 7,500 pounds.

TABLE 10.1 (CONTINUED)

6-inch (150-mm) Well Screen Specifications

	Open Area (percent)			Screen Transmitting Capacity (gallons per minute)			Collapse Strength (pounds/inch^2)		
	SS V-wire	PVC V-wire	PVC Slotted	SS V-wire	PVC V-wire	PVC Slotted	SS V-wire	PVC V-wire	PVC Slotted
Slot Size									
7	9.7	3.2	2.9	34.3	11.3	10.2	182	72.5	53
10	15.9	5.3	4.5	56.1	18.7	15.9	174	72.5	53
12	NA	10.0	5.0	NA	35.3	17.6	170	NA	53
15	22.1	14.3	NA	77.9	50.4	NA	NA	NA	53
20	27.9	18.2	6.0	98.1	64.2	21.2	153	68.2	53
30	37.1	21.7	8.0	130.9	76.5	28.2	136	65.3	53
40	44.2	NA	8.5	155.8	NA	30.0	122	62.4	53
50	50.4	NA	9.0	177.6	NA	31.7	111	59.5	53

Tensile strength: stainless-steel v-wire – 5,600 pounds; PVC v-wire 4,600 pounds; PVC slotted 11,000 pounds.

Data compiled for nominal 2-inch (50 mm), 4-inch (100 mm) and 6-inch inside diameter (150 mm) wells from Johnson Screen. 2004. www.johnsonscreens.com. JS-BR5230-0204.

Slot size is thousandths of an inch. Allowable flow rate based on entrance velocity of 0.1 ft/s (3.0 cm/s) for an extraction well. Allowable flow rate for injection wells should be decreased by factor of 2, or 0.05 ft/s (1.5 cm/s).

SS V-wire – 304 stainless steel (W60 model Free-FlowTM), vee-wire wrapped screen construction.
PVC V-wire – Schedule 40, polyvinyl chloride, vee-wire wrapped construction.
PVC Slotted – Schedule 40 polyvinyl chloride, horizontal mill-slot construction.

well construction, for several screen diameters and a range of slot-sizes. The designer should consult with the manufacturer to get the most current specifications.

Using Driscoll's maximum exit velocity of 1.5 cm/s (0.05 ft/s) for an injection well, it is possible to determine the screen transmitting capacity, Q_{stc}, by multiplying the open area for a given length of screen by the maximum exit velocity as shown in Equation (10.1):

$$Q_{STC} = A_{screen} \cdot V_{max} = F_{oa} \cdot L_s \cdot V_{max} \ (L/s)$$ (10.1)

where F_{oa} is the open area per unit length of screen (m^2/m or ft^2/ft), L_s is the length of the screen (m or ft) and V_{max} is the maximum exit velocity (cm/s or ft/s). Normally, open area is given by manufacturers in cm^2/m or inches^2/ft, so it is important to check units for dimensional consistency. **Table 10.1** provides the open area as a percentage, so it is necessary to compute the open area of the screen by multiplying the area of the cylinder ($\pi \cdot d_o \cdot L_s$, based on the outer diameter, d_o, of the screen and length) by the open area factor. Using the data from **Table 10.1** for a 50-mm (2-inch), 20-slot stainless-steel v-wire screen, and a 1.5-m (5-ft) screen length with an open area factor of 24.8-percent results in an open area of 584 cm^2 (0.65 ft^2). Substituting these values in Equation (10.1) yields a screen transmitting capacity of 0.94 L/s (0.03 ft^3/s or 14.7 gallons per minute [gpm]). Driscoll also recommends making a correction on the screen transmitting capacity to account for partial blockage of the screen openings by grains in direct contact with the screen. He recommends that the open area be reduced by a factor of 2 when calculating allowable flows in actual wells

(Driscoll, 1986). Applying the correction factor to Equation (10.1) leads to the maximum allowable injection rate, Q_{inj}, or maximum allowable extraction rate, Q_{ext}:

$$Q_{inj} \ or \ Q_{ext} = \frac{F_{oa} \cdot L_s \cdot V_{max}}{2} \qquad (L/s) \qquad (10.2)$$

where, V_{max} is 1.5 cm/s (0.05 ft/s) for injection and 3.0 cm/s (0.10 ft/s) for extraction. Thus, the maximum allowable injection rate for the example above would be 0.47 L/s (7.4 gpm). For extraction purposes, the screen could transmit 0.94 L/s or 14.7 gpm.

If the design injection rate were planned at 1.3 L/s (20 gpm), the screen length or diameter of the well could be increased to meet the exit velocity criterion. Because the allowable flow is proportional to screen length, a 3-m screen could accommodate twice the injection flow rate, if this were a match to remedy objectives. Alternately, a 100-mm (4-inch) diameter screen of 1.5 m (5 ft) length would increase the open flow area of the screen to 970 cm^2 (1.09 ft^2), since the open area factor for a 100-mm (4-inch) stainless-steel screen is 20.8%. As a result, using a 100-mm (4-inch) diameter screen would increase the allowable flow by the ratio of the open areas, 970 cm^2/584 cm^2, or 1.7 times the 50-mm (2-inch) rate of 0.94 L/s (14.7 gpm), to yield an allowable flow of 1.6 L/s (24.7 gpm).

The advantage of stainless-steel v-wire wrapped screen is illustrated by comparing the open areas with PVC wire-wrapped and PVC mill-slot construction as shown in **Table 10.1**. For a 50-mm (2-inch) diameter well screen, the open area for stainless-steel is approximately two times greater than PVC wire wrapped construction and 4 to 5 times greater than PVC mill-slot. This translates into a 2- to 5-fold higher allowable extraction or injection rates when using stainless steel for 50-mm (2-inch) wells. The flow advantage is maintained for all screen diameters, as indicated in **Table 10.1**. **Figure 10.2** presents allowable extraction rates for 50, 100 and 150 mm (2, 4 and 6 inch) diameter, 1.5 m (5 ft) screens using the information from **Table 10.1** and Equation (10.2). **Figure 10.3** shows the allowable injection rates for the same screens.

Although the flow areas for wire-wrapped PVC screens are closer to stainless-steel for larger slot sizes, the collapse strength and durability of the stainless-steel is much greater for all diameters. The only structural advantage for PVC mill-slot is its tensile strength, which is higher than for stainless-steel or PVC wire-wrapped screens. However, because well screens are not typically placed in tension, tensile strength does not provide a significant design advantage.

MATCHING SCREENS AND FILTER PACK TO FORMATIONS

Proper design of remediation wells requires grain-size (sieve) analysis of the formation so that a suitable filter pack and well screen can be selected. The primary purpose of the filter pack is to retain the fine grain-size fractions minimizing their transport across the well screen. It is common practice among environmental applications to use an engineered filter pack with a grain-size distribution appropriate to filter out the fine particulates. However, in many coarse-grained aquifers, it is possible to use formation collapse and aggressive well development techniques to develop a natural filter pack that effectively meets this goal. Regardless of which approach is used, it is critical to match the screen and filter pack to the formation.

Engineered filter packs are required in formations that consist of well-sorted, fine-grained materials, or interbedded facies with abundant fine-grained materials, since the required screen-slot size required to retain these materials will be too small to maintain hydraulic efficiency. Engineered filter packs are also required when formations do not naturally cave around the screen, providing the support and filtration necessary for hydraulically-efficient operation.

Although the convention used in classifying soils is based on sieve analysis plots with cumulative-weight-percent-finer or passing, the convention adopted by practitioners in well design

FIGURE 10.2 Summary of allowable extraction well flow rates for common well screen materials, based on 1.5-m (5-ft) screened intervals. Calculations were based on a maximum exit velocity of 3.0 cm/s (0.1 ft/s), using a formation blockage factor of 0.5. Data was compiled from Johnson Screens (www.johnsonscreens.com) for stainless steel vee-wire (SS V-wire), polyvinyl chloride vee-wire (PVC V-wire) and polyvinyl chloride mill-slot (PVC slot) screen materials.

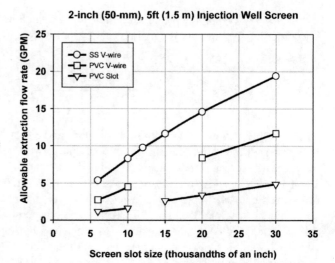

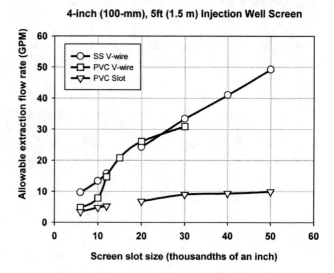

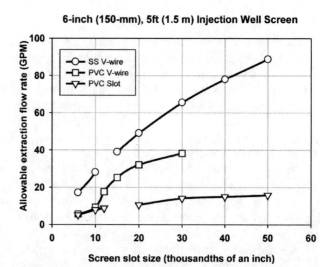

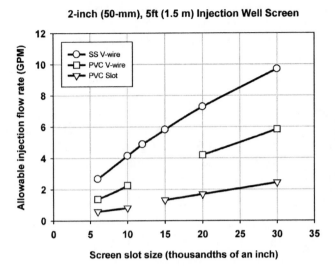

FIGURE 10.3 Summary of allowable injection well flow rates for common well screen materials, based on 1.5-m (5-ft) screened intervals. Calculations were based on a maximum exit velocity of 1.5 cm/s (0.05 ft/s), using a formation blockage factor of 0.5. Data was compiled from Johnson Screens (www.johnsonscreens.com) for stainless steel vee-wire (SS V-wire), polyvinyl chloride vee-wire (PVC V-wire) and polyvinyl chloride mill-slot (PVC slot) screen materials.

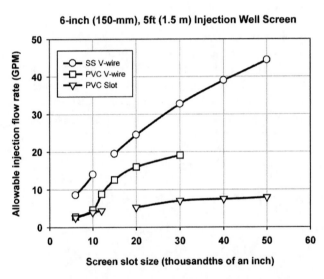

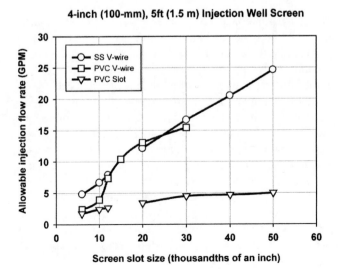

is to plot the sieve analysis as cumulative-percent-retained with grain size decreasing to the right. Rather than perpetuate the confusing mix of terminology, we advocate completing the analysis using the percent-finer terminology. For the seasoned well designer, this might also be confusing; however, we intend to make the process simpler, through the discussion that follows.

NATURAL FILTER PACKS

Natural filter packs can be used, if the screen slot size required to retain 40- to 70-percent of the formation is greater than 0.010 thousandths-of-an-inch, or 10-slot. For natural filter pack designs, Driscoll recommends that the screen-slot size should normally be designed to retain 30- to 60-percent of the aquifer material – that is, sized according to the sieve analysis grain size that allows 40- to 70-percent of the material to pass through the screen, or the d_{40} to d_{70} grain size from the cumulative percent finer plot (Driscoll, 1986):

Poorly-sorted aquifers, consisting of coarse-sands and gravel with finer grained sands, with d_{40} greater than 0.020-inch (0.50-mm). Applicable values for the uniformity coefficient, $C_u = d_{60}/d_{10}$, in these cases range between 4 and 6. The screen slot-size can be sized large enough to allow between 50 and 70-percent of the aquifer material to pass through the screen (the d_{50} to d_{70} range). This approach will require extra efforts in well development, as the fines will be conveyed through the screen during development. However, this design will enable use of the largest slot size and produce the highest yield well.

Poorly-sorted aquifers, consisting of medium to coarse grained sand with finer sand and limited silts and d_{40} greater than 0.010-inch (0.25 mm). Applicable uniformity coefficient values range from 4 to 8. The screen slot size can be sized large enough to allow between 40- and 60-percent of the aquifer material to pass through the screen (the d_{40} to d_{60} range). This design strikes a balance between development efforts and well-efficiency.

Well-sorted aquifer, consisting of medium- to fine-grained sands with silt. The screen slot-size is selected to allow between 40- and 50-percent of the aquifer materials to pass through the screen (the d_{40} to d_{50} range). This design accepts a reduced well flow capacity, in exchange for reduced well development effort. For fine-grained cases where the uniformity coefficient is less than 4 and d_{40} is less than 0.010-inch (0.25 mm), engineered filter packs are advisable to enable use of larger screen slots and improve hydraulic performance.

ENGINEERED FILTER PACKS

Filter pack materials should be selected from inert materials, preferably 95-percent or greater quartz. Well-rounded, well-sorted (C_u less than 2.5) quartz sands and gravels make the best filter packs. It is advisable to use filter packs with uniformity coefficients of 2 or less. Angular materials are prone to bridging and forming voids, while poorly sorted materials tend to sort hydraulically during emplacement, leading to a vertically-graded filter pack due to settling in the annular space.

The filter pack should be at least 2 inches (50 mm) thick to assure complete coverage for most well construction. However, there is not a benefit to building filter packs to greater dimensions, because thicker filter packs may slow the development process, without improving the well performance. A minimum thickness of 2 inches is generally required to build the filter pack properly and avoid bridging and voids. According to Driscoll, in theory, a thickness of three grains would have the filtration capacity to function properly, but it would be impractical to successfully construct the filter pack in this manner (Driscoll, 1986). Building on this research, many vendors have developed pre-packed screens that utilize sand packs between an inner- and outer-screen, or simply use synthetic media such as sintered polyethylene or comparable material to replace the filter pack. Pre-packed screens are much more common in monitoring well applications today and there are significant potential benefits from their application in remediation wells. Application of these pre-pack systems should be approached using the same design approach as used in conventional screen and filter packs.

The grain size of the filter pack should be selected based on the finest material in the formation, if the aquifer consists of interbedded, or graded formations that contain fine-grained silts or clays. If the hydrostratigraphy shows distinct grain-size distributions in the vertical profile, it is advisable to design a filter pack and screen slot-size for each viable unit. Driscoll recommends using casing to "blank-off" unfavorable sections of aquifers for water supply (Driscoll, 1986). We also recommend it for remediation wells, as extraction in these intervals will lead to sand and silt production and injection will short-circuit to the coarser-grained facies. Driscoll recommends the following steps in designing a filter pack (Driscoll, 1986):

1) Evaluate hydrostratigraphy of target interval(s) and perform sieve analysis for each interval. Evaluate d_{40} and uniformity coefficient using criteria above to determine if engineered filter pack is required.
2) If engineered filter pack is required, use the uniformity coefficient to determine the appropriate multiplier for d_{30}: If C_u is less than 4 and d_{40} is less than 25 mm (0.010 inch), multiply d_{30} by 4 to 6; If C_u is greater than 4 and d_{40} is greater than 25 mm (0.010 inch) use a multiplier of 6 to 10. The higher-range of multipliers is warranted when the formation contains abundant silt, or silt and clay stringers.
3) Use the cumulative grain-size plot (percent finer verus $\log_{10}$ grain size) for the formation sieve analysis and plot the multiplied range of d_{30} on the chart. Draw a smooth line from d_{10} through the calculated filter pack d_{30} and to d_{60} with the desired uniformity coefficient — typically 1.5 to 2 is desirable. The values for d_{10} and d_{60} which satisfy the uniformity coefficient slope through d_{30} can be directly calculated based on the geometry of the problem using Equations (10.3) and (10.4) below:

$$d_{60} = d_{30} \cdot \sqrt{C_u} \qquad\qquad (10.3)$$

$$d_{10} = d_{30} \cdot \frac{1}{\sqrt{C_u}} \qquad\qquad (10.4)$$

4) Plot the filter pack grain-size curves for the range of desired C_u on the chart and project the trend up to d_{90}. The filter pack d_{90} values can be calculated recognizing that the slope of the curve is the same between d_{60} to d_{90} as it is for d_{30} to d_{60} as shown in Equation (10.3):

$$d_{90} = d_{60} + [d_{60} - d_{30}] \qquad\qquad (10.5)$$

5) Plot the grain-size distribution for commercially available filter packs on the grain-size chart. Select the filter pack whose grain-size distribution falls within the bracketed range by the constructed filter pack plots. If more than one falls within the constructed filter pack range: use finer if the formation is likely to produce fines and development is at a premium; use the coarser if hydraulic efficiency is the main objective.
6) The well-screen slot-size is determined using the d_{10} for the commercially available filter pack. Pick the nearest smaller commercially available slot-size if fines are anticipated; pick nearest larger slot-size for maximum hydraulic efficiency.

EXAMPLE WELL DESIGN PROBLEM

The following example illustrates the design approach. **Figure 10.4** illustrates the grain-size distribution for a sample collected from a dune sand formation. The well of interest is intended to be used for injection at a planned design rate of 10 gpm in a 2-inch (50 mm), 5-ft (1.5 m) screen. Based on the formation's grain-size distribution, the calculated uniformity coefficient, C_u, is 2.1 and d_{40} is 0.012 inch (0.30 mm). The low C_u value and marginal d_{40} indicate that this formation is a candidate for an engineered filter pack.

To verify this interpretation, compare the allowable injection rate with the required slot-size for the formation. Use d_{40} as a guide to size the screen slots. The low value of C_u implies that a 10-slot (0.25 mm) screen would be required; however, from **Table 10.1** and Equation (10.2), the allowable injection rate for a 5-ft (1.5 m) screen would be approximately 4 gpm (i.e., screen transmitting capacity for an injection well divided by two for formation blockage).

From **Figure 10.4**, the d_{30} for the formation is 0.0096 inches (0.24 mm). Using Driscoll's recommendations, multiply d_{30} by a range between 4 and 6 to yield 0.038 to 0.058 inches (0.96 to 1.46 mm). Select C_u for the filter pack between 1.5 and 2 because of the formation's d_{40} and C_u. For a filter pack with C_u of 2, apply Equations (10.3) through (10.5) to solve for the calculated filter pack d_{10}, d_{60} and d_{90} for a factor of 4-multiplier as follows:

$$d_{10} = d_{30} \cdot \frac{1}{\sqrt{C_u}} = 0.038 \times 0.71 = 0.027 \, inches \, (0.68 \, mm)$$

$$d_{60} = d_{30} \cdot \sqrt{C_u} = 0.038 \times 1.41 = 0.054 \, inches \, (1.44 \, mm)$$

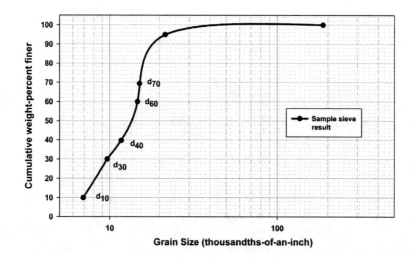

FIGURE 10.4 Example grain-size plot from a sieve analysis on a dune sand. Key sieve analysis parameters include d_{40} = 0.24 mm (0.0096 in) and C_u = 2.1. Based on these parameters, the required slot size for a natural filter pack would be 0.25 mm (0.010 in), or "10-slot."

$$d_{90} = d_{60} + (d_{60} - d_{30}) = 0.054 + (0.054 - 0.038) = 0.070\,inches \quad (1.78\,mm)$$

For a multiplier of 6 and C_u=2, we get d_{10}=0.041 inches (1.03 mm), d_{60}=0.082 inches (2.08 mm), d_{90}=0.105 inches (2.67 mm) for the calculated filter pack curve. The filter pack grain-size distribution curves are plotted on **Figure 10.5**.

To select the filter pack, plot the commercially available filter packs on the chart to see which one falls in the bracketed range. As shown on **Figure 10.6**, the #5 Global sand meets the design criterion. The screen slot-size is then selected using the d_{10} for the #5 Global sand, which is 0.038 inches (0.97 mm). Because the formation is well-sorted, select the smaller available screen slot-size, which would be 30-slot (0.80 mm). Looking back to **Table 10.1**, a 2-inch diameter, 30-slot wire-wrapped screen would allow an injection rate of 9.7 gpm; using a 4-inch (100 mm) screen would allow an injection rate of 16.7 gpm. The final selection should be based on the cost-sensitivity of operations to flow rate. We recommend that the 4-inch well is the best choice, as it provides greater flexibility in flow capacity, well-development and long-term operability.

INJECTION-OPTIMIZED WELLS

The design and construction practices described throughout this chapter were originally developed to support extraction well systems — water supply development and contaminant plume containment systems. Injection wells now represent a very large portion of remedial system well construction and there are several important considerations that apply to injection well design optimization.

Well safety — Injection wells may be operated at significant pressures, creating a potential health and safety hazard. In addition, *in situ* reactions can produce gases that can also pressurize the well head, even when injection is conducted under gravity feed conditions. We recommend that all injection wells be designed using pressure gages, pressure relief valves and well cap tethers that mitigate potentially dangerous conditions at the well-

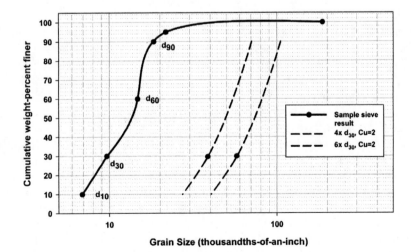

FIGURE 10.5 Filter pack design approach example. Using a d_{30} multiplier of 4 to 6 yields a filter pack d_{30} between 0.96 and 1.46 mm (0.038 to 0.058 in). The filter pack design grain size is bracketed using a uniformity coefficient, C_u, of 2, and plotted using Equations (10.3) through (10.5). The ideal filter pack for this example falls between the two dashed lines, representing (4 x d_{30}, C_u = 2) and (6 x d_{30}, C_u = 2).

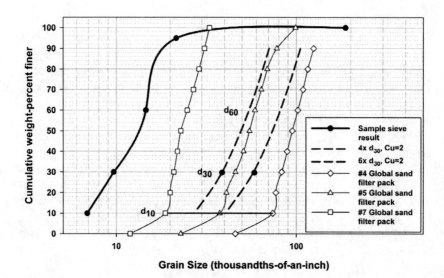

FIGURE 10.6 Selecting the filter pack and screen slot size. The plot illustrates grain size distribution curves for three filter pack materials: #4 Global sand ($C_u = 1.38$), #5 Global sand ($C_u = 1.33$) and #7 Global sand ($C_u = 1.33$). The #5 Global sand is the best match to the design criteria, as it falls within the bracketed range from the engineered filter pack calculations (Figure 10.5). The screen slot size is taken from the selected filter pack (#5 Global sand), using the d_{10} grain size from the plot. Based on the formation's d_{40} and low C_u, select the nearest smaller standard slot size, 0.80 mm (0.030 in, or "30-slot").

head. **Figure 10.7** illustrates the recommended injection well construction, including well head equipment for safety precautions.

Pressure tolerance — Injection wells must be designed to ensure structural integrity of the annular seal and to avoid unintentional fracturing of the aquifer matrix. Annular seals are particularly at-risk and the use of neat cement seal construction is an important element of a pressure-tolerant annular seal. Operating pressure limits are discussed in more detail in **Chapters 2** and **13**.

Filter pack — There is a common misconception that injection wells do not require filter pack design or the effort at development associated with extraction well construction. Wells that are used for intermittent or pulsed injection are subject to water hammer effects and post-injection pressure relaxation — both processes can induce back-flow into the well and create siltation. Therefore, it is critical to design the filter pack and well screen properly and to conduct well development. It is also important to avoid the temptation to utilize larger screen aperture than is appropriate for the formation and filter pack.

Reagent filtration — For injection wells that are used to deliver reagents to aquifers, it is essential to consider the influence of total suspended solids (TSS), air entrainment, by-products of the chemical reaction and chemical precipitates in the filter pack and on the well screen during the design process. Reagents with significant TSS will ultimately plug the formation over time, as will reagents that react *in situ* to form precipitates in the formation. We recommend using bag-filters or in-line filtration systems to minimize injection of filterable solids into formations, as TSS concentrations of 1 mg/L can lead to plugging and fouling in a relatively short time.

Washout — For reagents that react with formation minerals or that form precipitates, it is advisable to follow the reactants with clean water flushes to clear the well and filter pack. In some cases, stabilizers can be added to delay the reaction, or reduce near-well precipitation.

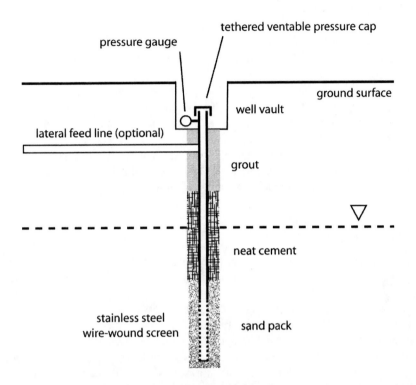

FIGURE 10.7 Well construction for a fluid injection well. It is very important to provide a high-quality well screen to allow thorough well development and low flow resistance. Neat cement provides a pressure-resistant well seal. Conventional grouting can be used above the neat cement seal. Safe well operation requires a pressure gauge and tethered, ventable pressure cap, with provision for safely handling vented fluids or gases.

Air entrainment and gas production — Care must be taken during injection not to entrain air in the fluids, as the air can block formation pores, reducing the effective hydraulic conductivity (refer to **Chapter 4**) and creating vapor lock conditions; air entrainment can also induce chemical precipitation or bacterial fouling, especially when injecting carbon solutions to stimulate anaerobic biological activity. Deep well reagent injections are particularly susceptible to air entrainment or water hammering. In these wells, it is helpful to deliver reagents through a drop line with a snubber or flat plate orifice at the end, to restrict flow and allow the drop line to fill with fluid at the desired flow rate.

Flow limitations — Extraction well performance is aided by gravity, whereas gravity works against injection wells — gravity drainage delivers groundwater to the vicinity of extraction wells, while an injection well pushes a continuously increasing mass of water through a resistive medium. As a result, most injection wells cannot accept injected water at the rate at which it can be withdrawn. Conventional extraction well hydraulic analysis typically fails to provide an adequate description of well performance under injection. A reasonable rule-of-thumb is that the short-term injection rate for any well operating in injection mode will be 1/3 to 1/2 of its capacity for groundwater extraction. Also, injection capacities tend to decrease during extended operation, particularly in cases where a rising injection water mound encounters lower permeability strata. We recommend testing injection capacity at planned flows and durations, to verify injection performance.

Injection pressure limits — Specific pressure limits must be established for each injection well, based on aquifer soil strength, effective stress and other factors described in Chapter 13. Pressure limit calculations are provided in Equations (13.44) to (13.46).

Screen length limitations — One important characteristic shared by injection, extraction and monitoring wells is that flows across the well screen are proportional to the aquifer matrix hydraulic conductivity and heterogeneities in the formation generate concentrations of flow to or from the most conductive strata. When injection well screens span multiple conductive strata, unintended concentrations of flow may occur. Shorter screened intervals reduce the risk of flow allocation problems and injection tracer tests are recommended to assure that injected reagents reach all the target zones. In our experience, screened intervals in the 1.5 to 3 m range (5 to 10 ft) are most likely to succeed and screened intervals of 6 m (20 ft) have generated significant flow allocation problems.

MONITORING-OPTIMIZED WELLS

The primary purpose of monitoring wells is to enable collection of representative samples and water level measurements from the aquifer of interest. Nielson recommends that monitoring well filter packs and screens should be designed using d_1, rather than d_{10}, of the engineered filter pack to minimize conveyance of fines (Nielson and Schalla, 1991). This approach makes sense if the only objective of the well is collection of silt-free water, or if well development is anticipated to be difficult. However, if the well is intended for hydraulic measurements or testing, using too fine a slot size will result in a well construction that affects the hydraulic performance of the well. We favor conventional approaches of well design instead, but understand additional efforts will be required during well development to achieve the same low-turbidity, silt-free groundwater samples.

One of the most important factors in designing monitoring wells is selecting the interval that will be monitored. We recommend that the decisions regarding well placement should be dictated by the hydrostratigraphy, not simply the depth interval. For this reason, we favor well construction techniques that enable collection of discrete interval samples. This can be accomplished using modular, multi-level well construction techniques such as the FLUTe™ multi-level system (FLUTe, 2007), Solinst's Multilevel Technology (Solinst, 2007), Westbay's Multilevel Monitoring System (Schlumberger, 2007) or conventional nested-well designs. The advantage of the modular systems is that they can be designed to monitor several discrete intervals through the installation of one borehole. The individual monitoring intervals are typically designed using pre-packed screens or filtered sampling ports that are isolated using various packer or annular-seal systems, and connected to sampling devices through dedicated tubing. These methods can yield high-quality groundwater samples; however, well development techniques are limited by the construction so it can be difficult to ensure a high degree of communication with the formation. Modular multi-level construction techniques also make it more difficult to obtain water level measurements directly, but the vendors continue to evolve the products to enable wider functionality.

While it is possible to install multiple nested wells within a single borehole, it is very difficult to ensure an adequate seal around the multiple casings without drilling large-diameter boreholes. In most cases, the best approach is to use separate borings for each of the wells in the multi-level nest, when conventional drilling and monitoring well construction methods are used. For deep installations, this approach is very costly, but it provides a high-degree of reliability in well performance, both from a sampling and hydraulic testing perspective.

The advent of direct-push, or CPT-driven microwells (38-mm or 1.5-inch or less), allows for the cost-effective installation of multilevel well nests using pre-packed screen construction. The primary disadvantage of microwells stems from the fact that the borehole is created by driving the rods through the formation, rather than drilling. In formations that contain clay layers or interbedded zones, this can lead to smearing of the borehole wall, which can be difficult to remedy by well development because of the well diameter. In silt or clay-rich formations, compression of the aquifer matrix reduces hydraulic conductivity along the probe track, decreasing hydraulic communication between the formation and the screen. However, in relatively permeable formations, this approach can be a suitable alternative to drilled borings or modular multi-level well construction.

The advent of permeable diffusion bags has enabled discrete interval sampling within longer monitoring well screens. In cases where the flow can be shown to be horizontal in a formation and the contaminants are compatible with the membrane materials, permeable diffusion bags afford the designer an opportunity to collect samples at 0.5 to 1.0 meter (1.6 to 3 feet) intervals across the vertical profile of the well screen. As the membrane technologies are advancing rapidly, the interested reader is advised to check with the manufacturers regarding contaminant compatibility and regulators regarding acceptance of this approach. The Interstate Technology & Regulatory Council (ITRC, 2007) is a good resource for further information.

10.3 WELL DEVELOPMENT

Proper well development is critical to the function of all well types, but current experience indicates that well development is not given sufficient attention. All too often we find the use of modest surging techniques and limited pumping, which results in wells that never run clear or have poor communication with the aquifer. We also find that when aggressive well development such as surging and pumping is deployed, it is used in fine-grained aquifers that tend to produce silt. Aggressive surging in fine-grained systems is not recommended, because it results in formation damage and blockage of the filter pack. There are numerous tools and strategies that can be used for well development, and there is not one approach that suits all formations. The designer should carefully consider the intended function of the well and the formation characteristics, factoring in past experience with a formation when it is available, to formulate the best well development strategy.

Driscoll's *Groundwater and Wells* (Driscoll, 1986) provides the most comprehensive approach to well-development. Another good reference is the ASTM "Standard Guide for Development of Ground-Water Wells in Granular Aquifers" (ASTM, 1998). Well development is required to remedy the formation damage that occurs during drilling and ensure a high-degree of hydraulic communication with the formation. The requirements for well development depend on the nature of the well completion, characteristics of the formation and drilling method used to install the boring.

Natural filter pack well completions require development to remove the formation fines from the annular space and create a graded filter pack adjacent to the well screen. The graded filter pack is created because the energy associated with surging and jetting dissipates with distance from the screen, enabling to most efficient removal of fines nearer the well, but the efficiency decreases with distance. Coupled with the filtering of fines that occurs in the formation, the natural filter pack becomes progressively finer (graded) away from the well. The tendency of increasing the grain size nearer the well increases the porosity and permeability of the filter pack and enhances the hydraulic connection with the formation.

Engineered filter packs are selected to provide comparable porosity and permeability to the formation, but it is necessary to remove the disturbed formation at the borehole wall to create the best connection with the formation. An engineered filter pack becomes well-graded through the development process, as the formation and filter pack fines are removed. Using too thick a filter pack counteracts the development process, as the energy associated with surging and jetting cannot be transmitted to the borehole wall to remedy the formation damage. When mud or drilling fluids are used, this can be particularly important as it is necessary to remove them from the formation to ensure optimal hydraulic performance. Driscoll recommend use of polymers and drilling fluid additives that decompose rapidly, to reduce the overall well development effort in these cases (Driscoll, 1986).

GRANULAR AQUIFERS

Well development in granular aquifers is typically completed using a combination of pressure surging or jetting to mobilize the fines and pumping to remove the fines from the well. Often, brushing and swabbing are used to mechanically remove the fines that are stuck in the screen slots. Many different methods or combinations of methods can be used to efficiently develop a well in a granular aquifer; however, the process will be most-efficient if the filter pack and well screen were designed appropriately.

The key objective in surging a well is to ensure that the pressure is transmitted along the entire length of the well screen. The most effective means of surging is accomplished using a drill rig equipped with a surge-block with multiple rubber flanges sized to the inner-diameter of the screen. The drill rods can be rapidly raised and lowered into the well, creating negative and positive pressure, which surges the filter pack, rearranging the individual grains, and mobilizing the fines. A weighted surge-block can also be mounted on a cable to achieve the same dynamic range of pressure required to mobilize the fines. Prior to surging, it is advisable to pump the well to remove any fine particles that have accumulated in the casing. Surging should progress from the top of the screen to the bottom, starting with relatively gentle surging pressure, and progressing to more aggressive pressures as the well develops. Surging techniques are not recommended when 1) the formation particles are angular, because sorting is inefficient, or 2) contains abundant silt and clay, as these particles tend to block the filter pack and screen. High-pressure jetting is recommended in these instances, and for finer-grained formations, as the method is more effective at dislodging particles that block the screen.

Pumping should be done frequently to ensure that accumulated fines are removed, rather than forced back into the formation. When pumping, it is advisable to raise and lower the pump through the well column to maximize removal of fines; this process also increases the efficiency of pumping, as the flow velocities are slightly higher near the pump intake and flow can be concentrated in particular segments of the screen. To gage progress, we recommend pumping at fixed rates or removing fixed volumes with each cycle of surging (or jetting). Logs should be maintained to characterize the water clarity (use of a tubidity meter is advised), drawdown during and after pumping, and water level recovery times. In addition to improved water clarity, the well should demonstrate less drawdown with pumping and quicker recovery times as development proceeds.

Pumping alone is often inadequate to dislodge particles that are blocking the screen, since the flow is unidirectional. For this reason, reverse circulation or back washing can improve the development process, by forcing water into the formation and then removing it. The oscillating pressure has the same effect as mechanical surging, but is less aggressive and generally less useful as a sole development technique. It is advisable to use a dedicated pump for well development, as the sand and silt removed during the development process have detrimental effects on the pump mechanism. Pumping should be initiated slowly and increased gradually as flow develops, to avoid sand-locking in the pump.

Jetting techniques are often used in stratified aquifers that contain abundant silts and clay, or highly angular grains. Jetting is used in combination with pumping and air-lift methods to remove the fines as they are dislodged from the filter pack and screen. According to Driscoll, jetting and air-lift methods are the most effective means of well development (Driscoll, 1986); however, we find that they are rarely used in environmental applications. Jetting tools typically utilize 4 or more hydraulic nozzles to deliver relatively high-pressure water in a concentrated section of the screen. The jetting tool is typically lowered into the well while continuously pumping or air-lifting the fines out of the well. The pumping or air-lift removal rate should always exceed the jetting rate to ensure effective removal of fines. In general, Driscoll recommends jetting at 1,380 kPa (200 pounds per square inch [PSI]) and entrance velocities of about 50 to 100 m/s (150 to 300 ft/s) for metal screens. PVC screens must be jetted with caution, as pressures over 700 kPa (100 PSI) can damage the integrity of the screen. This is a disadvantage of PVC screens, as less aggressive methods are required to avoid damage. The jetting tool should be rotated and moved continuously across the screen, making several passes to ensure that fine materials have been removed.

FINE-GRAINED AND STRATIFIED AQUIFERS

Surging methods are not recommended in aquifers that contain abundant fines, or consist of highly stratified facies with fines. As indicated above, surging can force fines into the formation, or mine excessive amounts from the formation, leading to plugging and formation damage, which results in poorly performing wells. We recommend jetting and air-lift/pumping methods in these cases because the technique focuses on removal of the fines from the screen and filter pack. Air-lift is usually preferred, as pumping capacity in these types of systems is limited due to low formation permeability. Because PVC screens are not compatible with aggressive jetting techniques, and because of the limited flow area for small slot sizes, stainless-steel wire wrapped construction is preferred in these types of aquifers.

10.4 WELL ECONOMICS

The selection of well construction materials and drilling practices are often driven by the immediate cost of installation, without consideration of full life cycle cost for the installation. Well construction choices affect the operability of wells through their full life cycle and we strongly recommend extending the economic analysis beyond the immediate construction costs, to include all of the costs that are associated with operating and maintaining a well. When this analysis is performed, it will normally be clear that the cost of higher-quality well construction is quickly paid back through better well performance, lower maintenance cost and a reduced well attrition rate.

There are several sources of cost in the operation of any well, whether the well is designed for a flow (injection or withdrawal) or monitoring function:

Drilling and materials — The drill rig and crew, attending hydrogeologist and well construction materials all contribute to the base cost for a well installation. The materials cost typically range from 5 to 10 percent of the total cost, depending on the depth of installation and type of formation.

Well development — The cost for well development is based on the drill crew and rig and attending hydrogeologist time. Well construction practices that allow more effective well development practices shorten the time on-hole and that, alone, may cover the cost of higher quality well materials.

Sampling — The cost of sampling is largely determined by the time required to staff and to obtain a quality sample from the well. Sampling cost in poorly-developed wells can be very high. For bailer sampling, recovery time is extended and it may be very difficult to obtain a low-turbidity sample. In some wells, even low-flow sampling methods cannot achieve low turbidity. Investment in higher-quality screen and well development can alleviate this recurring problem.

Injection flow rates — The flow rate for each injection well determines the crew time required to complete an injection event and may affect the number of wells required to achieve treatment area coverage.

Extraction flow rates — For an extraction well network, the flow rate that can be achieved from each well directly determines the number of wells required to create an effective capture. The cost of high-yield screens is easily offset by the expanded capture radius that is achieved.

Maintenance — Larger screen open area extends the time between required maintenance events.

Redevelopment — Stronger well screens are better able to withstand the rigors of redevelopment, if that becomes necessary (it is less likely with a higher-quality installation).

When all the well life cycle costs are accounted, the choice of higher-quality well materials and development practices is made clear. We recommend the use of wire-wound stainless steel construction, based on several advantages that lower the overall well life cycle cost:

- Larger open area gives higher flow capacities and improves development;
- Higher material strength allows the use of more effective well development tools for lower-permeability formations;
- More effective well development provides a more effective hydraulic connection between the well and the formation, giving a more accurate hydraulic response.
- More effective well development reduces the likelihood of false positive contaminant measurements, especially for metals, that are associated with fine particulates entering the sample.

Our experience suggests that the long-term operability of stainless-steel wire-wrapped screens more than offsets the higher construction cost for remediation related wells. We recommend stainless-steel well-screens as standard equipment in injection and extraction wells for this reason. We also recommend returning to stainless-steel for monitoring wells, a common practice in the past, for the same reasons. To avoid galvanic corrosion problems, the stainless steel screen sections can normally be coupled with PVC riser.

11 Design and Interpretation of Aquifer Characterization Tests

Conventional hydraulic testing methods were developed to provide answers that are best suited to water supply problems, not remediation hydraulics problems. The key issue is that most hydraulic tests average out the descriptive information like a damping filter: the long-term, steady-state tests encompass large aquifer volumes and are biased low with respect to the mobile porosity. Short-term, low-volume tests are better because we can isolate particular depositional elements during the test to characterize the hydraulic conductivity values and trends associated with facies changes in the aquifer. Ideally, hydraulic testing methods should be selected that enable remediation hydrogeologists to make hydrostratigraphy interpretations more quantitative. The goal is to quantify the hydraulic conductivity associated with particular facies and facilitate an integrated interpretation of plume geometry and migration pathways.

A major obstacle in recognizing the importance of the fine- and stratigraphic-scale heterogeneities has been the common viewpoint that larger-scale hydraulic testing methods provide a more realistic and representative estimate of hydraulic conductivity. While this is certainly true from the perspective of regional groundwater flow problems, where the primary objective is to understand the hydraulic budget for water supply problems, it is the spatial distribution and connectedness of the high-conductivity facies that control reagent distribution and contaminant transport at the remediation hydrogeology scale. The key to success is understanding how to use data obtained from both scales, so that important information is not lost or inappropriately averaged-out.

For example, the average of many slug tests at a site typically does not match the large-scale pumping test results. Most practitioners favor pumping-test derived hydraulic conductivity estimates over slug-test derived estimates, because of the perception that the larger-scale estimates are more accurate. However, limitations of the pumping-test approach were shown in Chapter 3, in the context of average conductivity for flow parallel to layers in simple systems – the pumping test under-estimates the highest values and over-estimates the lowest values. Slug-test results are often averaged across the site with limited regard to facies or facies trends, using harmonic and geometric means, which ultimately leads to a bias in the site-wide estimate, similar to the bias in estimates derived from pumping tests. Large-scale test procedures (e.g., pumping tests), or mathematical averages of small-scale test results, obscure critical site detail and tend to underestimate groundwater transport rates.

Slug tests and other small-scale hydraulic tests should be correlated to facies, to highlight the site-specific relationship between facies trends and hydraulic conductivity distributions. The resulting interpretations will honor the broad range of hydraulic conductivity and emphasize the fact that the well-sorted facies demonstrate hydraulic conductivities that are several orders of magnitude higher than the poorly-sorted facies. Applicable test methods include grain-size analysis, slug tests, CPT (**cone - penetrometer**) dissipation tests, short-term mini pumping tests and transient hydraulic interference tests.

Ultimately, hydraulic testing methods still fall short as a reliable means of evaluating transport behavior. The primary limitation of hydraulic methods stems from our inability to accurately map the mobile and immobile porosity comprising the three-dimensional architecture of aquifer systems between our measurement locations. Hydrostratigraphic methods provide the best means of interpolating the structure where we do not have data, but we are still challenged with developing quantitative mapping of hydraulic conductivity. Hydraulic tomography is an emerging technology that extends the application of transient interference testing to a three-dimensional testing and monitoring network. Current applications of that technology appear to offer a means to better interpret the facies boundaries that control flow in aquifers and show promise for improving our ability to map the structure and distribution of hydraulic conductivity.

We recommend that practitioners strive to make hydrostratigraphic interpretations more quantitative, by integrating facies mapping, small-scale hydraulic conductivity estimates and plume distributions when developing conceptual site models to describe contaminant transport and reagent distribution. Ultimately, direct measurement of transport behavior through tracer testing, as discussed in Chapter 12, provides the most reliable means of predicting injected reagent behavior.

11.1 METHODS FOR ESTIMATING HYDRAULIC CONDUCTIVITY

There are numerous methods available for field testing aquifers to obtain estimates of hydraulic conductivity. Most of the methods were developed for broad aquifer characterization and many are not well suited to detect the structural details that control transport through the formation. The test methods also are limited in their characterization of hydraulic conductivity extremes:

High conductivity systems — Aquifers (or facies within them) with hydraulic conductivities greater than 10^{-2} cm/s (30 ft/day) often exhibit very rapid drawdown/displacement responses during pumping and slug tests. In addition, the magnitude of the response is often relatively low, making it difficult to separate the hydraulic response from noise associated with background stresses such as distant pumping, tidal influences and barometric pressure variations. As a result, hydraulic testing in high-conductivity systems requires special provisions to generate sufficient hydraulic stress to ensure a high signal-to-noise ratio in the hydraulic response and that instrumentation is capable of resolving the rapid changes in the drawdown/displacement.

Low conductivity systems — Hydraulic testing in aquitards or poorly developed hydrofacies is simpler, in theory, because the drawdown/displacement responses are slow and the magnitudes are relatively large, enabling accurate characterization of the recovery curve. For very low conductivity formations with hydraulic conductivity values less than 10^{-6} cm/s, it is very difficult to monitor long enough to adequately characterize the response curve. Laboratory permeameter[1] testing methods are workable for these low-conductivity aquifer segments, as the approach allows for multiple samples to be collected and analyzed so that the variability can be properly evaluated.

Our focus in this section is to guide the practitioner in selecting the most appropriate testing method to correlate hydraulic conductivity with hydrostratigraphy. **Table 11.1** summarizes the hydraulic testing and hydraulic conductivity estimation methods available to meet this objective. We provide more detailed guidance on the application of these methods in the sections that follow. There are many excellent reference books available that provide the theoretical derivation of the boundary value problems and analytical solutions for the hydraulic testing methods. Among the

---

[1] Although the laboratory methods typically provide estimates of vertical hydraulic conductivity, rather than horizontal, it is the vertical hydraulic conductivity that is most important for aquitards. This is due to the fact that aquitards are typically invoked as vertical barriers to contaminant migration and because hydraulics induce vertical rather than horizontal flow through these hydrostratigraphic units.

TABLE 11.1

Summary of hydraulic testing and hydraulic conductivity estimation methods. We recommend small-scale, transient testing approaches that enable correlation of facies trends with conductivity estimates rather than large-scale averaging approaches

Test Method	Applicability	Comments
Grain-size analysis and Kozeny-Carman Equation	Discrete, small-scale estimates of horizontal conductivity in aquifer matrix only; not suitable for cohesive, plastic soils or cobble-sized distributions.	Well-suited for correlation of facies trends with conductivity; samples can be collected during boring installation and may be analyzed in the field.
Laboratory Permeameter Testing	Vertical hydraulic conductivity of discrete samples.	Best suited for aquitards and other low-conductivity strata; samples can be collected during boring installation.
CPT Dissipation Testing	Less than 10^{-3} cm/s.	Can be collected during CPT sounding with limited added cost; best suited for silts and clays.
Slug Testing	Applicable to all matrices: high-conductivity systems require specialized testing and data acquisition methods; low-conductivity system require specialized analysis if complete recovery not obtained.	Can be conducted during vertical aquifer profiling or monitoring wells; easily repeated.
Interference Testing and Hydraulic Tomography	Similar to slug testing, but requires additional equipment to monitor slugged well and monitoring wells; requires specialized data acquisition methods.	Extends scale-of-measurement for conventional slug testing; Hydraulic tomography requires sophisticated simultaneous solution. Potential to map three-dimensional facies boundaries and conductivity.
Mini-Pumping Tests	Short-duration (4 to 8 hrs), step-test approach provides estimates of horizontal and vertical conductivity (when appropriate monitoring network is installed). Aquifer storage parameters when multiple radial distances monitored and high-frequency data acquisition methods used.	Test can be scaled to evaluate localized or average behavior; multiple tests can be completed to correlate facies trends with quantitative estimates. Can be performed as injection or withdrawal to estimate injection/extraction capacity.
Conventional Pumping Tests	Long-duration (24-hr plus) tests provide best estimate of water budget and long-term injection/extraction well performance. Provides large-scale average estimates of vertical and horizontal hydraulic conductivity, aquifer storage, and classic type curve response. Can assist in identifying boundary conditions such as leakage through aquitards, recharge from surface water, and flow barriers.	Best suited for water supply, hydraulic containment, or recirculation-type *in situ* system design. Not suitable for transport assessments as volume-averaging filters out important information.

applied texts, are *Aquifer Testing* by Dawson and Istok (1991) and *The Design, Performance, and Analysis of Slug Tests* by Butler (1988). The interested reader is also encouraged to review the detailed derivations provided by Bear in *Dynamics of Porous Media* (Bear 1988).

GRAIN-SIZE ANALYSIS

Grain-size analysis has not gained the credibility it deserves as a means of estimating hydraulic conductivity — particularly for establishing a link between facies and hydraulic conductivity. This is primarily due to mis-application of the approach for estimating conductivity of cohesive soils and improper sampling and analysis across facies boundaries. As discussed in Chapter 3, the Kozeny-Carman formula provides a reliable means of estimating hydraulic conductivity from grain-size distribution data, provided that the formation does not contain abundant fines, which result in cohesive or plastic behavior, or cobble-sized grain-size distributions.

Proper sample collection and sieve analysis is required to evaluate effective grain diameter and sorting — the key is to collect representative samples that enable us to map facies variability within hydrostratigraphic units. Grain-size conductivity estimates should be organized by facies within hydrostratigraphic units, to show the trends and correlations, not to estimate averages.

For cohesive silts and clays, Shelby tube sampling and permeameter analysis provide a useful estimate of *vertical* conductivity, which can be important in evaluating the integrity of aquitards as flow barriers. Because aquitards and low-conductivity facies act as reservoirs for contaminant mass, it is important to obtain accurate estimates in these depositional elements.

CONE-PENETROMETER (CPT) DISSIPATION TESTING

CPT methods provide a qualitative estimate of hydraulic conductivity when conventional soundings are performed via soil behavior type classification. The qualitative estimates of hydraulic conductivity are demonstrated through the continuous pore-pressure response that is logged during each sounding. This makes CPT an ideal tool for discriminating between the mobile porosity associated with the well-developed facies and the immobile porosity associated with poorly-developed silt and clay depositional elements.

CPT dissipation test methods also provide quantitative testing of low-conductivity formations with a limited effort. As discussed in Chapter 9, formations with hydraulic conductivity values greater than 10^{-3} cm/s do not build pore pressures greater than the hydrostatic pressure as the tool is advanced, because the formation is capable of readily dissipating fluid pushed ahead of the probe. In contrast, silts and clays exhibit strong negative- and positive-excess pore-pressure deflections. In silts and clays, where the pore pressure deflections are significant, the probe advance can be stopped to allow monitoring of the hydrostatic pressure recovery rate, providing estimates of hydraulic conductivity. Silts and over-consolidated clays typically demonstrate a dilatory pore pressure recovery response, which is characterized by both increasing and decreasing pressure trends. Normally-consolidated clays typically exhibit monotonically decreasing trends. In silts and clays, the time to achieve complete pore pressure recovery can be slow, so specialized methods of analysis have been developed by researchers including Burns and Mayne (1998) and Mayne (2002), to account for the specific behavior and truncate the recovery time while still providing accurate hydraulic conductivity estimates.

Hydraulic conductivity estimates can be obtained for higher-conductivity formations, by performing slug-tests in the CPT rods when outfitted for groundwater sampling, just as it can be done during vertical aquifer profiling when using direct-push methods. We provide recommendations for slug testing in the section below. The interested reader is also referred to the works of Butler (1998) and Butler et. al (2001) for additional guidance and analytical solution methods.

11.2 SLUG TESTING

Slug testing in high-conductivity systems has received a significant amount of attention in recent years, as researchers have demonstrated the importance of inertial effects in these systems (Butler, 1998; Bohling, et al., 2002; Zurbuchen, et al., 2002; Bohling, et al., 2007). This research has led to the understanding of under-damped, oscillatory displacement/recovery responses, enabling much more accurate estimation of hydraulic conductivity in high-conductivity systems. As late as the mid-1990s it was common for practitioners to dismiss the oscillatory response as interference in the well due to over-aggressive placement of the slug — sometimes called the "sploosh effect," — or attributed to movement of the pressure transducer in the well during the test.

When the response is monotonically decreasing, we find that practitioners often under-estimate hydraulic conductivity from slug tests by seeking to match the wrong portion of the recovery curve. Most of the conventional methods are derived such that the designer should seek to match the linear portion of the recovery curve, which constitutes 90 percent of the change in head; however, we often see slug tests analyzed by matching 90 percent of the data based on time, not recovery. Analysis of the simpler response is also improved by comparing the initial displacement to the theoretical displacement, which is obtained by calculating the volume of the slug compared to the annulus and well screen. This provides a quality check on the test and provides a frame of reference for the initial displacement.

Improvements in data logging and acquisition equipment, as well as the use of pneumatic slug testing methods, have facilitated dramatic improvements in hydraulic conductivity estimates in recent years. In addition, researchers have extended slug-testing methods from "one well slugged and one well monitored" to "one well slugged and several monitored", also known as interference testing, enabling rapid and cost-effective testing of larger aquifer volumes, without "filtering-out" the heterogeneities. The primary difference between conventional slug testing and interference testing is the requirement for synchronized, high-frequency data acquisition rates — 20 to 100 Hz (1 sample every 0.05 to 0.01 seconds).

These interference testing methods are now being extended to hydraulic tomography testing by performing "many intervals and wells slugged - many intervals and wells monitored." Under this approach, discrete intervals within each well are isolated using packers to enable slug testing on the scale of feet or meters at each well or boring, while pressure transducers are deployed at multiple intervals in many wells to capture the wave-forms associated with propagation and attenuation of pressure in the formation. When combined with sophisticated simultaneous inversion methods (Yeh and Liu, 2000; Zhu and Yeh, 2005, 2006), the method may be capable of providing three-dimensional mapping of hydraulic conductivity in natural aquifers. However, the literature does not yet provide examples of success at the field-scale. Reported trials are presented in the context of model calibration for hydraulics, or sand-box transport models, but we are unaware of field-scale transport results at this time.

Interference testing and hydraulic tomography methods hold significant promise for resolving the three-dimensional aquifer architecture at the remediation hydrogeology scale, because they are tailored to resolving the high-frequency, transient responses. It is important to note that hydraulic tomography methods suffer from the same limitations that seismic and electrical geophysical methods do. Interpretations can be ambiguous because the observable response is based on the product of the feature thickness and the contrast in material properties (characteristic seismic wave velocity or electrical resistivity), which results in many non-unique solutions. The scale of resolution is also influenced by that contrast in material properties and the spacing of sensors. This limitation has been overcome in reservoir geophysics by acquiring true velocity and electrical resistivity data and calibrating the response and interpretation accordingly. At a minimum, the geophysical methods constrain the geometry of facies bodies and aid in defining the reservoir architecture at the appropriate scale for petroleum exploration and production. We expect that the current hydraulic tomography methods will provide useful information for interpreting facies geometry and hydraulic properties, at intermediate scales that are inaccessible through conventional technologies.

The key question is whether the tools can be used to economically resolve the aquifer architecture and hydraulic properties at the scale that controls contaminant transport and reagent distribution – the remediation hydrogeology scale. The methods require specialized monitoring equipment that is capable of collecting synchronized, high-frequency head measurements at many locations simultaneously. The process has to be repeated at numerous locations in order to establish the cross-connections between each tested location and the surrounding formation. The required instrumentation and analytical tools have been available in geophysical applications, such as seismic and electrical methods, for many years. We look forward to the results of field-scale applications of this emerging technology.

CASE STUDY – SLUG TESTING IN HIGH-CONDUCTIVITY AQUIFERS

Figure 11.1 presents the hydraulic response of a slug test that was performed in a high-conductivity glacial outwash aquifer, consisting of medium-grained sand and gravel mixtures. This test was performed recognizing that the conductivity of the system was very high, so the field team was prepared to synchronize the initiation of the test with a pressure transducer capable of logging the data at 0.1 second (10 Hz) intervals. As shown, the hydraulic response exhibits a classical under-damped, oscillatory recovery pattern that is characteristic of high-conductivity formations. The data were analyzed using AQTESOLV® and the KGS solution method after Hyder et al. (1994), which indicated that the hydraulic conductivity was 0.13 cm/s. Not surprisingly, this estimate was nearly one order-of-magnitude higher than prior estimates, derived from conventional pumping tests which indicated a mean hydraulic conductivity of 3.9×10^{-2} cm/s. As we will see in Chapter 12 (Case Study 3), the permeability in some portions of this aquifer may be even higher than indicated by analysis of the oscillatory response, because tracer testing performed at this location indicated transport velocities of several meters per day.

If the field crew had not anticipated the oscillatory response, they might have used a lower sampling frequency and that would have filtered out the true hydraulic response. We often see slug tests completed using a log-scale sampling strategy, collecting samples every 0.1, 1, and 10 seconds. Were this the case for high-conductivity dataset, it is likely that the hydrogeologist would have dismissed the test results, because the characteristic response of the recovery curve is difficult to resolve and the quality appears questionable. The lower panel on **Figure 11.1** shows the same data set sampled using a log-scale strategy. Based on the quality of the recovery curve, it is easy to see why the oscillatory response is often mischaracterized as a "sploosh-effect" or transducer movement, when the sampling frequency is inadequate to resolve the response.

11.3 PUMPING TESTS

Successful implementation of pumping tests in high-conductivity systems requires the use of data logging pressure transducers that are capable of recording head measurements at one-tenth of a second (0.1) intervals, or less. Frequent and high-quality flow measurements should also be recorded, preferably logged digitally, as flow rates sometimes vary with drawdown in the well as a function of the pump curve. In water supply testing, the convention is to use a log-scale sampling frequency strategy, such that samples are acquired at 0.1, 1, 10, 100 second-intervals, allowing the large-scale trends and characteristics of the drawdown response to be resolved for comparison to the appropriate type-curve (confined aquifer, leaky aquifer, delayed yield, etc.). Our experience suggests that the log-scale strategy is appropriate for water supply, but it filters out the early-time, transient responses that are required to identify the heterogeneities that control reagent distribution and contaminant transport in aquifers. There are no longer barriers to the high-resolution testing and analysis needed to observe aquifer heterogeneities:

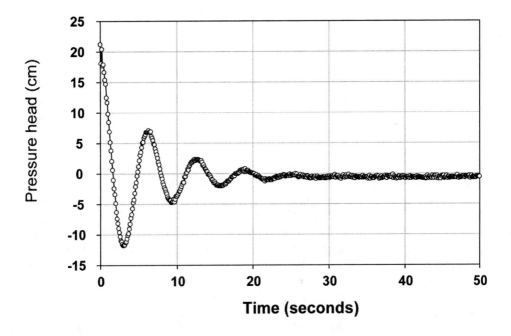

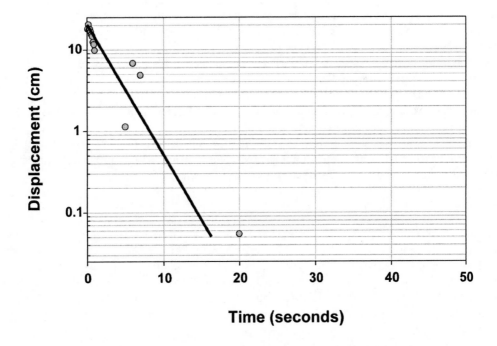

FIGURE 11.1 Example of an under-damped, or oscillatory response for a slug test, performed in a high-conductivity aquifer in Northeastern Canada. Inertial effects associated with rapid fluid displacement in formations with hydraulic conductivity greater than 10^{-2} cm/s can result in the oscillatory pressure response, as seen in the upper panel, which was logged at 10 samples per second, or 10 Hz. If the transducer logging frequency is set too low, the oscillatory response will not be detected. The lower panel shows the same data set, sampled using the log scaling of observation times - the oscillatory response is not observed at the lower sampling frequency.

- Data loggers and PC workstations enable high-frequency data collection and analysis without any significant effort or cost.
- Aquifer analysis tools such as AQTESOLV® enable the practitioner to evaluate pumping tests with variable flow rates, eliminating the need to average pumping rates and allowing the designer to separate these influences on the drawdown response from the heterogeneities that are so important in remediation hydraulics.
- The more sophisticated analytical tools also enable accurate estimates of aquifer storage through the use of simultaneous solution of monitoring data acquired at several radial distances.

Because the objective in remediation hydraulics is to capture and resolve aquifer heterogeneities (rather than to obtain the best average estimate), there is greater value to be derived from short-duration pumping tests — we call them mini pumping tests — than from long-term testing. To maximize the value of the hydraulic testing result, the tests should be completed using several different pumping rates, as in a step-drawdown test, and the monitoring strategy should emphasize synchronized, high-frequency data collection (head and flow rate) as discussed above. In addition to reducing pumped volumes (which adds considerable cost at remediation sites due to treatment and disposal requirements), a major advantage of the mini tests is that many individual tests can be performed, allowing the designer to better correlate hydraulic conductivity to facies and account for the variability across the remediation site.

Pumping test analysis methods should be selected based on the type of aquifer and considering the potential influence of boundary conditions. One of the most important steps in developing a plan for pumping tests is to have a well developed conceptual site model. The conceptual site model frames our expectations so that we can identify important features of the aquifer system that will affect and control the behavior of our tests. The designer should develop estimates of the magnitude and rate of drawdown propagation, based on the conceptual site model, to ensure that the monitoring network geometry (both horizontal and vertical spacing) is adequate to resolve hydraulic responses to the test. For example, a test conducted in a relatively thick, stratified system should utilize a three-dimensional network of piezometers and monitoring wells spanning several radial distances, directions, and stratigraphic intervals. Ideally, such a system should also be stressed by extracting or injecting in specific hydrostratigraphic intervals, if the goal is to establish the degree of communication between units. Classification of the aquifer enables selection of the most appropriate analytical solution for interpreting the data and will ensure that monitoring data meet the assumptions associated with the analytical solution.

CASE STUDY – PUMPING TESTS AND HETEROGENEITY

Almost all groundwater remedial actions rely on the use of traditional aquifer testing methods, either extracting groundwater from a well and observing the response at other wells or single well slug tests, to characterize the hydraulic properties of aquifers. In this section we examine the ability of a pumping test to determine the hydraulic properties of an aquifer. The aquifer represented on **Plate 7** is again the focus of our analysis. **Plate 7** presents the horizontal hydraulic conductivity distribution that would be expected to result from sediment deposition by a braided alluvial channel.

As introduced in Chapter 3, the synthetic aquifer is 6 meters (20 feet) thick, confined, and groundwater flow is from the left to right (west to east), driven by an average hydraulic gradient of 0.0025 ft/ft. No-flow boundaries are present along the top and bottom (north and south) of the aquifer. The model extends 300 feet from east to west and 400 feet from north to south. The aquifer was discretized using a 1 foot by 1 foot grid, consisting of 120,000 cells. Groundwater flow was simulated using MODFLOW (McDonald and Harbaugh, 1988; Harbaugh, et al., 2000). The objective was to predict the flow rate and groundwater velocity through the system and, more

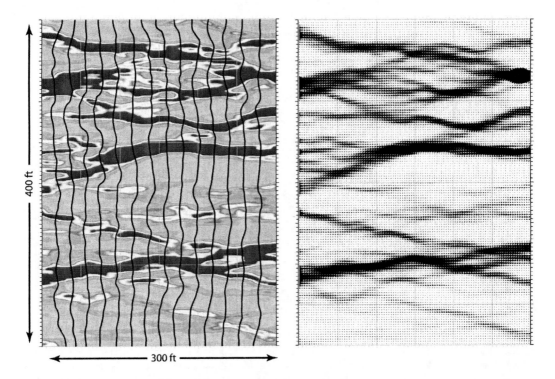

PLATE 7 Heterogeneous porous medium, synthesized to analyze the detectability of high-flow zones using water elevation contours. The relative hydraulic conductivities are shown on the left diagram, ranging over three orders of magnitude (dark blue is 1,000-fold higher conductivity than the light green shaded areas). The grid area is 300 x 400 feet and the water elevation contours, shown in black, are at 0.5-foot intervals. Groundwater flows are highly concentrated in this heterogeneous system, as shown by shading on diagram at the right. In this setting, roughly 70% of the flow occurred in 25% of the cross-sectional area, and more than 90% of the flow occurred in 50% of the cross-section. These flow concentrations are not discernible in the groundwater elevation contours. *(See Plate section for color version of Plate 7.)*

importantly, to determine whether the presence of heterogeneities and anisotropies in the formation could be detected in pumping test results. As described in Chapter 3, the groundwater surface elevation contours, built from monitoring well networks, would not reflect the anisotropies of such an aquifer.

A pumping test was designed with 6 observation wells installed as shown on **Figure 11.2**. All wells were screened across the entire aquifer thickness. The observation wells were aligned with the pumping well in the direction of groundwater flow (left to right) and perpendicular to groundwater flow (top to bottom). The orientation and close spacing of the observation wells was selected to permit quantification of hydraulic properties in the direction of groundwater movement, as well as a qualitative assessment of the anisotropy of the aquifer. The aquifer test was 24 hours in duration, 8.4 hours of pumping at a uniform rate of 1.3 L/s (20.5 gpm), followed by 15.6 hours of recovery.

Figure 11.3 shows the water level response in wells X+10 (10 feet east of pumping well) and Y-10 (10 feet south of pumping well). These two wells have very similar response curves, with drawdown starting nearly instantaneously with the start of pumping (consistent with confined aquifer conditions). Drawdown stabilized after approximately 0.1 days (8,640 seconds) as it extended to the boundary of the aquifer (in this case the constant head boundary condition along the right side of the model domain). The drawdown levels were similar in the nearest off-channel and in-channel

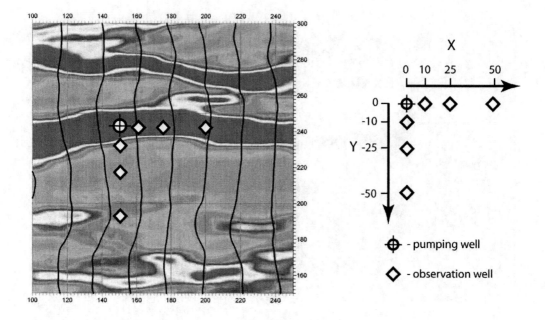

FIGURE 11.2 Layout for pumping test analysis of an anisotropic porous medium. The test area is a subset of the synthetic aquifer domain created for Plate 7. The pumping well was placed in a high-conductivity zone and two lines of observation wells were laid, one in the high-conductivity zone at X+10, X+25 and X+50 feet, and a similar set was placed perpendicular to the high-conductivity zone at Y-10, Y-25 and Y-50 feet.

wells (Y-10, with 2.88 feet of drawdown and X+10, with 2.72 feet of drawdown, respectively). The observed drawdown in each well was analyzed using AQTESOLV®, to estimate hydraulic properties. The analysis showed that early-time data, before 0.001 days (86 seconds), diverged significantly from the prediction line (**Figure 11.3**). After that time, there was close agreement between observed (dots) and predicted responses (gray curves) with estimated transmissivity values of 510 and 540 ft²/day at wells Y-10 and X+10, respectively, for observations later than 0.001 days. The same pattern was observed in the remaining wells, as shown on **Figure 11.4**. Standard practice for pumping test analysis uses the later-time data and an analyst might conclude (incorrectly) that the aquifer is homogeneous isotropic and is only slightly more permeable in the x-direction. We will examine the importance of the early time data for the detection of anisotropies, below.

Figure 11.5 shows a comparison of the transmissivity estimates with distance from the pumping well in the northing (Y) and easting (X) directions, which are roughly perpendicular and parallel to the known anisotropy associated with the high-conductivity channels. The plots for the in-channel (X-series) wells and off-channel (Y-series) wells were developed from the late-time transmissivity estimates shown on **Table 11.2**. The influence of volume averaging on the transmissivity estimate is highlighted on the graphs by comparing the trends along each axis to the transmissivity of the channel, the domain average and the off-channel average. The transmissivity estimates for the monitoring points nearest the pumped well (X+10 and Y-10) are comparable, at 540 ft²/day and 510 ft²/day, but biased-low with respect to the domain average of 840 ft²/day.

As shown, the transmissivity of the channel wells decreases slightly with distance from the pumped well to about 475 ft²/day at X+50. Overall, the channel well transmissivity values are about one-half the domain average transmissivity of 840 ft²/day, but an order of magnitude less than the true transmissivity of the channel of 6,100 ft²/day. The poor correspondence between the true channel transmissivity and estimated values along the channel reflects the influence of the low-

X + 10

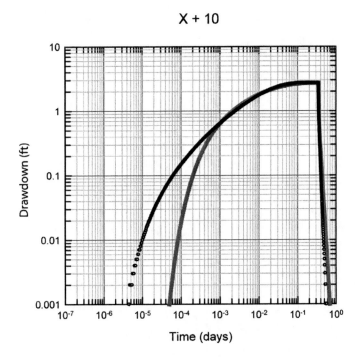

Y - 10

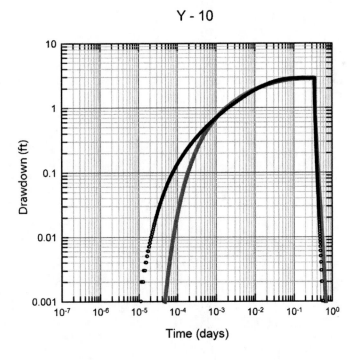

FIGURE 11.3 Examples of pumping test results in the synthetic domain. The well located at Y-10, shown on the bottom, is located in a low-conductivity zone and the X+10 well lies in the high-conductivity zone. Automated curve fitting forces a match to the late-time data (by convention). This abandons important data for each of these curves that would have disclosed the presence of much higher hydraulic conductivities near the respective observation points.

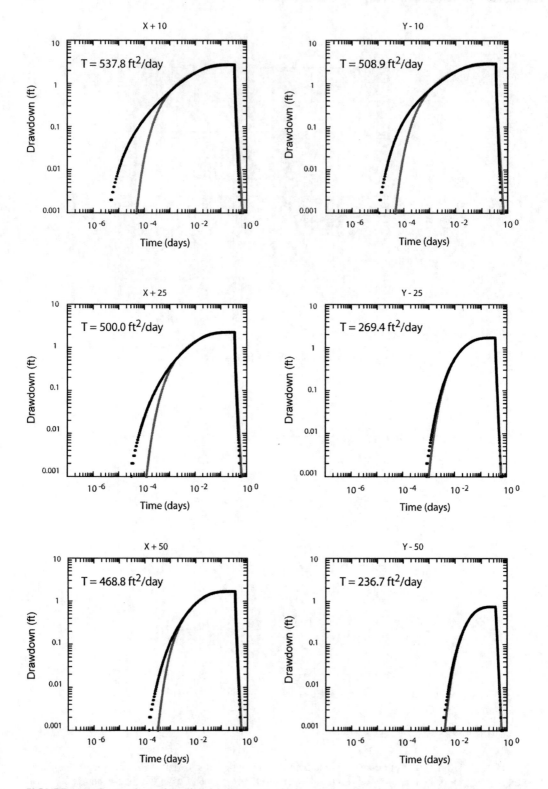

FIGURE 11.4 Comparison of predicted and "observed" drawdown response at observation wells associated with the pumping test on the synthetic aquifer. The test layout is shown on Figure 11.2. Early-time drawdown deviated from predicted values in all wells, while late-time data closely fit predicted drawdown.

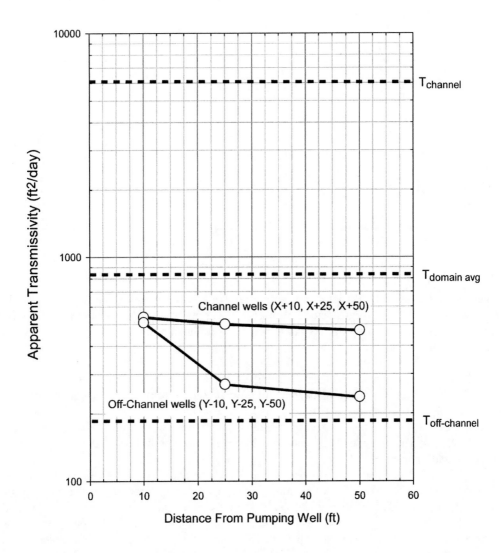

FIGURE 11.5 Apparent transmissivity as a function of distance from the pumping well, for observation wells located in the high-permeability channel and in the low-permeability aquifer matrix adjacent to the channel. The transmissivity in the channel was 6,100 ft^2/day and the average transmissivity for the domain was 840 ft^2/day. The off-channel average transmissivity was 190 ft^2/day. The well layout is shown in Figure 11.2.

TABLE 11.2
Summary of transmissivity estimates for simulated pumping test case study in an anisotropic aquifer. The actual site condition values were derived from the point values in the model. Pumping test estimates were derived using the conventional late-time fits from the simulated drawdown data.

Well	Actual Site Conditions		Pumping Test Estimates	
	T [ft²/day]	K [ft/day]	T [ft²/day]	K [ft/day]
MW-1	6580	329	--	--
X+10	7260	363	537.8	26.9
X+25	7500	375	500.0	25.0
X+50	3250	162.5	469.8	23.5
Y+10	290	14.5	508.9	25.4
Y+25	46	2.3	269.4	13.5
Y+50	242	12.1	236.7	11.8

conductivity features in the volume-averaging, much like the equivalent hydraulic conductivity for flow parallel to layers in Chapter 3. However, it is important to note that the transmissivity estimates derived from early-time curve matching were much more accurate than the late-time values compared here.

The transmissivity values for the off-channel wells decrease from about 510 ft²/day at Y-10 to approximately 240 ft²/day at Y-50, approaching the average off-channel transmissivity of 190 ft²/day. While the estimates for the off-channel wells are biased slightly high compared to the average transmissivity, the results are much more useful for evaluating flow conditions than the results for the in-channel wells. In simple terms, the off-channel trends reflect an averaging process that is dominated by the low-conductivity features, similar to the equivalent hydraulic conductivity perpendicular to layers.

So, how well did the pumping test characterize the aquifer? The hydraulic conductivity distribution shown on **Plate 7** varied from 1 ft/day (light green shading) in the finest-grained sediments to 1,000 ft/day (deep blue shading) in the coarsest-grained sands. The hydraulic conductivity locally at each well and the estimated parameters from the pumping test are summarized in **Table 11.2**. Comparison of site conditions with pumping test estimates showed that the pumping test:

- consistently under-estimated the hydraulic conductivity in the high-conductivity zones;
- over-estimated the conductivity in the low permeability zones;
- under-estimated the average hydraulic conductivity of the aquifer by roughly 100 percent.

For both well sets, the apparent transmissivities poorly matched the actual values, both for the vicinity of the testing and for the aquifer domain, as a whole. What is surprising in this example is that, even in a highly controlled analytical experiment, a single pumping test was unable to recognize that $K_x{:}K_y$ was more than 300:1, and that the hydraulic conductivity near the pumping well was more than 300 ft/day.

The failure to recognize high conductivity zones will result in under-estimation of transport velocities and improper design of *in situ* remedies. Unfortunately, the pumping test inaccurately indicates that groundwater velocities are slow (0.3 to 0.7 ft/day) which might have encouraged selection of an *in situ* reactive zone technology with periodic injections. Such a system, if implemented, would be a challenge to operate, as the true groundwater flow velocities are 12 to 28 times faster (up to 8.2 ft/day); the high groundwater velocities would quickly dilute and flush reagents out of the treatment area.

The standard analytical approach, which pushed the curve fitting to later-time data, discarded information that was critical to understanding this formation. From **Figure 11.4**, we can see that the divergence of early-time data from the analytical line was significant, particularly for the channel wells. To assess this further, we plotted the estimated transmissivity for well X+10, as a function of results tabulated over successively longer intervals, beginning with the earliest data. **Figure 11.6** shows the break points in pumping time that were used for the analysis. The results were tabulated in **Figure 11.7**, which shows the estimated transmissivity, as a function of the pumping time. The transmissivity estimated from the first seconds of pumping came the closest to matching the actual

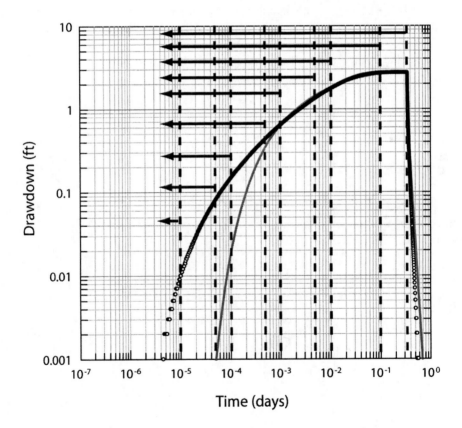

FIGURE 11.6 Pumping test results for the X+10 observation well, with data cutoff times that were used for the analysis given in Figure 11.7 indicated by dashed lines. For example, the left-most dashed line indicates the cutoff for data earlier than 10^{-5} days; the dashed line on the far right indicates "all data." The pumping test results were analyzed using datasets from progressively longer time intervals, beginning with only data from earlier than 10^{-5} days and ending with the entire dataset. The results of that analysis are shown in Figure 11.7.

transmissivity of the channel area. This suggests that the transients, which are often discarded or discounted, contain critical data for the characterization of an aquifer, particularly in cases of heterogeneous, anisotropic formations. This analysis also supports the premise that single well slug tests provide critical information on near-well hydraulic properties, which cannot be obtained using pumping tests.

The pumping test provided only a fair estimate of the total groundwater flow through the heterogeneous anisotropic aquifer (only accurate to within an order of magnitude). Because the test generated an under-estimate of the actual value for total groundwater flow through the unit, it may be considered as a conservative estimate for water supply development purposes. However, the groundwater velocity estimates were orders of magnitude below the "actual" values and the pumping test would provide no value in discerning mass transport velocities or localization within the formation.

11.4 MONITORING APPROACHES – WHAT A MONITORING WELL SEES

Our accumulated experience teaches us that solute concentrations and groundwater velocities can vary dramatically over very short vertical intervals. We have observed orders-of-magnitude ranges of solute concentration and hydraulic conductivity, all within the span of a 5-foot monitoring well screen. In some cases, the high flow and high concentrations coincide, while in others, they do not. As discussed in Chapter 7, the maturity of the plume determines the relative partitioning of contaminant mass between the mobile and immobile segments of aquifers. The degree of connectedness between the mobile and immobile segments dictates the plume behavior during advancement of the plume and remediation through diffusive mass exchange. Advective transport

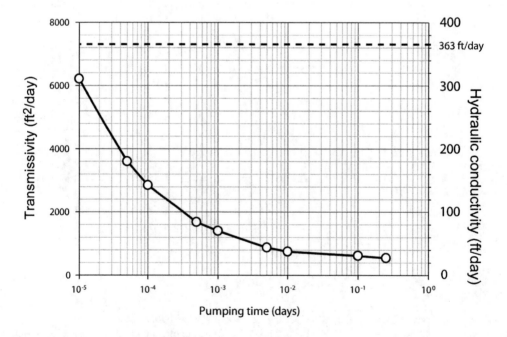

FIGURE 11.7 Importance of early-time data in correctly assessing hydraulic parameters in an anisotropic, heterogeneous aquifer matrix. The dashed line indicates the "actual" hydraulic conductivity for the observation well (X+10).

dominates in the mobile segments, but is tempered by diffusive interaction with the immobile reservoirs, where mass transport is dominated by diffusive mechanisms. The superficial result is that plume transport velocities appear to be retarded compared to the advective flow rates, as discussed in Chapter 7. While it is true that the mass transfer between the mobile and immobile porosity segments of the aquifer reduces the apparent plume migration rate, the important implication for remediation is that a significant component of mass in the system is stored in the immobile porosity reservoirs comprising the lower-K facies in the aquifer.

We have provided examples of these observations in Chapters 7 and, now that we've trained ourselves to be cautious in reading monitoring well results, these patterns of striking heterogeneity appear to be the rule, rather than the exception. Each method of monitoring well sampling carries a bias and it's necessary for groundwater scientists to understand these sampling biases, as well as the variability of aquifer systems, to develop an effective sampling strategy for any site. In the following sections, we discuss the most common methods of monitoring well sampling, describing the biases of each and how they might be overcome or harnessed.

SAMPLING METHODS

The primary objective in groundwater sampling is to collect a "representative" sample of the groundwater in the formation. Definition of what a "representative" sample is still remains a contentious issue in the literature and regulatory environment; however, the United States Environmental Protection Agency (USEPA) guidance documents available through the Resource Conservation and Recovery Act (RCRA) and Comprehensive Environmental Response, Compensation, and Liability Act (CERLA) provide prescriptive methodologies that are designed to reduce sampling error and variability. While there are many acceptable methods for collecting groundwater samples, the approaches are generally described by the three categories below. Our focus is to evaluate the sampling methodology based on the hydraulics of the sampling process. The interested reader is encouraged to review appropriate regulatory guidance documents to ensure methods are selected to meet project-specific data quality objectives and state or federal requirements.

Conventional Bailing and Pumping Methods — The conventional approach to groundwater sampling relies on pre-sample purging by bailing or pumping three or more well volumes from the well. The basic idea is that removing the stagnant water from the well draws groundwater from the formation, enabling collection of an appropriate sample. The key distinction between conventional bailing and pumping methods is that the purge volume is somewhat arbitrary — there is little consideration to the flux through the well, except in low-yield cases, where purging causes the well to run dry. In these cases, the well is allowed to recover prior to sample collection. Because the purging methods involve pumping or bailing to remove relatively high-volumes of water from the well, the groundwater contributions are distributed throughout the formation in the well-screen in proportion to the hydraulic conductivity. The high flow rates associated with pumping, or bailing methods, mix the groundwater in the well, homogenizing the sample. For this reason, we designate conventional bailing and pumped sampling methods as conductivity or flux-proportional sampling methods.

Low-Flow Sampling — Low-flow sampling methods, also known as minimal-drawdown sampling, utilize low-flow rates (100 to 500 mL/minute) to collect groundwater samples and minimize the entrainment of sediments during the process. Typically, low-flow methods utilize real-time monitoring and stabilization of field geochemical parameters including pH, oxidation-reduction potential, dissolved oxygen and turbidity to determine when an appropriate purge volume has been completed.

The flow rate and drawdown are monitored to ensure that a discrete sample can be obtained. The question of whether low-flow sampling can provide a representative sample from a discrete interval in the monitoring well screen remains a contentious issue. From our perspective, the hydraulics of the well provide a simple answer — if the flux through the target interval is relatively high compared to the sampling rate, we believe that a discrete sample can be obtained; however, if the flux is low compared to the sampling rate, in-well mixing will result in flux-proportional averaging of the measured concentration. This topic will be discussed more in this section and in Chapter 12, where we present results of tracer studies.

Permeable Diffusion Bags — Permeable diffusion bags utilize compound-specific membranes to allow groundwater flowing through the well to equilibrate with the fluid in the bag, thereby eliminating the need to purge before sampling. While permeable diffusion bags are applicable to a broad range of contaminants, it is important to select the membrane to match the compound of interest.

Permeable diffusion bags are ideally suited to sampling in longer well screens, provided that the flow in the aquifer is horizontal. Vertical flow within the well limits the utility of the method, as concentrations will appear uniform because of the equilibration process, even though there might be stratification of the plume. Similarly, low-conductivity formations might also lead to uniform concentrations across long well screens as diffusion is the driving force for sampling by this method.

Ultimately, case studies documented by the ITRC (2007) indicate that the discrete sampling approach provided by permeable diffusion bags reveals that plumes can be highly stratified in formations. Provided that the conditions are suitable and compounds are compatible, we recommend permeable diffusion bag sampling because the discrete samples can be correlated with the facies in the hydrostratigraphic units.

CASE STUDY – SAMPLING IN A HETEROGENEOUS FORMATION

Based on our experiences in collecting data from tracer tests and remediation system monitoring, we have come to the consensus that how a well is sampled has a significant impact on the result. In addition, the results of this case study example will also illustrate that position of the sampling location relative to the mobile and immobile porosity segments of the aquifer has a controlling influence on resulting observations.

In this case study example, we examine solute transport simulations conducted in a synthetic stratified aquifer, based on the data derived from high-resolution conductivity testing at the Borden aquifer. The transport models were set up to simulate the advance of a clean-water front leaving an *in situ* reactive zone barrier. The simulations were designed to evaluate the concentration trends that would be observed as the aquifer cleans up, using three different sampling methods: 1) continuous readings from a fixed-elevation probe, 2) low-flow sampling, and 3) high-flow purging to represent conventional bailing or pumping methods. The model parameterization is discussed in the following paragraphs. The results of the simulations are evaluated in the following subsections.

MODEL PARAMETERS

Transient groundwater flow and solute transport simulations were completed using MODFLOW (Harbaugh, et al., 2000) and MT3DMS (Zheng and Wang, 1999). The model consisted of 65, 5-cm thick layers using the vertical hydraulic conductivity distribution associated with right panel shown on **Figure 3.11** and Core 16 on **Figure 3.12**. The conductivity values ranged from 7×10^{-4} cm/s to 1×10^{-1} cm/s, with a geometric mean of 3.0×10^{-2} cm/s and standard deviation of 2.6×10^{-2} cm/s. The model was treated as an unconfined aquifer with a specific yield of 0.2 and a total porosity of 0.35. The model domain was 1,000 cm (200 cells) in the horizontal dimension, parallel to flow.

A horizontal hydraulic gradient of 7.2×10^{-4} cm/cm was applied across the model using constant head boundaries at each end. Based on the gradient and porosity, this yielded a superficial groundwater velocity for the model of 5.3 cm/day for the geometric mean conductivity and 7.1 cm/day for the harmonic mean. Taking this superficial perspective on velocity, the transit time across the 1,000-cm model domain would range between 140 and 190 days. However, particle tracking results obtained from each of the 65-layers led to an average velocity of 8.6 cm/day, with a range between approximately 1 and 18 cm/day[2]. Therefore, the range of computed advective transit times ranged between 56 and 1,000 days.

The transport models were set-up using an initial bromide concentration of 1,000 μg/L throughout the entire model domain to simulate the advance of an upgradient, clean-water front that would exit an *in situ* reactive zone barrier at time zero. Clean-water with a zero bromide concentration was allowed to perfuse the model through the constant heads at the left-hand side of the model for a period of 480 days. The model explicitly simulated only advection and diffusion, by prescribing zero values for all dispersivity coefficients. The molecular diffusion coefficient of bromide was set at 1.84×10^{-5} cm^2/s. This modeling approach simulates the effects of immobile and mobile porosity explicitly through the diffusive interchange of mass between the lower- and higher-conductivity segments in the aquifer. The solution was obtained using the TVD and conjugate gradient method to minimize numerical errors and numerical dispersion. This led to a computed mass balance error of less than 0.08 percent for the models in this case study.

Transport Modeling Results

Figure 11.8 shows the time-lapse advance of the clean-water front from left to right across the system at the following times: 3, 35, 78, 132 and 199 days. The darkest gray-scales at the left correspond to clean-water front with a zero bromide concentration, while the light gray-scales to the right correspond to the initial bromide concentration of 1,000 μg/L. The brightest whites correspond to the leading edge of the advancing front, where concentrations range between 0 and 1,000 μg/L. The shape of the advancing clean-water front highlights the relative transport rates within the 65 layers comprising the model. The locations of the sampling intervals evaluated in this analysis are shown on the figures as white boxes.

The transport observation package was used to generate bromide concentration breakthrough curves for each of the 65 model layers at 700 cm from the origin of the clean-water front. The entire column of the model at this location was treated as a fully penetrating well by prescribing a hydraulic conductivity of 10,000 cm/s and using a porosity of 1.0. The bromide breakthrough curves were evaluated in detail at four model layers to evaluate the simulated sampling results at 280 cm, 215 cm, 125 cm and 45 cm model elevation (from base). These intervals were chosen to highlight the transport behavior in the fastest (mobile) and slowest (immobile) velocity segments compared to the overall average behavior in the system: the 280-cm interval is situated in a band of the highest conductivity; the 215-cm interval is in the thickest segment of low conductivity; the 125-cm interval is placed in an interbedded region of the model; and the 45-cm interval is located in a relatively thin low-conductivity band. As a result, we can examine the influence of sampling methods on the observed breakthrough curve behavior and gain insight into the role of mobile and immobile porosity on the relative rate of clean-up in the synthetic stratified system.

---

[2] The low-end result is higher than would be computed using the lowest hydraulic conductivity values, because, during sample extraction, the particles tended to short-circuit out of the low-K layers to take the path of least resistance.

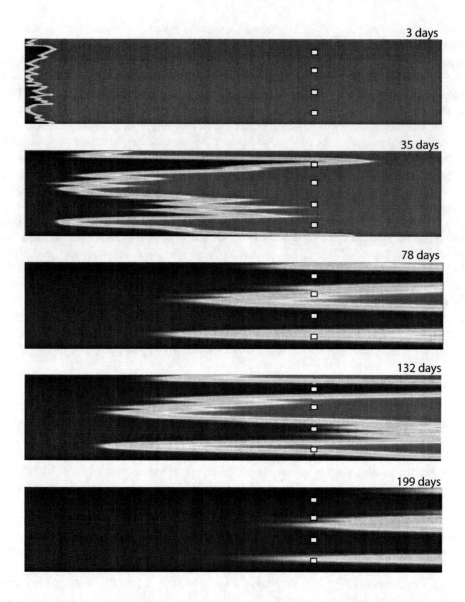

FIGURE 11.8 Time-lapse of simulated clean-water front migration past a vertical array of low-flow sampling points (shown by white rectangles). The 1,000 x 325-cm model domain was uniformly charged with 1,000 ug/L bromide then flushed by clean-water from the left hand side for 480 days. Diffusive mass transfer of solute stored in lower-conductivity strata slowed the advance of the clean-water front and led to asymmetric breakthrough curves.

The sampling methods were simulated using the following calculation approaches:

Continuous probe sampling — Breakthrough curves were computed for the duration of the model using the simulated results from each of the four locations at 3 day intervals at 280, 215, 125 and 45 cm elevations in the model domain. Because the probe sampling results correspond to concentrations within each 5-cm cell, these results are the best representation of the "true" bromide concentrations in each particular layer or aquifer segment.

High-volume purge sampling — High-volume purging was simulated to represent a fully-penetrating (3.25-m long) monitoring well screen. The breakthrough curves for each of the 65 model layers were combined using a groundwater-flux-weighted average (flux-proportional average) to simulate a conductivity-proportional sample that would result when sampling using bailers or conventional pumping methods. Because the observed concentration is dominated by the aquifer segment with the highest flux, the utility of the observation depends on whether the majority of the mass is co-located with flux.

Low-flow sampling — 5-cm-long wells were used to simulate 1 hour of sampling every three days for the duration of the transport model. The individual flow rate for each well was set at 0.75% of the total flow through the vertical profile, or a combined rate for the 4 simulated wells of approximately 3 percent of the instantaneous flux rate through the model. Concentrations were recorded prior to sampling, at the initiation of sampling, mid-way through sampling and at the end of the sampling period to characterize the potential changes in concentration related to sampling. The low-flow sampling was evaluated using a separate flow and transport model; however, the comparison of the breakthrough curves with- and without pumping indicates that sampling has a negligible effect on overall transport rates through the model.

Comparison of Breakthrough Curves

The relative transport rates in each of the four sampling intervals can be evaluated by comparing the breakthrough curves (concentration versus time trends) for each probe sample with the flux-proportional sample as shown on **Figure 11.9**. To simplify the language, we refer to the discrete sample interval as the **probe sampling**, and the high-volume purge as **flux-proportional sampling**. The flux-proportional sample represents the average concentration across the vertical profile of the aquifer, or the result of high-volume purging and sampling across a 3.25 m (10.6 ft) well screen. Because the probe and flux-proportional sampling breakthrough curves depict the concentration trends at the same distance from the clean-water source, but at different depths for each sampling location, the relative placement of each curve on the plot provides a relative measure of the migration rate, or clean-up rate at each interval.

This analysis can be completed two different ways using the breakthrough curve chart: by selecting a target concentration on the y-axis and reading the time from the x-axis at the intersection of the individual curve, or by selecting a time from the x-axis and reading the concentration from y-axis at the intersection of the curve. Using the first approach and selecting 200 µg/L as our target concentration for comparison, we find that the 280-cm probe breakthrough occurs first at approximately 42 days, followed by the 125-cm interval probe at 70 days, the flux-proportional sample at 100 days, the 215-cm interval probe at 160 days and the 45-cm interval probe at 182 days.

Because the clean-water front represents a continuous source with zero concentration, the center-of-mass can be estimated from the chart by selecting the concentration, C, that corresponds to 50 percent of the initial concentration, C_o of 1,000 µg/L, or 500 µg/L. The average transport velocity, V_{avg}, indicated by each curve can be estimated by dividing the distance traveled, x, by the time to reach $C/C_o = 50$ percent, t_{50}:

$$V_{avg} = \frac{distance}{time} = \frac{x}{t_{50}} \left(\frac{cm}{s} \right) \tag{11.1}$$

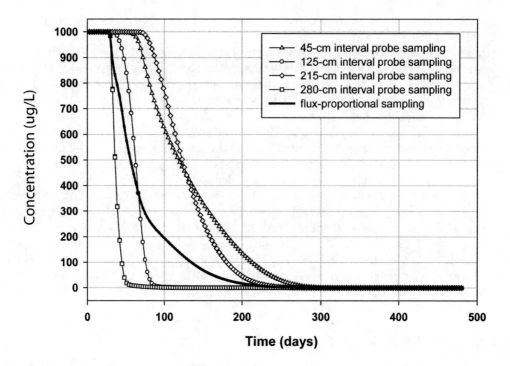

FIGURE 11.9 Comparison of clean water breakthrough curves, based on simulated probe and flux-proportional sampling methods. The simulated probe sampling results represent the temporal concentration trends observed at each of the 5-cm cells at 45, 125, 215 and 280-cm vertical intervals in the model, 700 cm downgradient from the clean water source. The flux-proportional sample is the flux-weighted concentration average from the 65 cells comprising the vertical profile of the aquifer at 700 cm. Comparison of the individual breakthrough curves to the flux-proportional sample indicates that the migration rates at the 280-cm and 125-cm intervals are greater than the average for the system; migration rates through the 45-cm and 215-cm intervals are slower than the average.

For the 280-cm interval probe, we find the t_{50} equals approximately 36.8 days from **Figure 11.8**. Substituting 36.8 days and 700 cm into Equation (11.1) yields an average clean-water front velocity of approximately 19.1 cm/day. Similarly, we find the average transport velocity of 11.1 cm/day for the 125-cm interval probe, 5.9 cm/day for the 45-cm interval probe and 5.7 cm/day for the 215-cm interval probe. The average transport velocity for flux-proportional result is 12.7 cm/day.

Notice that the slopes of the individual breakthrough curves flatten out with time, leading to significant asymmetry in the shape. The reason for the change in the slope of the breakthrough curves is diffusion of mass from the low-conductivity strata, or immobile reservoirs within the system. This behavior is consistent with the mass transport discussion in Chapter 7 and the back-diffusion process illustrated on **Plate 10**. The changing slope of the breakthrough curves reflects the fact that diffusion is a time-dependent process. Although the high-conductivity segments (280 and 125 cm intervals) of the aquifer show rapid decreases in concentrations, diffusive mass transfer from the low-conductivity strata, such as the 215-cm and 45-cm intervals, continues to provide mass that affects the overall concentrations in the system. This result is shown by the change in slope of the flux-proportional sampling breakthrough curve, which reflects the decreasing diffusive flux rates over time.

Another convenient perspective is offered by treating the breakthrough times for successively decreasing concentrations as relative measure of the change in the mass-transfer from the immobile to the mobile segments of the aquifer. **Figure 11.10** shows the average velocity of concentration-fronts between 900 and 5 μg/L for the four sampling locations and flux-proportional sampling. In this case, the term average velocity is used because the travel distance of 700 cm is divided by the time for breakthrough of each threshold concentration on each breakthrough curve. The average concentration-front velocities are highest initially and decrease with time, as indicated by the changing slopes.

It is important to note that the concentration-front velocities would be constant if there was no diffusive mass transfer between the mobile and immobile segments of the aquifer. Under conditions of no diffusion, the graphs plot as a series of vertical lines based on the hydraulic conductivity of each layer. The higher-conductivity intervals (280 and 125 cm) show the greatest change in slope for the discrete sampling. Notice that the flux-proportional sampling curve is bracketed by the fastest interval (280 cm) and the slowest interval (45 cm). This trend suggests the relative importance of the high-conductivity strata during the early advance of the clean-water front and the low-conductivity strata as a continuing source of mass during the later times in the conductivity proportional sample.

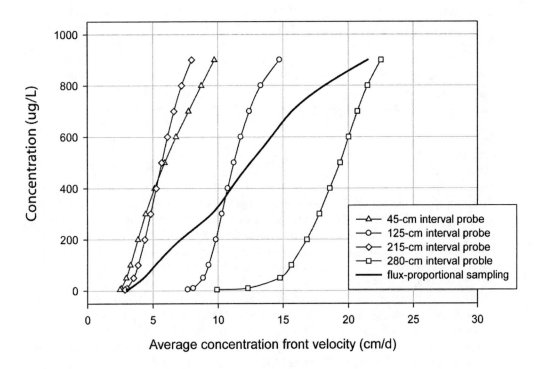

FIGURE 11.10 Comparison of concentration front breakthrough velocities for discrete interval probe and flux-proportional sampling. The concentration front breakthrough velocities decrease over time, as diffusive mass transfer from the low-conductivity strata (215 and 45 cm) provides a continued source of mass to the faster flowing strata (280 and 125 cm). The flux-proportional sampling curve illustrates how the high-conductivity strata dominate early-time behavior and lower-conductivity strata affect late-time behavior, as the clean-water front migrates through the system.

EVALUATION OF SAMPLING METHODS

The breakthrough curves for each of the four intervals of interest are shown using a consistent convention to enable analysis of sampling at each location and comparison among the graphs. The simulated low-flow sampling results are depicted using small circles. The results for the discrete interval probe are shown using a thick gray line to illustrate the concentration trends for each 5-cm cell corresponding to the location of interest. In addition, we show the breakthrough curves for the adjacent layers — these are shown using an upward pointing triangle for layer above in the model and a downward pointing triangle for the layer below in the model. The flux-proportional sampling results are shown using squares. The breakthrough curves illustrate the concentration that would be observed from sampling using each of these methods during the advance of the clean-water front.

Comparison of sample-specific breakthrough curves at each of the four locations indicates that the measured concentration depends on the sampling method. There is a significant degree of variability in the low-flow breakthrough curves compared to the discrete interval sample, which corresponds to a continuous probe reading. The bandwidth of breakthrough curves for the cells above and below the interval of interest illustrates the vertical stratification of the plume at each interval. It might seem counterintuitive, but the magnitude of the concentration fluctuation for low-flow sampling is not controlled solely by the concentrations in the adjacent layers, but by the concentration of the interval that contributes the majority of the flux for the sample volume. Because the simulation represents a dynamic exchange of mass between the mobile and immobile segments of the aquifer, the behavior also depends on the temporal evolution of the plume.

45-cm Interval — As shown on **Figure 11.11**, the magnitude of concentration fluctuation is more than 100 µg/L at the 45-cm interval. The probe breakthrough curve coincides with the peaks of the low-flow sampling breakthrough curve, while the minima deflect toward the interval 5-cm below the sampled interval. The lower panel on **Figure 11.11** shows a zoomed-in view of the breakthrough curve between 75 and 125 days. This result indicates that the low-flow sampling induces flow from the higher-conductivity layer beneath it, which has lower concentrations as indicated by its earlier breakthrough curve. This example matches the behavior we have seen in numerous tracer studies, and is discussed further in Chapter 12. The magnitude of the concentration fluctuations dampen with time, as the difference in concentration above and below this interval decreases with the advancing clean-water front.

Closer examination of the concentration distribution in the vertical profile across the model shows the influence of sampling on the transient vertical concentration distribution. The left panel on **Figure 11.12** depicts the concentration contours before sampling at 78 days; the right panel shows the configuration of contours during sampling at 78 days. Notice the deflection of concentration contours toward the sampling intervals in the right panel. Because the 45-cm sampling location is placed near the middle of a relatively thin, low-conductivity layer, pumping induces flow from above and below the sampling point. Although higher-concentration water is drawn downward from above the sampling location, there is greater flux from the lower-concentration water from below, resulting in a reduced concentration during low-flow sampling (See **Figure 11.12**).

125-cm Interval — The influence of sampling methods in the interbedded zone corresponding to the 125-cm interval (shown on **Figure 11.13**) is similar to the thin low-conductivity layer at 45-cm. Notice that the range of concentrations in the adjacent cells is much narrower and that the transient range of concentrations more closely approaches the adjacent layers. The contrast in hydraulic conductivity among the layers in the interbedded zone at 125 cm is comparable to the 45-cm strata; however, the variation occurs on the scale of 5 to 10 cm instead of 25 cm in the 45-cm case.

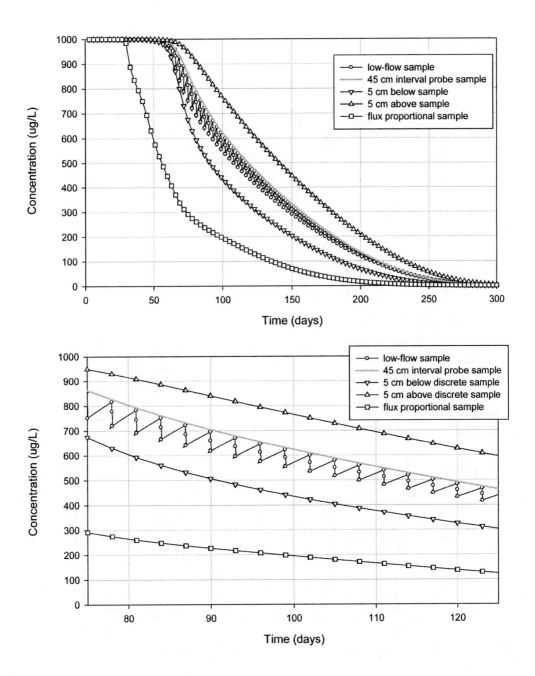

FIGURE 11.11 Effect of sampling methods on breakthrough curves for the 45-cm interval. The lower panel shows a zoomed-in view to illustrate the concentration fluctuations induced by 1 hour of low-flow sampling every 3 days.

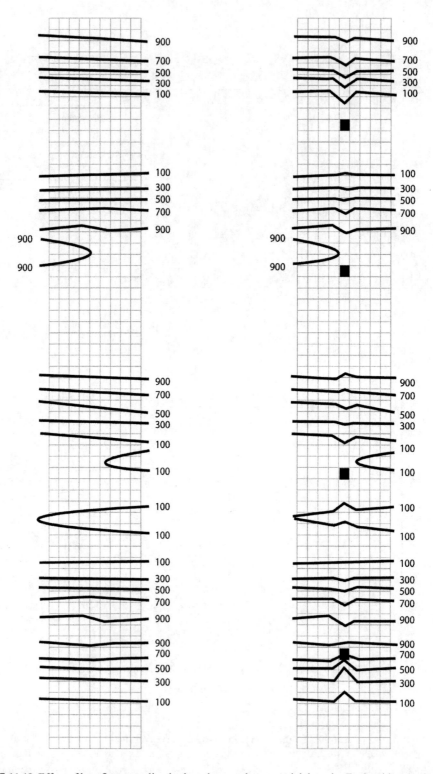

FIGURE 11.12 Effect of low-flow sampling in the solute wash-out model domain. Each grid square represents 5 cm². Contour lines on the left panel show concentrations prior to low-flow sampling. On the right, low-flow samples have been collected from the four darkened grid locations and the solute concentration contours have been disrupted by cross-over from adjacent intervals.

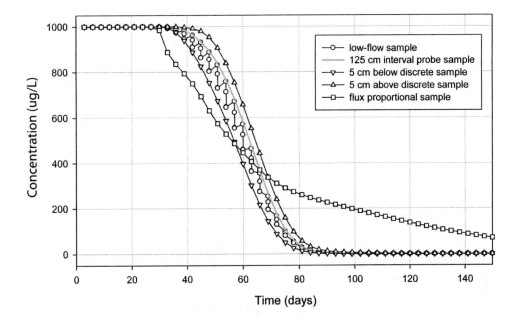

FIGURE 11.13 125-cm interval breakthrough curves, illustrating the influence of the sampling methods on observed concentrations. The 125-cm sampling interval is located in an interbedded zone in the model, so low-flow sampling induces flow from low-concentration strata below the sampled interval.

The upper-limit of the low-flow concentration peaks corresponds well with the probe sampling result. During sampling the minima approach the concentration of the higher-conductivity layer beneath it, as shown on **Figure 11.13**. Notice that the flux-proportional sampling breakthrough curve crosses the probe and low-flow sampling curves between 57 and 69 days; whereas, the 45-cm interval breakthrough curves occur later in time. The relative position of the discrete interval breakthrough curves at the 125-cm interval reflects a higher degree of connectedness with the mobile porosity segments of the aquifer than the thicker low-conductivity strata at the 45-cm interval. While we make no attempt to quantify the mass-transfer coefficients in this case, the behavior is consistent with discussion in Section 7.2 (See Figure 7.10).

215-cm Interval — **Figure 11.14** shows the breakthrough curves for the thickest low-conductivity strata in the model. As expected, the flux-proportional sampling breakthrough curve occurs much earlier than the low-flow and discrete interval sampling results. This is expected because the hydraulic conductivity in this interval is much lower than the average and because the thickness of the combined strata reduces the rate of mass transfer out of the immobile reservoir. Notice that the range of transient concentration fluctuations associated with low-flow sampling is relatively low, even though the discrete sampling result for the layers above and below show more than 400 μg/L range, when the separation is greatest at about 135 days. In this case, the probe sampling breakthrough curve nearly bisects the fluctuations from the low-flow sampling. The narrow range of concentration fluctuations is the result of the localized flux-proportional sampling that occurs during low-flow sampling. Because the 215-cm sampling interval is situated near the middle of the thick, low-conductivity strata, the observed concentrations decrease toward the lower-

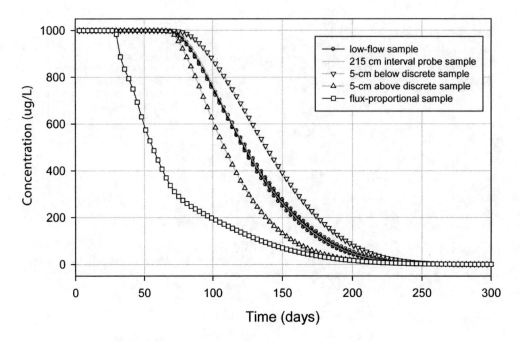

FIGURE 11.14 Comparison of breakthrough curves for various sampling methods at the thick, low-conductivity stratum surrounding the 215-cm interval.

concentration, higher-conductivity strata above the sampled interval. The deflection of the contours during sampling is also shown on **Figure 11.1**2. After sampling, concentrations rebound toward the lower-conductivity, higher concentrations due to the influence of diffusion.

280-cm Interval — **Figure 11.15** shows the sampling breakthrough curves for the highest-conductivity region in the model at 280 cm. The low-flow and probe sampling results precede the flux-proportional curve, reflecting the high-rate of clean-water flux through this mobile porosity segment of the aquifer. During the first 40 days, the low-flow sampling concentration fluctuations decrease under the influence of the higher conductivity, low-concentration strata. After about 50 days, low-flow sampling increases the concentration above the low-levels that are measured by probe sampling, as slightly higher concentrations are drawn in from the layer above. **Figure 11.12** shows the deflections of the contours from above during sampling at this location. **Figure 11.13** shows a zoom-in of the breakthrough curves between 40 and 120 days, illustrating the cross-over between the probe and low-flow sampling results.

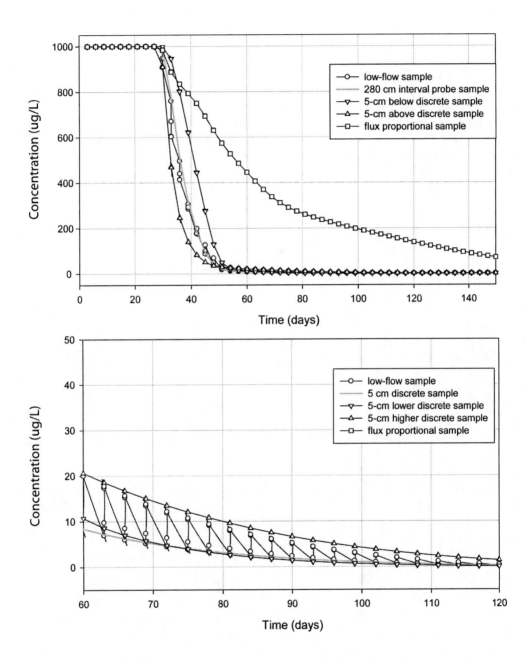

FIGURE 11.15 Breakthrough curves for different sampling methods at the 280-cm interval. The upper panel illustrates the influence of low-flow sampling in this higher-conductivity band. During the first 40 days of clean water migration, sampling induced flow from the lower-concentration strata, reducing the low-flow concentration. The lower panel highlights the behavior from 60 to 120 days, when diffusive flux of mass results in higher concentrations during low-flow sampling.

SUMMARY OF CASE STUDY RESULTS

The transport simulations provide a practical example of the controlling influence that mass transfer between the immobile and mobile segments of the aquifer has on clean-up time. The modeling results indicate the method of sampling has a significant impact on the observed concentrations in a moderately heterogeneous system because of plume stratification and the mass transfer from the immobile segments that act as reservoirs for contaminant mass. Therefore, the measured concentrations depend on the location of the sampling device and the volume of water removed compared to the prevailing flux in the system. These results are consistent with the modeling work by Cosler (1997, 2004) that described the potential influence of plume and hydraulic conductivity stratification on groundwater sampling results and rate-limited mass transfer associated with mobile-immobile domain processes. We expect that Varljen, et al. (2006) came to the contrary conclusion because their work considered steady-state particle tracking, rather than transient transport simulations.

12 Tracer Study Design and Interpretation

> **"There's no truth like tracer truth"**
>
> *James F. Quinlan[1]*

If you want to understand how injected reagents and contaminants behave in an aquifer, inject some reagent and pay close attention. Tracer studies are the most powerful tool we have to unmask the heterogeneous structure of groundwater flow and we now view tracer studies as a standard element for all reactive zone remedial designs. In the past three years, we have harvested a rich base of tracer study data, often with stunning results — three, four, possibly five discernable tracer peaks passing a series of 1.5-meter long well screens following a single tracer injection pulse, representing a five-fold range of groundwater velocities, all of them exceeding the average groundwater velocity determined by hydraulic testing at the site. The analysis of these tracer studies is causing us to re-think much of our sense of fine-scale aquifer structure and groundwater and contaminant velocities, and is one of the main motivations for developing *Remediation Hydraulics*.

We use tracer studies to support remedial investigations and designs, determining injection-volume/coverage relationships, reagent dilution ratios and solute velocities in aquifers. Because they link so closely to remedial system designs, tracer studies are probably the most cost-effective of the possible pre-design data collection activities. Tracer studies can be designed to determine the following:

Volume-radius relationships — When we revised our conceptual aquifer mass transport model, reducing the expectation of transverse dispersivity, there was a very important impact on reagent injection: reagent injections must span the entire contaminant profile in the y-z plane to obtain complete reagent coverage, as depicted in **Figure 5.20**. Transverse dispersivity isn't going to boost the reagent spread much beyond what we achieve on the day of injection. Tracer studies allow us to estimate the transverse coverage that can be achieved at a radial injection well, supporting the design of injection well spacing and injection volumes.

Average and mobile fraction groundwater velocities — Tracer studies provide a basis for estimating average groundwater velocities, which we expect to match the results obtainable through aquifer pumping testing. Mass transport for contaminants and reagents is normally dominated by mobile pore space fractions that are dramatically more conductive than the formation average would indicate. Tracer studies can give us a picture of the range of actual velocities in a formation, from which we can estimate reagent transit times through treatment zones and velocities for the clean water front following source removal actions.

---

[1] Research geologist, Mammoth Cave National Park (1973 – 1990) and Adjunct Professor, Eastern Kentucky University until his death in July, 1995.

Small-scale aquifer behaviors — The movement and dilution of injected reagents and contaminants are determined largely by aquifer characteristics that are too small to characterize with many test methods. Tracer studies can unmask important details of fine-scale and stratigraphic-scale aquifer structure that cannot be observed easily with other aquifer characterization tests.

Safe injection pressures — Fluid injection pressures can easily exceed a formation's structural capacity, in shallow systems with small effective stress and at sites with significant depth to the groundwater surface, where pressures exerted in the screened interval far exceed gauge pressures at the ground surface. Tracer injections provide an opportunity to test the formation response to fluid injection, supporting design of the full-scale reagent injection network. Occasionally, formation responses observed during tracer studies will cause reconsideration of the viability of an injection-based remedy.

Reagent dilution ratios — Injected reagents are diluted through diffusive interaction with immobile groundwater, as the injected fluid travels along the mobile pore space pathways. Tracer studies give us a basis for estimating the dilution that occurs in transport and to adjust the solution strength for injected reagent, accordingly.

Solute velocities and washout rates — The velocity of solute movement in the subsurface is typically much faster than would be derived from average groundwater velocities. Tracer studies allow us to quantify maximal velocities and the pace of reagent washout from the injection zone, providing a basis to determine the injection repeat frequency for barrier-type treatment systems.

Above all, tracer studies consistently remind us of the distinction between hydraulics and transport. While aquifer testing clearly demonstrates hydraulic connectedness over large areas for most formations, tracer studies show that transport is constrained much more narrowly, suggesting that the flow systems we have studied comprise a multitude of parallel laminar flow streams with limited transverse horizontal or vertical interactions.

12.1 TRACER STUDY DESIGN AND LAYOUT

Typical approaches of low-volume, low signal-to-noise tracers that were common a few years ago have given way to large-volume, high-signal-to-noise tracers, in response to the development of large-scale, reagent-dependent *in situ* remedies. There are currently two principal modes of tracer evaluation used to support *in situ* remedial design:

Transport studies — Injected tracers can be tracked as they migrate from the injection zone to downgradient locations, carried in flowing groundwater. Tracer concentration breakthrough curves (concentration versus time) are developed from observations at one or more distances from the injection zone. These studies require a number of observation wells aligned on the flow path, which is often not known for the study area.

Washout studies — Tracer concentration-depth profiles can be recorded in the injection well at regular intervals following injection, to identify strata of higher groundwater flow. The injected tracer volume can be balanced by extraction from the same well, inserting a volume just large enough to replace groundwater in the well. Alternatively, it may be useful to observe concentration versus depth after routine reagent injection, to identify asymmetries in the distribution of injected reagents.

Most tracer studies associated with contaminant distribution studies and treatment system designs are set up for transport analysis, with an injection location and multiple downgradient observation points. **Figure 12.1** introduces a standard transport tracer study layout with a group of dose response wells at a short radius, along with arcs of tracer transport observation wells at

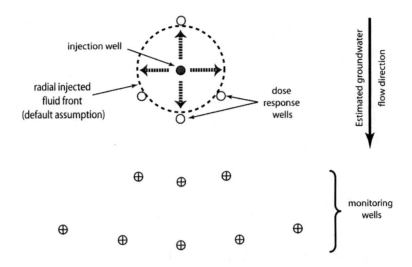

FIGURE 12.1 Tracer study well setup. Dose response wells are placed at the planned injection radius, to observe development of the tracer cloud in real-time. Monitoring wells are placed at two or more distances along the presumed groundwater flow path, with lateral replication to cover variation in groundwater flow axis, relative to the presumed direction.

multiple radial distances from the injection point. It is not uncommon for injected tracers to be "lost" in the subsurface and the design approach we present is intended to maximize the likelihood of obtaining the intended measurements.

Tracer injection well — The tracer injection well should be optimized for injection of a large fluid volume with little resistance. We recommend the use of wire-wound screens, well packing selected to match the formation characteristics, and a complete well development. Well injection pressures must be managed to avoid fracturing the aquifer formation (refer to **Chapters 2 and 13** for more information on injection well design and operation).

Dose response wells — The dose response wells are constructed as typical monitoring wells and are placed to measure the volume-radius relationship, a direct measure of the aquifer mobile porosity. The dose response wells are monitored in real time for the appearance of tracer, often using a visible dye as the indicator, superimposed on a quantitative tracer. In an optimal design, two or three dose response wells are constructed: one along the expected groundwater flow axis and two at right-angle positions, relative to the groundwater flow axis. Each of these wells must be developed to be certain that it is hydraulically connected and as responsive as possible to aquifer flow conditions.

Tracer observation wells — It is important to understand that, until the tracer study is well underway, the groundwater flow path is typically an unknown at this scale and the flow axis may deviate significantly from the fall line of the groundwater elevation contour map. Consequently, it is wise to install multiple transport observation wells at desired observation distances, to assure that the tracer cloud passage is observed. We recommend covering a 60-degree arc with at least one set of observation wells, to assure capturing passage of the tracer cloud. Otherwise, it is possible to monitor the observation network for an extended period, without ever observing the tracer mass. Tracer flow is likely to follow unpredictable pathways (relative to the elevation contour map) and to travel faster than the average groundwater velocity estimates. As with dose response wells, each of these wells must be properly developed to be certain that it is hydraulically connected and as responsive as possible.

Salt washout — The special case of salt washout is often focused on an injection well, alone, although it can be coupled with downgradient observation wells, especially in fractured bedrock systems (the very low injection volumes of strictly-administered salt washout studies makes the downgradient capture of the tracer plume somewhat improbable). **Figure 12.2** shows a salt washout tracer setup.

Most of the errors and failures we observe in tracer study design and implementation occur when the project team thinks they already know the aquifer behavior and the tracer study is merely a required exercise. Tracer studies often expose unanticipated aquifer behaviors and the project team must be alert and adaptive to capture their full potential to deliver insights on the subsurface.

12.2 TRACER PROFILES AND BREAKTHROUGH CURVES

When tracer is injected into flowing groundwater, there are two possible modes of observation: tracer profiles and breakthrough curves. A tracer profile is a plot of tracer concentration versus distance from the injection well, at a specified time after the tracer injection. **Figure 12.3** shows a conceptual tracer profile as it develops from the initial (idealized) "square" tracer pulse at t = 0. Tracer movement occurs primarily in groundwater flowing through the mobile pore space, at a velocity, V_{mobile}, that exceeds the average (superficial) velocity for the formation. However, diffusive tracer mass exchange between the mobile and immobile pore space slows the overall

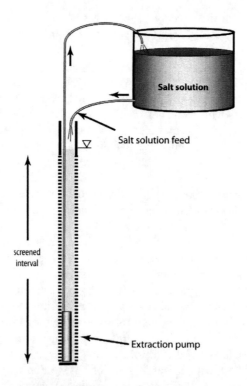

FIGURE 12.2 Salt washout tracer setup. Groundwater from the well bore is extracted in balance with salt solution returned at the top of the well bore.

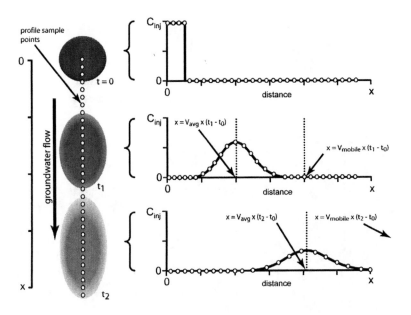

FIGURE 12.3 Conceptual development of a groundwater tracer concentration-distance profile in an aquifer with 50 percent mobile fraction. The center of tracer mass travels at the average groundwater velocity. It is impractical to collect sufficient samples to build tracer concentration-distance profiles.

tracer movement to the average groundwater velocity for the formation (V_{avg}). At t_1 in **Figure 12.3**, water that initially carried tracer has moved to a distance $x = V_{mobile} \times (t_1 - t_0)$, but the center of tracer mass has traveled only to a distance $x = V_{avg} \times (t_1 - t_0)$. At t_2, diffusion along the flow axis and between the mobile and immobile pore spaces spread the tracer distribution and decreased the peak concentration. The center of tracer mass continues to travel at the average groundwater velocity.

The center of tracer mass travels at the average groundwater velocity, regardless of the mass transfer characteristics of the formation or the tracer compound. If the rate of tracer mass transfer between the mobile and immobile pore spaces is high, the tracer distribution will be Gaussian, as shown in **Figure 12.3**. If mass transfer is limited by the tracer diffusivity or by geometric limitations to mass exchange, the tracer distribution will be more log-normal, with the peak concentration moving at a velocity closer to the mobile velocity and an extended tailing at lower concentrations. As the tracer pulse travels along the flow axis, diffusion pushes tracer ahead and behind the original pulse, so the theoretical square tracer pulse will break down, regardless of diffusive interaction that may occur between tracer in the mobile and immobile porosities.

The number of sample points required to construct an accurate tracer profile renders the concentration-distance curves impractical. Fortunately, tracer breakthrough curves provide a very effective alternative to profiles. As shown in **Figure 12.4**, high-resolution tracking of tracer movement can be accomplished through frequent measurements at a limited number of observation wells. Tracer behavior is normally tracked using breakthrough curves.

Tracer breakthrough curves are compilations of tracer concentrations versus time at fixed observation points along the groundwater flow path. These curves behave like a mirror of the tracer profiles, with the highest velocities (earliest arrival times) on the left side of each curve. **Figure**

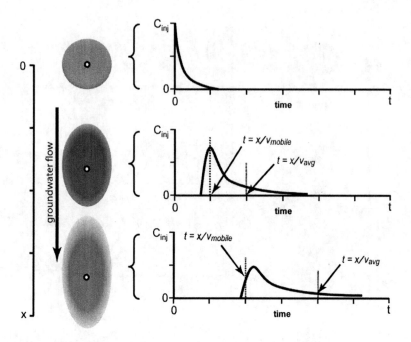

FIGURE 12.4 Conceptual model of tracer breakthrough curve (concentration-time) behavior in a dual or multi-porosity aquifer matrix. The center of tracer mass travels at the average groundwater velocity and when diffusive mass transfer is limited between the mobile and immobile pore fractions, the peak concentration travels near the mobile fraction velocity. The tracer breakthrough curve becomes strongly log-normal.

12.4 shows a tracer breakthrough curve that might arise in a system with 50 percent mobile pore fraction and a limited mass transfer rate. In this case, the peak initially travels at a velocity near the groundwater velocity in the mobile pore space (V_{mobile}), while the center of tracer mass travels more slowly, at the average groundwater velocity (V_{avg}). Over time, modeling indicates that the peak velocity slows to match the average groundwater velocity, but the difference in arrival times for the peak concentration and center of mass are "burned in" and persist (see Chapter 5 for more detailed discussion of peak and center of mass velocities).

Tracer behavior provides useful insight into mass transfer between mobile and immobile pore spaces. As we introduced in Chapters 5 and 6, aqueous-phase diffusion drives a notable mass exchange between closely-divided mobile and immobile pore spaces in the aquifer matrix. Migration into and out of the immobile pore space slows the movement of tracer mass, relative to groundwater velocities in the mobile pore spaces, and shapes the tracer breakthrough curves. **Figure 12.5** shows extremes of tracer behavior. When mass exchange between mobile and immobile pore spaces is limited, the breakthrough curve is strongly log-normal and the peak concentration arrives quickly, traveling near the mobile-fraction groundwater velocity, as shown on the left side of the figure. The normally distributed breakthrough curve, on the right of the figure, arises when there is a high rate of mass exchange between mobile and immobile pore spaces. In both cases, the center of tracer mass travels at the average groundwater velocity.

The tracer mass exchange rate can only be discussed in relative terms, because the tracer behaviors we observe in the field are the result of several factors that affect our single observation, including: 1) tracer diffusivity (thermal or chemical), 2) surface–to–volume ratios for the mobile-immobile contact and 3) multiple mobile velocities. Diffusivities for low-sorption chemical tracers

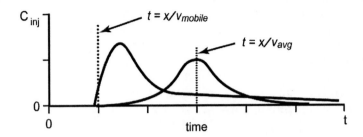

FIGURE 12.5 Extremes of tracer breakthrough behavior. The strongly log-normal curve, on the left, arises in a dual (or multi) domain system, when tracer mass transfer between the mobile and immobile pore spaces is limited by the tracer diffusivity or by the geometry of the domain structure. The normally-distributed curve, on the right, arises when the tracer diffusivity is high and the domain geometry favors tracer mass transfer. In both curves, the center of mass arrives at the average groundwater velocity. The peak of the log-normal curve travels near the groundwater velocity in the mobile pore space. X is the distance traveled by the tracer to the observation point.

cover a fairly narrow range; however, there is a very large contrast between chemical and thermal diffusivity and we can obtain simultaneous high diffusivity (temperature) and low-diffusivity (chemical) results from a single tracer injection. We examine this approach in Case Studies 3 and 5, below. We know that the surface-to-volume ratios for the mobile-immobile contact must have an influence on mass transfer, but we have no way of quantifying this factor in the field. Finally, we know that high resolution measurements of hydraulic conductivity lead us to expect multiple conductive pathways, operating in parallel, over the 1.5 to 3-m intervals commonly sampled for tracer breakthrough curve development. Each tracer breakthrough curve is likely to be a composite of parallel tracer flows and they are often difficult to untangle. Several of the case studies, shown below, provide evidence of multiple parallel breakthrough curves.

12.3 TRACER VOLUMES AND ESTIMATES OF MOBILE POROSITY

The injection volume required to reach the dose-response well is determined by the mobile porosity, one of the key formation characteristics we wish to quantify through the tracer study. The equation that relates injected volume to radius of injection is:

$$Volume = \pi \cdot r^2 \cdot h \cdot \theta_m \tag{12.1}$$

where r is the radius of the dose response well placement, h is the thickness of the injected interval (nominally, the screened interval thickness) and θ_m is the mobile porosity. For most conductive aquifers, the total porosity is assumed to be 30 to 40 percent and our observations suggest that the mobile pore space is normally 10 percent or less, conceivably reaching a maximum value of approximately 20 percent of the total aquifer matrix. It is important to prepare sufficient tracer volume to reach the dose response wells if the mobile porosity is in the 15 to 20 percent range. The tracer solution is injected steadily, logging injection flow rate, cumulative injected fluid and concentrations observed at the dose response wells. Groundwater elevation responses should be observed across the network, to assess mounding near the injection well and to assess the hydraulic continuity across the study area. It is acceptable to continue injections after the concentration plateaus in the dose response wells, noting that the tracer cloud diameter is calculated from the total injected volume and the mobile porosity extracted from the dataset.

The mobile porosity, θ_m, is determined by observing the buildup of tracer concentrations in the dose response wells. **Figure 12.6** shows expected pattern of tracer concentration increase that occurs as a function of injected tracer solution volume. Injected volumes and tracer concentrations are logged frequently during the injection process. Concentrations typically plateau in the dose response wells at levels slightly below the injected solution concentration.

We calculate the mobile porosity from the volume injected to reach 50 percent of the maximum observed value, as shown in **Figure 12.6**. The mobile porosity is then calculated from Equation (12.1), rearranged to isolate θ_m:

$$\theta_m = \frac{Vol_{inj\,50}}{\pi \cdot r^2 \cdot h} \tag{12.2}$$

where $V_{inj\,50}$ is the volume injected at the time the observed concentration reached 50 percent of its maximum value. Tracer concentrations at the dose response well are expected to be near the injected concentration and concentrations at the dose response well that are significantly below the injected concentration may indicate a problem in tracer delivery (Case Study 3 includes analysis of a low breakthrough concentration). There are several possible explanations for low dose response well tracer concentrations:

- Arrival of small tracer volumes via high-conductivity pathways can result in low tracer concentrations in the dose response wells. In these cases, continued tracer injection is expected to generate higher levels in the dose response wells.
- Structural failure of the aquifer matrix can lead to anomalously high or low dose response well behavior. The clearest indication of structural failure of the aquifer matrix arises

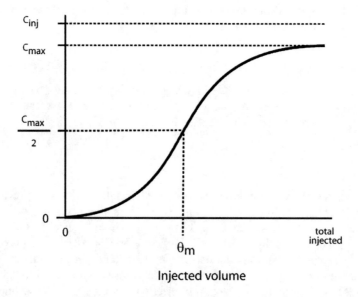

FIGURE 12.6 Injected tracer volume and the dose response well concentration — picking the volume to represent the mobile porosity. Maximum concentration observed at dose response wells is typically somewhat lower than the injected concentration.

when high-strength tracer reaches a dose response or observation well during the injection phase and no washout is observed during transport monitoring. It is important to note that formation failure should be noted by pressure and flow monitoring during the tracer injection.

- If the dose response well is placed too far from the injection well, diffusive exchange between the mobile and immobile pore spaces dilutes the tracer. This effect is observed in Case Study 5, where dose response wells were located at varying radial distances from the injection well.

Very few sites yield the "ideal" tracer distributions that allow easy analysis. It is not uncommon to use the results of an initial tracer injection as the basis for a conceptual mass transport model, then a second, more comprehensive, test can be used to validate the model.

12.4 CALCULATING THE CENTER OF MASS

Determination of groundwater velocity from tracer breakthrough curves requires estimation of the center of mass. The time of arrival of the center of mass can be calculated as shown by the general formula, Equation (12.3),

$$t_{com} = \frac{m_1 \cdot t_1 + m_2 \cdot t_2 + ... + m_n \cdot t_n}{m_{total}} \tag{12.3}$$

where the mass for each time interval is the area beneath that segment of the concentration-time curve and the time represents the midpoint of the interval between observations. If the tracer is present in the groundwater background, it is necessary to obtain an estimate of the background and subtract that value from observations during the tracer study.

When samples are collected at a constant frequency over time and the rate of concentration change is low, we can assign the time of sample collection to represent each mass (concentration) observation in the center of mass calculation and introduce little error. A constant sample collection frequency is shown in the breakthrough curve given in **Figure 12.7**. However, field data are rarely collected at regular intervals and concentration changes between observations can be quite large, especially over the relatively long time spans that are required to observe most tracer pulses. In these cases, estimates can be improved by calculating a center of mass for each sample in the series of observations.

Figure 12.8 shows a series of concentration observations taken at variable intervals. To estimate the center of mass for the sample collected at time t_2, we first calculate the mass in the second half of the interval prior to sampling, $mass_{pre}$ (lightly shaded area in **Figure 12.8**), and the first half of the subsequent interval, $mass_{post}$ (the darkly shaded area in **Figure 12.8**).

$$mass_{pre} = \left[t_n + \frac{(t_n + t_{n-1})}{2} \right] \cdot \frac{1}{2} \left[C_n + \frac{(C_n + C_{n-1})}{2} \right] \tag{12.4}$$

$$mass_{post} = \left[\frac{(t_n + t_{n+1})}{2} + t_n \right] \cdot \frac{1}{2} \left[C_n + \frac{(C_{n+1} + C_n)}{2} \right] \tag{12.5}$$

The center of mass for the sample collected at time t_n can be calculated as shown in Equation (12.6):

$$t_{CM_n} = \frac{mass_{pre_n} \cdot \left[\dfrac{t_n}{2} - \dfrac{(t_n - t_{n-1})}{4}\right] + mass_{post_n} \cdot \left[\dfrac{t_n}{2} + \dfrac{(t_{n+1} - t_n)}{4}\right]}{mass_{pre_n} + mass_{post_n}} \qquad (12.6)$$

In the example shown below, the center of mass for the observation at t_2 shifted to an earlier point in time when Equation (12.6) was applied, compared to the result obtained using Equation (12.3). When concentration changes between sample collections are large, or when sample interval variation is large, the time shift for center of mass can be significant.

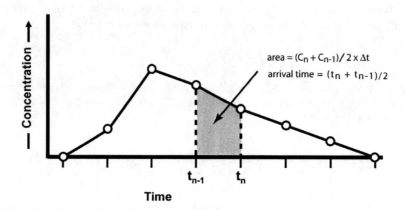

FIGURE 12.7 Calculating center of mass from samples collected at constant time intervals. This method uses the average of samples at the beginning and end of the interval to represent the concentration and the center of the interval to represent the time of arrival for the estimated mass for entry into Equation (12.3).

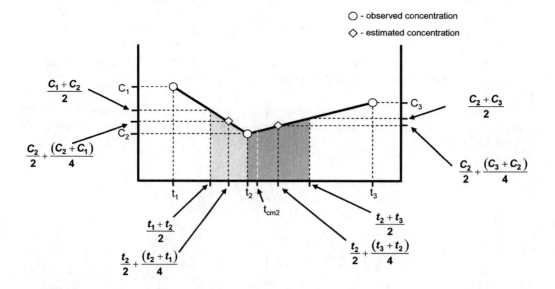

FIGURE 12.8 Graphical representation of the variables used to estimate center of mass from samples collected at unequal time steps. The center of mass can also be calculated for each sample result, shown here for the sample collected at t_2.

To illustrate the calculation of center of mass, we selected a simple dataset from recent studies in the Gaspur Aquifer of the Los Angeles Basin in the Western United States. Hydraulic testing at the site indicated the average groundwater velocity is between 9 and 15 cm/day. Total organic carbon from a first-time carbohydrate injection was used as the tracer and concentrations were observed at a groundwater monitoring well located 12 m downgradient from the injection location (when degradable organic carbon is first introduced to an aquifer, bacterial populations are low and the carbon travels through the formation with only limited degradation, so it can be used as a quasi-conservative tracer). The objective of the test was to determine velocities of the solute peak and the center of mass.

The aqueous-phase organic carbon content was measured 29 days prior to the one-time injection, to develop a baseline value. As shown in **Table 12.1**, the baseline value of 4.4 mg/L total organic carbon was subtracted from all observations to show only the incremental organic carbon from the injection in the column labeled "baseline-adjusted TOC." Equation (12.3) was used for the first calculations on the Gaspur formation tracer. There were 7 sample collection dates, which result in 6 curve intervals, such as those shown in **Figure 12.7**. The intervals between samples, arrival times and interval average concentrations were calculated as depicted in the figure. Areas under the curve were estimated for each interval, then were multiplied by the respective arrival times and totaled. As shown in Equation (12.3), the sum of arrival time — interval area products $(1,967,046 \text{ days}^2 \cdot \text{mg/L})$ — was divided by the total area under the curve $(14,400 \text{ days}^2 \cdot \text{mg/L})$, to obtain the arrival time for the center of mass, 137 days. Based on the 12-m distance from the tracer injection to the observation well, the center of mass arrival represented an average groundwater velocity of 9 cm/day, consistent with hydraulic testing at the site.

The same data were analyzed using the method depicted in **Figure 12.8**, to account for large concentration changes and variable intervals between sample collections. Equation (12.4) was used to calculate the area under the concentration-time curve, prior to each sample collection, and Equation (12.5) was used to calculate areas under the curve following each sample collection. **Table 12.2** summarizes the data analysis using this approach. The interval areas x arrival times are higher for this method than those shown in **Table 12.1**, due to the use of adjusted arrival times for each interval and the fact that this method breaks the area into 7 intervals, rather than the 6 of the earlier method.

The resulting center of mass arrival time was 125 days, shown on **Figure 12.9**, lagging the peak arrival by approximately 54 days. The extra calculation effort associated with the method of **Figure 12.8** provided only a modest change for the center of mass arrival, suggesting that the simpler method of **Figure 12.7** is workable, even for the highly variable dataset.

For the tested segment of the Gaspur formation, we concluded that the average groundwater velocity was 10 cm/day, in the range expected from previous hydraulic testing. To analyze solute mass transport in this dataset, recall from Chapter 3 the relationship between average groundwater velocity and the velocity in the mobile fraction:

$$V_{mobile} = \frac{V_{avg}}{\dfrac{\theta_{mobile}}{\theta_{total}}} \tag{12.7}$$

For any lognormal breakthrough curve,

$$V_{mobile} > V_{peak} > V_{avg}$$

In this example, the peak traveled at 17 cm/day, indicating that the mobile porosity is no greater than 58 percent of the total porosity (which would occur if the peak traveled at V_{mobile}).

TABLE 12.1.

Center of mass calculations for TOC tracer example, using Equation (12.3).

Elapsed Time (days)	Interval (days)	Arrival time (days)	TOC (mg/L)	Baseline-Adjusted TOC (mg/L)	Interval Average TOC (mg/L)	Interval Mass (mg/L x days)	Interval mass x arrival time
-29			4.4	0.0			
	71	6.5			2.65	188	1,223
42			9.7	5.3			
	8	46			5.25	42	1,932
50			9.6	5.2			
	9	54.5			5.4	48.6	2,648
59			10	5.6			
	12	65			115.8	1,390	90,324
71			230	226			
	63	102.5			133	8,379	858,847
134			44	40			
	197	223.5			22.1	4,353	1,012,072
331			8.6	4.2			
					Totals:	14,400	1,967,046

TABLE 12.2.

Center of mass calculations for TOC tracer example, using Equation 12.6.

Elapsed Time (days)	TOC (mg/L)	Baseline-Corrected TOC (mg/L)	Pre-sample area estimate (mg/L x days)	Post-sample area estimate (mg/L x days)	Combined Interval Area (mg/L x days)	Adjusted Sample Collection Time (days)	Interval area x arrival time
-29	4.4	0.0	0.0	47.0	47.0	- 10.5	- 494
42	9.7	5.3	141	21.1	162	27.9	4,520
50	9.6	5.2	20.9	23.9	44.7	55.7	2,490
59	10	5.6	24.8	364	388	72.7	28,208
71	230	226	1,024	5,642	6,665	85.6	570,524
134	44	40	2,712	3,029	5,741	144	826,704
331	8.6	4.2	1,285	0.0	1,285	282	362,370
				Totals:	**14,332**		**1,795,310**

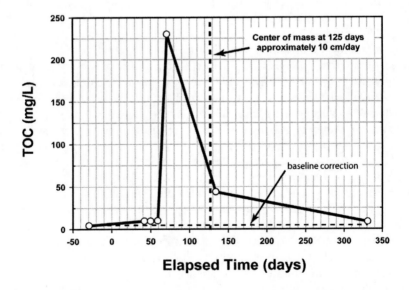

FIGURE 12.9 Tracer concentration behavior 13 m downgradient from a carbon injection well in the Gaspur Aquifer, Los Angeles basin. There is significant separation of the concentration peak and the center of mass, indicating significant immobile porosity and limited mass transfer between mobile and immobile pore water fractions.

12.5 TRACER SELECTION CRITERIA AND AVAILABLE TRACERS

The tracer we select has a significant bearing on the aquifer characteristics that can be observed. Here are the most important considerations in tracer selection:

Signal-to-Noise — Signal-to-noise tops the list of tracer selection issues, because it is possible to quickly lose a tracer signal against a variable background signal. Diffusive equilibration of solutes among high- and low-conductivity aquifer strata causes significant dilution of reagent pulses and a high starting concentration assures that the tracer signal will be measurable for an extended distance in the subsurface. Solute dilution is one of the important attributes of reagent mass transport that we wish to assess through tracer studies and a high signal-to-noise tracer is necessary to obtain that quantitative result.

Negative tracer — It is important to keep in mind the possibility that a tracer can be lower concentration than the groundwater, a "negative" tracer. One example is in high-solute or brackish groundwater: in that setting, it is difficult to obtain a signal contrast between groundwater and an injected salt-based tracer. A low-solute (clean water) tracer that gives a low-conductivity response is a very workable approach. Use of tracers in northern latitudes offers a chance to use low-temperature as a tracer during winter months and this approach is shown in Case Study 3.

Reactivity — When injected solutes react chemically with the aquifer matrix, tracer concentrations represent the combined effects of dilution and reaction, confounding our analysis. As a result, tracer studies designed to assess groundwater flow or contaminant migration patterns use chemically inert, non-sorptive tracers.

Mass transfer behavior — Diffusive equilibration effects are determined by the geometry of the porous medium and by the diffusivity of the selected tracer. Whether tracer studies are set up to mimic a reagent distribution or contaminant transport, it is important to understand the relative diffusivities of the selected tracer and the emulated compound. It is often possible to use the remedial reagents, themselves, as tracers and this is a desirable approach.

Multiple tracers — It is possible to superimpose two or more tracers, which provides an opportunity to compare breakthrough patterns as a function of mass transfer characteristics — the higher mass transfer tracer tracks with the average groundwater velocity and the lower mass transfer efficiency initially tracks with groundwater velocity in the high-conductivity strata. These separations can sometimes be observed in breakthrough behavior, as shown in the Case Study 3 results, below. Reagents can also be combined with properly selected non-reactive tracers to evaluate *in situ* reagent demand and reaction rates to aid in selecting appropriate solution strength for remedy implementation.

Automated data logging — Selection of a tracer that can be continuously logged with a down-hole instrument provides a cost-effective method for developing high-resolution breakthrough curves. However, the sensor patch on these instruments is very small and tracer movements may miss the sensor. The use of these down-hole instruments should be accompanied by a periodic disruption of the in-well volume to capture a periodic measurement of the "average" well bore concentration, or by a manual vertical profiling as described in **Figure 12.10**. Case studies 2 and 3 demonstrate how that periodic disruption of the well volume has provided invaluable insights to aquifer structure. Salt solutions (electrical conductivity signal), bromide ion (specific ion probe) and temperature-contrasted water (thermocouple) are examples of tracers that can be tracked with automatic data logging. We currently deploy automated data logging, combined with periodic grab sample collection (as a cross-check and to disrupt well bore stratification), whenever possible.

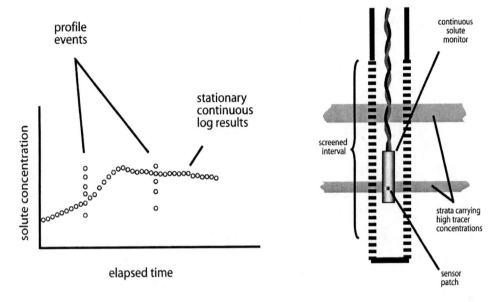

FIGURE 12.10 Combination profiling — using a continuous-recording probe to obtain periodic solute concentration profiles.

Laboratory analysis availability — Laboratory analysis is the primary means of tracer quantification for TOC or low-concentration fluorescein. Laboratory analyses can also be used to cross-check field instrumentation for tracers used in automated data logging applications. The ability to conduct automated data logging, coupled with laboratory confirmation, gives a potential advantage to bromide as a tracer, although field probe stability has been a problem for some of our bromide tracer studies.

The most common tracers currently in use are soluble salts of sodium or potassium, including sodium chloride, sodium bicarbonate and potassium bromide. Each of these produces a strong electrical conductivity signal with low reactivity following injection. **Figure 12.11** shows the conductivity signal for a sodium chloride solution, for a range of concentrations that can be used in tracer applications. Automated field logging is available for electrical conductivity, making the soluble salts a preferred tracer for most applications. The conductivity measurement responds reliably to total dissolved ions and the probes appear to be stable over extended deployment. Bromide and other ion-specific probes are also available and they can improve the signal-to-noise ratio when the total dissolved ion concentration background is high. Current field experience suggests that frequent recalibration may be required for these probes, as is discussed in Case Study 5, below.

Fluorescing dye tracers, such as fluorescein, provide a very high signal-to-noise ratio when their concentrations are quantified in laboratory analyses and they are preferred in applications where large dilution is expected – karst formations and long-distance tracer applications in porous media. We have also used fluorescing dyes as visible tracers, to observe initial tracer breakthrough in dose response wells. In most new applications, we are using soluble salts, monitored with a conductivity or ion-specific probe to track the dose-response well, in the manner shown in **Figure 12.6**. In earlier dose-response determinations, we stopped the tracer injection shortly after first arrival, but we abandoned that approach and replaced it with a larger-volume injection that continues until tracer concentrations stabilize, as in **Figure 12.6**.

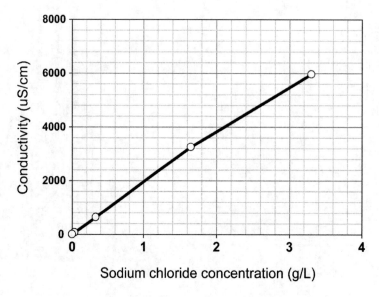

FIGURE 12.11 Relationship between sodium chloride (NaCl) concentration and electrical conductivity in water at 20° C.

Reagent injections can also be tracked effectively as soluble tracers, as long as the reactivity can be taken into consideration. Dissolved organic carbon, for example, behaves as a conservative tracer when it is injected into a formation at levels that far exceed aquifer equilibrium concentrations (this works until bacteria populations expand in response to the larger pool of available carbon, limiting a degradable organic carbon tracer to a one-time use at any site). Permanganate can be measured with a conductivity probe or using a field permanganate ion determination and can serve as a tracer to track the effectiveness of injections, as described in Case Study 1.

Temperature is one of the most effective and underutilized tracers we have at our disposal — we have very reliable, automated data logging for temperature, it is easy to formulate, and has a very effective signal-to-noise ratio. Temperature offers a very high diffusivity and its breakthrough curves can be compared to chemical tracers which have much lower diffusivities. We recommend adjusting chemical tracer injections to temperatures that contrast with groundwater temperatures by 5 to 10 °C. That induces only a limited density difference (which can be used to offset the density contrast induced by dissolved solids content) and provides a signal-to-noise ratio exceeding 100:1 for most applications. Case Study 3 and 5 utilize temperature as a tracer.

Table 12.3 provides listing of tracer compounds commonly used in groundwater studies, along with several of their important attributes. **Table 12.4** provides formulation recipes for several common tracers, each based on a 1,000-liter batch volume. It is important to note that the tracer solution adds total dissolved ions to the solution, thus adding to its electrical conductivity. **Table 12.4** identifies the expected addition to electrical conductivity. If the background electrical conductivity is high, the tracer may not be effective (low signal-to-noise). It is always critical to measure background levels for any tracer under consideration.

TABLE 12.3

Common subsurface tracers and their attributes.

Tracer	Visible Response	Signal-to-Noise at Typical Tracer Injection Strength	Conductivity Response	Compound-Specific Probe Available	Diffusivity relative to TCE	Reactivity	Sorptivity	Comment
Bromide	No	1,000	Yes	Yes	similar	low	low	Measurable in real time with field probe; lab cross-check advised
Organic dyes[1] – qualitative	Yes	100	No	No	similar	low	low	Can be confounded by contaminant fluorescence; real-time indicator for injected fluid arrival
Organic dyes[1]	No	>500,000	No	No	similar	low	med	Highest available signal-to-noise ratio, sorptivity dependent on dye type[3]
Temperature	No	10	No	Yes	very high		high	Often overlooked
Salt solutions	No	100	Yes	No	similar	low	low	NaCl or Na_2HCO_3 to generate conductivity signal
Sucrose	No[2]	10,000	No	No	similar	low	low	Microbial consumption rate is initially low with injected solutions far in excess of background DOC levels

[1] Includes fluorescein, rhodamine and other fluorescing organic dyes that can provide either a visual signal, at low signal-to-noise ratio, or a UV-spectrometer signal, at a very high signal-to-noise ratio. Special consideration should be given in instances where discharge to surface water is possible.

[2] Molasses is a common source of sucrose and it can be strongly colored (brown to pale straw).

[3] Rhodamine tends to sorb more strongly than fluorescein. However, both will sorb to organic material, with studies showing up to 20% loss over time due to sorption therefore will be less conservative than salt tracers.

TABLE 12.4

Formulations for common tracers.

Tracer	Available Formula	Target concentration (mg/L)	Batch Volume (L)	Tracer mass (kg)	Approximate Electrical Conductivity (uS)	Comment
Bromide	NaBr (77.65% Br)	200 as Br⁻		0.257	minimal	200 mg/L gives minimum workable signal-to-noise; not usable for conductivity
		2,000 as Br⁻		2.57	4,000	
	KBr (66.44% Br)	200 as Br⁻	1,000	0.301	minimal	High concentrations risk density-driven flow
		2,000 as Br⁻		3.01	4,000	Potassium salt preferred to avoid sodium addition to groundwater
Organic dyes[1]	Fluorescein: $C_{20}H_{12}O_5$ ------- Rhodamine:	40	1,000	0.0400	~ 0	40 mg/L provides strong visible signal, without sacrificing effectiveness as a quantitative tracer
Salt solutions	NaCl	2,000 as NaCl	1,000	2.00	3,800	Alternatives to sodium salts can be considered for potable water supplies
	$NaHCO_3$	8,900 as $NaHCO_3$	1,000	89	26,000	High concentrations risk density-driven flow
Sucrose	$C_{12}H_{22}O_{11}$	200 as TOC	1,000	0.200	~ 0	Purchased as baking sugar
Molasses	complex mixture	4,500 as TOC	1,000	10.0	1,400	Approximately 45% organic carbon, by weight. Conductivity signal is usable in groundwater with low dissolved solids

[1] Fluorescein, rhodamine and other fluorescing organic dyes that can provide either a visual signal, at low signal-to-noise ratio, or a UV-spectrometer signal, at a very high signal-to-noise ratio. Special consideration should be given in instances where discharge to surface water is possible.

12.6 CONTROLLING TRACER DISTRIBUTION

The ideal tracer cloud is circular in plan view, confined to the injection well screened interval and uniformly distributed vertically and horizontally. It is unreasonable to expect a perfect tracer distribution, although we can manage the tracer injection process to come as close to the ideal as is possible. The process variable with the greatest influence over the distribution quality (other than the aquifer, itself) is the injection flow rate; variables that exert lesser influence include tracer fluid density and surface tension.

Tracer Injection Flow Rate — Finding an optimal tracer injection rate is the most challenging aspect of tracer study design. If the injection rate is too low, the resulting tracer cloud will be distorted into a thin ellipsoid, elongated along the groundwater flow axis. If the injection rate is too high, fracturing occurs and the bulk of the tracer mass is deposited outside the interval of interest, often flooding into the vadose zone or onto the ground surface. We recommend the following strategy to set the tracer injection flow:

Pressure-restrict the injection flow rate — Set the tracer solution injection flow at the rate that can be achieved at the maximum safe formation injection pressure (determined according to **Equation 13.44**). This provides a maximal safe flow. If a high-quality injection well construction practice is used, it is unlikely that this flow rate would be too low to establish transverse tracer coverage.

Tracer Solution Composition — The tracer solution composition influences its fluid properties, with the potential to bias tracer movement in the subsurface. It isn't possible to add chemical tracers without increasing the solution density, relative to low-solute groundwater, but if densities are too high, the injected fluid could sink following its injection. Contrasting temperature injections can effect density, viscosity and surface tension, and some organic compounds can effect surface tension. Although it isn't possible to construct tracer solutions that exactly match all groundwater fluid properties, the following guidelines are helpful in formulating tracer solutions:

Limit tracer concentration — High tracer concentrations can increase the solution density or alter its surface tension, causing the fluid to sink or to flow differently than groundwater. For aquifers with low total dissolved solids, we normally use tracers in the range of 400 to 4,000 mg/L total dissolved solids.

Be aware of temperature impacts — Temperature contrasts between injected solutions and the aquifer can be very useful when we wish to use temperature as a high-transfer-rate tracer (see Case Study 3). However, a large temperature contrast can also alter fluid properties, relative to groundwater, biasing the injected tracer movement through the aquifer.

Tracer mounding — The aquifer accommodates injected fluid through an increase in aquifer volume and, if injection pressures are maintained at safe levels, fluid accommodation occurs through mounding of the injected fluid into the overlying vadose zone. The aquifer pressures should be logged before, during and following tracer injection, to track development and dissipation of the water mound. We expect the mound to be a mixture of groundwater and injected tracer which should dissipate laterally, under the influence of gravity. Some portion of the injected tracer is lost to the water mound, so our estimates of mobile porosity are slightly lower than actual, because we generally assume that the entire injected tracer volume is available to reach the dose response wells. It is possible to correct for tracer mounding, if sufficient data on water elevations has been collected and if an effort is made to confirm that tracer has migrated into the overlying strata. Without this information, it is sufficient to understand that the assumed volume of the tracer cloud is likely to be an over-estimate.

12.7 TRACER CASE STUDIES

The following case studies were drawn from recent field studies, to illustrate the results that can be obtained using a variety of tracer strategies. Each of these studies provided important insights into the subject aquifer structures and their patterns of groundwater flow and solute transport. A very interesting result, repeated at several sites, was the passage of multiple distinct solute peaks through very narrow vertical intervals. These observations were consistent with our discussion of aquifer fine-scale structure in Chapters 3 and 6, and were made possible by the use of continuous data logging instruments, combined with transient disturbances of the stratified flow.

CASE STUDY 1

The first case study is drawn from a permanganate injection pilot study in the Eastern United States. The site is underlain by fractured sandstone that has been contaminated by chlorinated solvent and the study was undertaken to map the distribution of injected permanganate solution, in support of a full-scale oxidation treatment. **Figure 12.12** shows the study area layout and a cross-sectional diagram of the formation. Each of the monitoring wells was outfitted with a data logger that recorded electrical conductivity and groundwater level in the well bore. Eight 2,300-liter batches of 4-percent permanganate were injected over a 9-day period and a 10th injection comprised a limited volume of permanganate solution and stock tank rinseate that was gravity-fed into the formation. Each injection was completed during a standard 8-hour work shift.

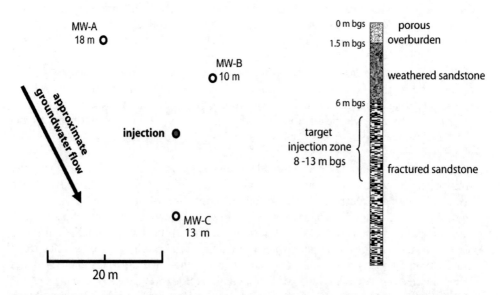

FIGURE 12.12 Layout for injection and observation wells in Case Study 1, which used a continuous electrical conductivity probe to track coverage and persistence of sodium permanganate injected in a fractured bedrock aquifer in the Eastern United States. The cross-section is depicted on the right side of the diagram. Scales are approximate.

The hydraulic pulse associated with injections was observed at each of the observation wells and was independent of the arrival of oxidant, as shown in **Figure 12.13**. This reinforces the fact that aquifer hydraulic connections can extend over ranges much greater than mass transport and that a strong hydraulic connection cannot be taken as an indication that mass transport will occur between two observation points.

The permanganate propagation, as indicated by electrical conductivity, was first observed in MW-C, located 13 m downgradient from the injection well, during the first 2,300-L permanganate pulse. Permanganate concentrations increased at MW-B during each of the first four injections. Fifty-percent of C_{max} was reached at MW-B after less than 4,600 L of solution was injected. Using Equation (12.2) and assuming the aquifer thickness was 5 m, the mobile porosity was less than 0.29 percent. Based on the value calculated from observations at MW-C, the mobile porosity was

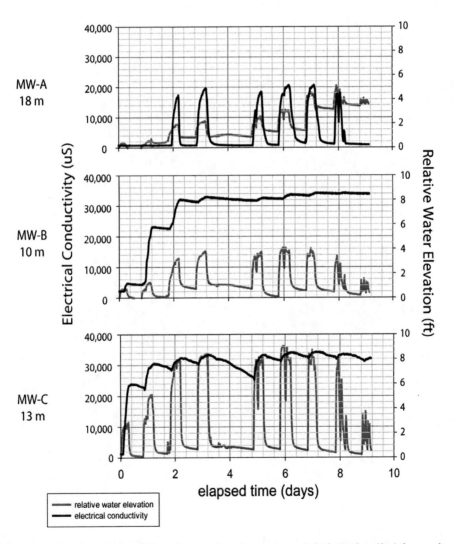

FIGURE 12.13 Electrical conductivity and water elevation response (relative to baseline) for monitor wells located 18 m (MW-A), 10 m (MW-B) and 13 m (MW-C) from a sodium permanganate injection well in Case Study 1. Each of 9 injection episodes caused a spike in the water elevation. At the upgradient edge of the injection area (MW-A), oxidant was washed out of the well between injection events. Inside the injected radius, permanganate persisted between injections.

less than 0.14 percent. These values are in the range commonly observed in fractured bedrock formations.

Based on these observations, those of Case Study 4 and numerous other field sites, we note that fractured bedrock systems are often highly volume-sensitive, due to the low mobile porosity (low secondary porosity). Small volumes of injected reagent, recharge or contaminated groundwater can generate large hydraulic or chemical concentration responses. In this case, the very low mobile porosity makes *in situ* chemical oxidation using permanganate solution a very cost-effective remedial strategy, which will be carried into a full-scale design.

CASE STUDY 2

The second case study site is also in the Eastern United States, in a fine-grained alluvial system comprised of silt, sand and clay mixtures. The test injection program was designed to determine the volume-radius relationship and solute velocities for injected reagents, to support the design of an *in situ* enhanced reductive dechlorination remedy. **Figure 12.14** shows the study site layout — a dose response well (MW-A) was located approximately 6 m downgradient from the tracer injection well, along the expected groundwater flow axis. The screened interval for injection and observation wells was 1.5 m. Tracer observation wells were located on the expected flow axis at an 8-m radial distance (MW-B) and two additional observation wells were placed at the 8-m radial distance, approximately 30 degrees of arc to either side of the expected flow axis, providing coverage in case the groundwater flow was not aligned with the elevation gradient.

The primary tracer was bromide at 2,200 mg/L, with a visible dye tracer, fluorescein, superimposed at approximately 40 mg/L. 7.7 m³ of the dye solution was injected while the field crew collected periodic samples from the dose response well, MW-A. The samples were compared

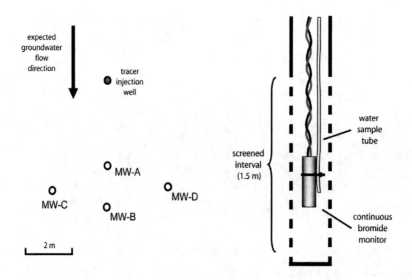

FIGURE 12.14 Tracer injection and monitoring well layout and monitoring setup used in Case Study 2. Bromide concentrations were monitored continuously using a bromide-selective ion probe. 3.85-liter (1 gallon) grab samples were collected for fluorescein analysis at variable intervals. Grab sample collections caused disruption of groundwater in the monitoring well, indicated by significant short-duration changes in bromide concentrations.

to visible dye standards carried by the field crew, to obtain a rapid approximation of the volume-radius relationship and to provide an estimate of the mobile porosity, θ_m.

Bromide was monitored with data logging ion-specific probes suspended at the mid point of each 1.5-m screened interval. A sample tube was attached to each of the specific ion probes, as shown on **Figure 12.14**, providing for collection of groundwater grab samples for bromide laboratory analysis and visible dye checks. The sample collection process disrupted stratified flow through the monitoring wells and the transient concentration response provided a means of observing the passage of multiple tracer peaks.

The bromide ion probe response dramatically exceeded laboratory analytical results throughout the study (many readings exceeded the injected bromide concentration by more than two-fold), which calls attention to the need for bench scale testing of probe response linearity and drift, before deployment in the field. It also highlights the value of grab samples to validate probe response. Tracer study teams should consider using a simple salt tracer (NaCl, or equivalent) that can be measured using a conductivity probe, if high-concentration tracers are needed. Although the ion probe did not provide usable absolute concentrations, the probe data are useful to show relative concentration responses that indicate stratification of tracer in the screened intervals and the impact of transient disruption of the well bore.

The injection process was monitored by grab sampling at the dose response well, MW-A, for the visual dye tracer. Samples were compared against a set of standards carried by the field crew. **Figure 12.15** shows the dye response at the dose response well as a function of the injected volume. The dose response well was located 3 m from the injection well and the screened interval was 1.5 m. The dye response was first noted after 6 m^3 of tracer injection; however, the dye response reached only 10 percent of the injected dye tracer response when the tracer fluid supply was exhausted. Based on the available data, the mobile porosity, θ_m, exceeded 0.19 [based on Equation (12.2) and the fact that more than 7,900 L was required to achieve 50-percent of the injected material]. Fifteen days after the injection process was completed, dye concentrations at MW-A peaked at 40 mg/L, ten-fold higher than concentrations observed during the injection process. This suggests that the

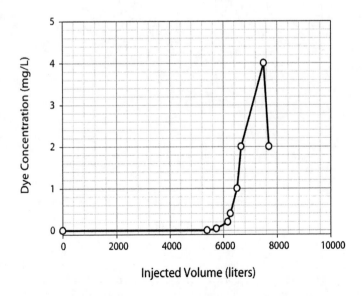

FIGURE 12.15 Dose response well dye concentrations during tracer injection for Case Study 2.

injected fluid was likely approaching 50-percent breakthrough and the mobile porosity was not likely much larger than the 0.18 approximation.

Figure 12.16 shows the bromide ion probe response curves for each of the monitoring wells. The bromide ion probe indicated concentrations at the dose response well, MW-A, that reached slightly more than double the injected concentration. A limited number of grab samples were collected for laboratory analysis of bromide and for visual dye comparison to prepared standards. The general pattern of the bromide probe was confirmed by these samples, but this serves as an example of limitations on some of the field instruments commonly deployed and the need to be certain of instrument response across the entire range of intended observations. The center of tracer mass arrived in 49 days at 3 m distance and in 60 days at the 4.6-m distance. These data translate to average groundwater velocity estimates of 6.1 and 7.7 cm/day, respectively. Note that both curves remained above baseline levels at the time data collected ceased and groundwater velocities were somewhat over-estimated.

The active sensor patches for data-logging probes cover only a small surface area, so if the groundwater flow is laminar, the probe responds only to a very thin layer of groundwater passing the monitoring well. Each grab sample collection disrupted this laminar flow and the results can be seen in the tracer response curves. **Figure 12.17** shows curve segments from **Figure 12.16**

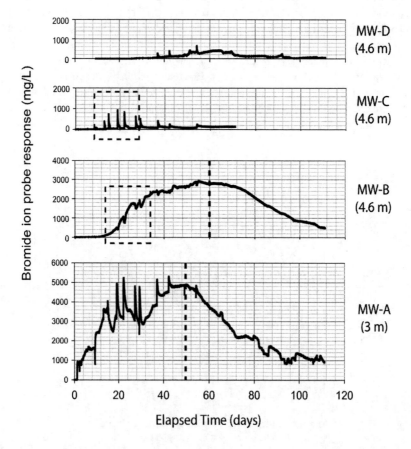

FIGURE 12.16 Bromide concentration trends at monitor wells downgradient from a tracer injection. Bromide concentration spikes and dips occurred during grab sample withdrawal. Figure 12.17 provides a detailed examination of these disruptions for MW-B and MW-C (areas indicated by dashed-line rectangles). Arrival times for the center of mass at MW-A and MW-B are indicated by vertical dashed lines.

(dashed areas) that were expanded to illustrate this point. The probe response in MW-C was stable at very low levels until a grab sample was collected, then an immediate, sharp increase in response was recorded. The probe response subsided over the following 24 hours, resuming the low-level response until the next grab sample disruption. In this case, it appears that low-bromide groundwater was flowing past the monitoring well at the elevation of the probe sensor patch, while a high-concentration layer of groundwater was flowing past the well at a different elevation. The grab sample process disrupted the stratification, increasing the observed probe response. Cleaner groundwater flushed the well at the probe sensor depth, maintaining the low response until the next grab sample collection. There are multiple possible explanations for the observations, but all scenarios must include both clean and high-bromide groundwater flowing past the well, simultaneously.

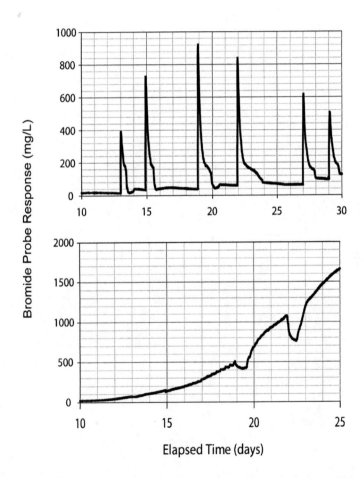

FIGURE 12.17 In each of these curves, grab samples caused a disruption in bromide concentration trends logged by the data logging probe. Top panel: Each grab sample collection from MW-C caused a dramatic spike in bromide concentrations that washed out over the following 24-hour period. This is an indication of a high concentration stratum flowing through the well but not contacting the bromide probe (refer to Figure 12.14 for the well monitoring setup). Bottom panel: Each grab sample collection from MW-B caused a decrease in bromide concentrations that lasted for approximately 24 hours, after which the preceding bromide trend was restored. This is an indication of a high concentration stratum passing through the well at the level of the bromide sensor, offset by lower bromide concentrations temporarily diluting groundwater at the probe elevation.

Case Study 3

Tracer injection studies were conducted in association with an aquifer sparge trial in a highly productive gravel and sand aquifer in Eastern Canada. The study layout, shown in **Figure 12.18**, consisted of 14 aquifer sparge wells arranged in three rows (2 deep, one shallow), 4 overlying soil vapor extraction wells and more than 20 groundwater monitoring wells. At the time of the tracer studies, the sparge wells were operated in 4 groups, each group receiving air for 15 minutes and idle for 45 minutes. **Figure 12.18** shows the plan view layout and screened intervals for each well set — deep sparge wells were set at the base of the aquifer and monitor wells used for the tracer

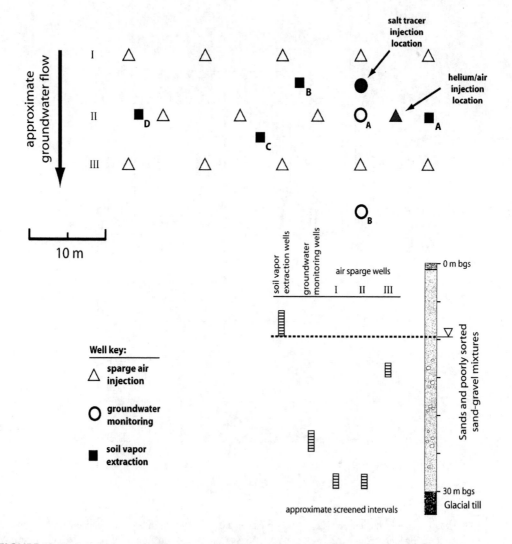

FIGURE 12.18 Layout for tracer Case Study 3 — an aquifer sparge site in Eastern Canada. Aquifer sparge wells are arranged in three ranks, perpendicular to groundwater; two ranks are placed deep in the formation and the third is shallow. Four soil vapor extraction wells overlie the 14-well aquifer sparge system. A helium tracer was injected into a deep zone air sparge well (indicated above) and helium appearance was tracked in each of the soil vapor extraction wells. A salt and temperature tracer was injected into a deep zone groundwater monitoring well (indicated above) and tracer passage was tracked in a subset of site monitoring wells, shown above.

study were set in the lower half of the aquifer (only a subset of monitoring wells are shown in the layout, for simplicity). A gas-phase (helium) tracer study was conducted to confirm that the bulk of injected air reaches the vadose zone in the vicinity of injection. An aqueous-phase tracer study (salt/temperature) was conducted to measure transport groundwater velocities through the sparge treatment zone.

Helium was injected, in lieu of air, for one 15-minute air injection at a deep sparge well on the edge of the pattern. Then, each of the four soil vapor extraction wells overlying the sparge network was monitored for the appearance of helium. **Figure 12.19** shows the helium concentrations observed in each of the four vapor extraction wells for 2,900 minutes following the 15-minute injection period. Most of the injected gas was observed in the nearby vapor extraction well and approximately 50 percent of the injected helium mass was recovered during the monitoring period, indicating that injected air is arriving near the injection location. Helium arrival began 10 hours after the injection, indicating that the injected gas followed a complex pathway to the vadose zone, ruling out a well seal failure and indicating that large-channel formation (piping) was unlikely to have occurred.

One of the objectives of pulsed-mode aquifer sparging is to cause cyclical vertical groundwater displacement, inducing bulk water mixing to improve mass transfer rates from groundwater to the injected sparge air. Continuous-recording water elevation probes were installed in several monitoring wells interspersed in the sparge well network and a typical response is shown in Figure **12.20**. This figure covers 180 minutes, three complete cycles of the sparge well groupings. Piezometric elevations varied over a 4.5-cm range, following a repeating cycle that tracked the sparge well operation cycles. A large vertical displacement is preferred (up to 30 cm), and this data led the project team to reallocate air flows with the goal of increasing vertical displacement.

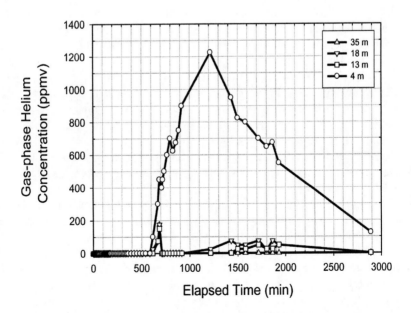

FIGURE 12.19 Gas-phase helium concentrations observed in an aquifer sparge injection tracer test. Helium was injected into an aquifer sparge well during the interval 0-15 min, and concentrations were observed in 4 overlying soil vapor extraction wells positioned from 4 to 35 meters from the sparge well location.

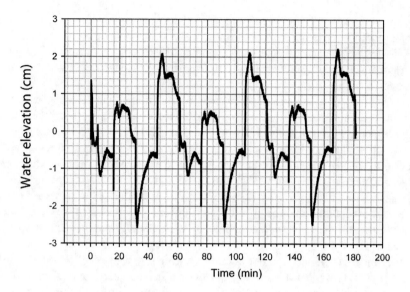

FIGURE 12.20 Cyclical water table elevation changes associated with cyclical variations in tracer concentrations shown in Figures 12.21 through 12.23. The elevation changes were driven by operation of an aquifer sparging system comprising four air injection well sets, each operating for 15 minutes out of every hour.

The groundwater velocity was assessed with a salt and thermal tracer study. A 3,000 mg/L (as NaCl) solution was injected at the salt tracer injection location noted on **Figure 12.18**. The solution was injected at 1 to 2 °C (injections were conducted in mid-winter) and the groundwater temperature was approximately 7 °C, generating a strong contrast between tracer and background temperatures. The air sparge system was stopped during tracer injection and restored to normal operation after the solution injection was completed. Electrical conductivity and temperature were logged in a series of wells arrayed along the expected flow path, including MWs A, B and C shown on **Figure 12.18**. Data logging was initiated when the sparge system was halted (t = 0). Clean water was injected from 940 to 1,070 minutes to calibrate flows, then tracer injection was initiated.

Tracer was injected at 14.8 L/minute until the electrical conductivity signal began to increase in MW-A, located 3.5 m from the injection well. Breakthrough was observed after 6,440 L of solution had been delivered (1,505 minutes). If the tracer cloud remained within the 1.5-m injected interval and was radially symmetrical, the estimated mobile porosity was at least 11 percent, based on Equation (12.2).

Figure 12.21 summarizes the results obtained at MW-A, 3.5 m downgradient from the salt tracer injection. The upper panel of the figure shows each of the electrical conductivity observations, which were logged at 1-minute intervals. Unplanned sparge system shutdowns impacted the tracer signal for short periods centered at 2,500 and 2,700 minutes and for a more extended period, from 3,200 to 3,800 minutes. The lower panel provides lines connecting the data points, focusing on the interval between 1,000 and 3,000 minutes. From this data, distinct tracer strengths can be seen associated with each position in the 4-part sparge cycle. This appears to be associated with the piezometric elevation changes induced by the sparge air injection and indicates that the very minor water movements inside the monitor well, associated with this cycling, are changing the water exposed to the conductivity sensor patch. Observations during the period from 1,500 to 2,000 minutes indicate that groundwater strata bearing two distinct tracer concentrations and at least

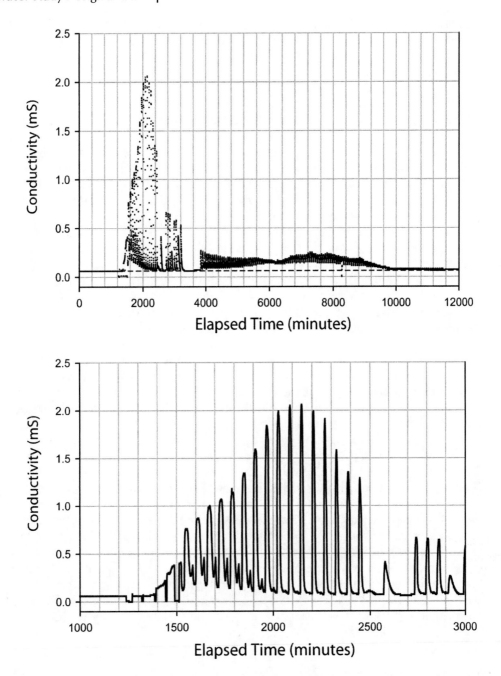

FIGURE 12.21 Results of sodium chloride tracer testing in a high-conductivity, heterogeneous aquifer in Quebec. The test was conducted in the vicinity of an active air sparging treatment. The upper graph shows 1-minute-interval data logger records for electrical conductivity as individual points and the lower graph shows conductivity detail for the period from 1,000 to 3,000 minutes. Tracer was moving past the 1.5-m long observation well in multiple conductive layers, with at least four resolvable concentration peaks. The sparge system consisted of four groups of sparge wells, each operating for 15 minutes each hour. Two of the sparge well groups can be seen pushing separate conductivity peaks into the conductivity monitor during the 1,500 to 2,000 minute period. The larger peaks are seen at 60-minute intervals; the lesser peaks also occur at 60-minute intervals, offset from the larger peaks by 30 minutes. The sparge system was turned off during tracer injection, and for three other periods, centered on 2,500, 2,700 and 3,500 minutes.

one clean-water stratum were passing sufficiently close to the sensor to be moved into contact by the relatively small vertical movement generated by the sparge system. Moreover, the groundwater velocity through the well bore was sufficient to clear the fluid within 15 minutes, in contrast to the 24-hour clearing times observed in Case Study 2, above. This is some of the most striking data we've seen in tracer studies and suggests that at least three tracer peaks are traveling at several distinct velocities, all within the 1.5-m vertical zone sampled by the monitoring well. This is the type of stratified movement anticipated in the modeling of detailed Borden Aquifer conductivity data, shown on **Figure 5.2** and it would have been missed if there wasn't a repetitive disturbance of the water column to expose the probe sensor patch to the distinct water masses. Based on experiences in Case Studies 2 and 3, we strongly recommend including vertical profiling or mixing of the well bore, to assure that distinct water masses can be detected as they pass the monitoring location. These observations also lead to renewed concerns regarding what a monitoring well "sees" and how our well construction and sampling practices control results.

The temperature and electrical conductivity curves for MW-A are compared in **Figure 12.22**. Solids lines were superimposed on the raw data tracks to outline the main peaks of temperature and conductivity passing the observation point. Note that the chemical peak is strongly log-normal, with peak arrival much earlier than for the thermal peak, which is much more Gaussian. Because thermal diffusivity is several orders of magnitude greater than chemical diffusivity, the "mass transfer" is much greater for the thermal peak. The higher transfer rate leads us to expect a more Gaussian curve, with the peak velocity more closely aligned with the average groundwater velocity.

Centers of mass were calculated for each of the entire datasets and the results are shown on **Figure 12.22**. Each of the centers of mass was affected by peaks arriving later than 6,000 minutes. Based on the entire dataset (all peaks), the average groundwater velocity was 180 cm/day based on the chemical tracer and 132 cm/day from the thermal tracer. The small chemical peak passing between 1,500 and 2,000 minutes traveled at greater than 1,000 cm/day and the major chemical peak (arriving at 2,150 minutes) traveled at 780 cm/day. These results are all indicating extremely high groundwater velocities, particularly for porous media.

The electrical conductivity signal remained strong when the tracer pulse arrived at MW-B, 7 m downgradient from the injection location. Results for this location are shown in **Figure 12.23**. The first peak was observed at 4,800 minutes (3,295 minutes after the tracer injection was completed), traveling at 305 cm/day. At least five distinct peaks passed the MW-B observation location with an overall center of mass arriving at 9,300 minutes, indicating an overall average velocity of 129 cm/day. The temperature peak was broad and flat at this location, with a maximum change of only 0.3 °C. High thermal diffusivity and the resulting signal attenuation limit the viability of temperature as a tracer to relatively short time spans in groundwater. The electrical conductivity peak observed at MW-C, approximately 14 m downgradient from the injection, was also broad and flat and was not monitored through its full development.

CASE STUDY 4

A salt washout study was performed in fractured sandstone of the Upper Trifels Formation in Southwestern Germany. The primary objective of the study was to confirm the relative washout rates from two conductive zones that were identified during the well construction. A further objective was to observe the passage of tracer, if possible, in downgradient monitoring wells. In most cases, especially porous media, the very low volume of tracer solution infused into the source well dramatically reduces the likelihood of intersecting the tracer cloud with downgradient wells. However, in bedrock with very small fracture porosities, it is possible to observe tracer migration (although not a certainty).

Case Study 4 used a solution replacement approach, slightly modified from that shown in **Figure 12.2**, above; the study layout is shown in **Figure 12.24**. The well fluid volume was 140

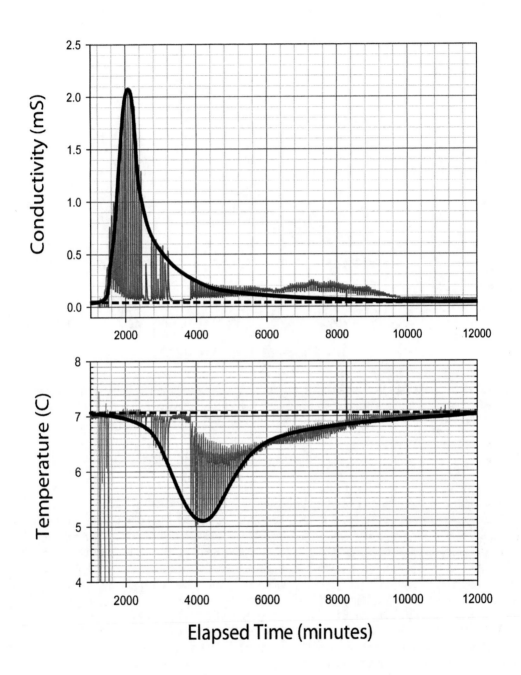

FIGURE 12.22 Results of sodium chloride tracer testing in a high-conductivity aquifer in Quebec, part 2. Chloride injection was conducted during cold weather and the injected fluid temperature (approximately 0° C) contrasted strongly with the aquifer temperature (approximately 7° C). Temperature serves as a second tracer, with a higher transfer coefficient than for salt-driven electrical conductivity (thermal diffusivity is greater than chemical diffusivity in aquifer matrices). The primary tracer pulse is outlined for each graph to highlight the comparison; dashed lines indicate baseline values for aquifer temperature and electrical conductivity.

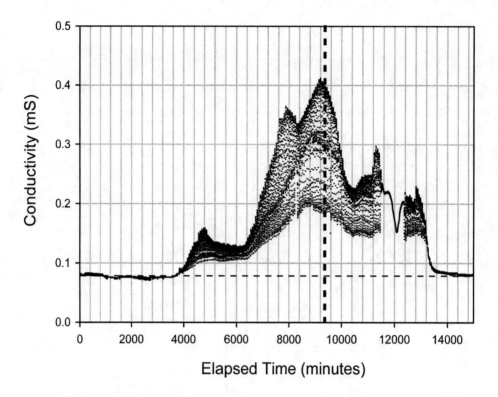

FIGURE 12.23 Case Study 3, observations at MW-B, 7 m downgradient from a salt tracer injection well. The background electrical conductivity for this dataset was 0.08 mS (horizontal dashed line). The center of mass arrival was calculated as 9,300 minutes, indicated by the vertical dashed line.

liters, so an equal volume of salt solution was prepared using 12.5 kg of sodium bicarbonate. A groundwater extraction pump placed at the bottom of the well extracted water at 3 liters/min. The extracted groundwater was replaced by water drained at an equal rate from the salt solution tank, maintaining a stable water elevation in the well. Groundwater extracted from the bottom of the well was pumped to disposal and the extraction-replacement process continued until the 140-L well volume was replaced by salt solution, approximately 50 minutes of pumping. **Figure 12.25** shows depth to groundwater for the salt infusion and nearby monitoring wells. Water elevation disturbances were noted during the infusion equipment placement and removal, but the infusion did not generate hydraulic responses in neighboring wells. Conductivity was then measured at 0.33-m vertical intervals to create hourly profiles for each of the monitoring wells.

Salt washout from the infusion well is summarized in **Figure 12.26**. Each curve comprises a set of electrical conductivity measurements taken at 0.33-m intervals. The first curve was constructed from measurement collected 1.3 hours after the infusion was completed. By this time, low-conductivity groundwater began to displace the salt solution from the uppermost elevations in the well. The displacement progressed over the following 7.8 hours, entirely washing the salt out of the upper half of the infusion well, while the lower half of the well remained near the initial salt concentrations. This indicates: 1) the flow through this well is primarily in the upper half, and 2) flow velocities in the upper portion are sufficient to clear out the well bore in less than 10 hours. This information is primarily qualitative. More valuable information was collected from downgradient observation wells.

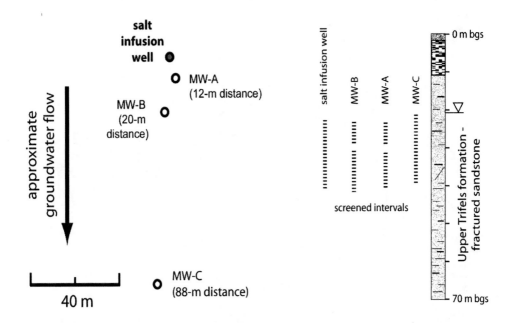

FIGURE 12.24 Salt tracer study layout. The study site lies in Southwestern Germany and the wells are constructed in the Upper Trifels Formation, a fractured sandstone. Screened intervals were established in sections of the bore hole that showed significant groundwater flux during geophysical borehole testing.

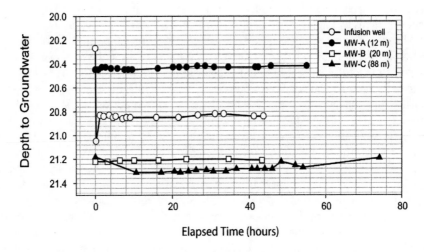

FIGURE 12.25 Results of groundwater elevation monitoring during salt infusion tracer study. The study layout is shown on Figure 12.24.

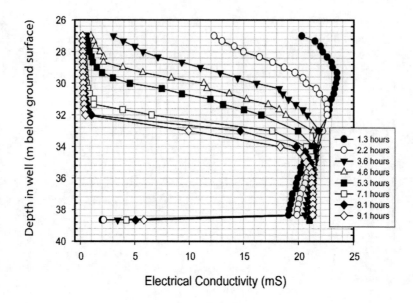

FIGURE 12.26 Salt tracer study — salt washout phase. Salt solution washes out of the infusion well in response to groundwater flow through the well. Variations in flow with depth are reflected in the rate of washout across the depth intervals. In this case, the dominant flow zone is in the uppermost 5 meters of the well.

MW-A, located 12 m downgradient from the infusion well, responded very quickly to the migrating tracer. **Figure 12.27** shows the arrival and washout of salt from the upper half of this well and the arrival of tracer in the lower half of the well. **Figure 12.28** shows the washout of tracer from the lower half of MW-A. The entire salt tracer pulse passed this location in less than 20 hours, indicating an average groundwater velocity exceeding 12 m/day. **Figure 12.29** shows the average conductivity values from each measurement episode, to develop a tracer breakthrough curve for MW-A. The center of tracer mass passed this location 7.6 hours after the tracer infusion was completed, indicating an average groundwater velocity of 38 m/day. This velocity is fast, but well within the range of values observed in fractured crystalline bedrock. At this location, cores indicate very small secondary fracture porosities, consistent with a low-volume, high-velocity system of the type indicated by the tracer results. It is also important to note that the electrical conductivities at MW-A peaked at one-quarter of the initial tracer strength in the infusion well. This is likely an indication of diffusive tracer interaction with primary porosity of the sandstone, which is consistent with the log-normal breakthrough curve.

A small electrical conductivity signal was also detected by the project team at the monitoring well 88 m downgradient from the infusion well and is shown in **Figure 12.30**. This signal is very small compared to the background electrical conductivity and it requires corroborating data for confirmation. If validated, the data indicate an average tracer velocity approaching 50 m/day.

These data provided the project team with valuable design support information at a relatively low cost. This is a low-volume, high-velocity system with an indication of primary porosity in the sandstone matrix. Although there may be contaminant mass stored in the porous matrix at this location, it is likely that a persistent source mass lies upgradient and that the most cost-effective approach will be to locate and treat those source zones, first. It is possible that source removal will lead to a rapid decline in contaminant concentrations in the low-volume, high-velocity sector of the

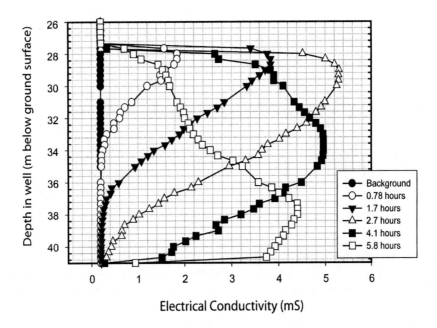

FIGURE 12.27 Salt tracer study — early-stage salt passage at a well 12 m downgradient from the salt infusion well (MW-A). The fastest electrical conductivity response was observed in the upper portion of the well, and tracer passage in the upper zone was nearing completion 5.8 hours after completion of the infusion process. The first tracer arrival was observed in the upper level less than 1 hour after the salt infusion was completed, 12 meters from the observation location.

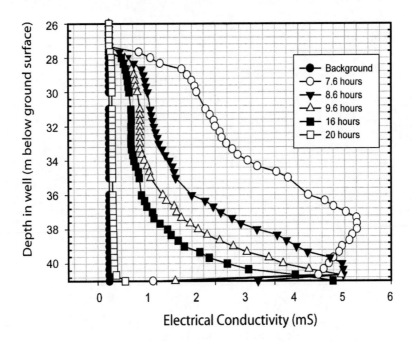

FIGURE 12.28 Salt tracer study — late-stage salt passage at a well 12 m downgradient from the salt infusion well (MW-A). The fastest electrical conductivity response was observed in the upper portion of the well, and tracer passage in the upper zone was nearing completion 5.8 hours after completion of the infusion process. The first tracer arrival was observed in the upper level less than 1 hour after the salt infusion was completed, 12 meters from the observation location.

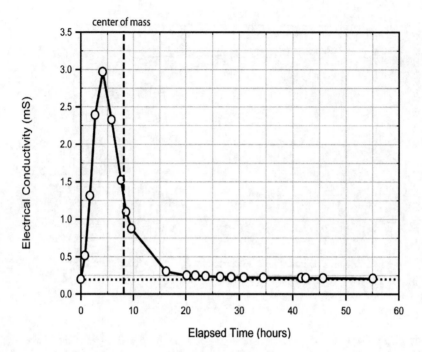

FIGURE 12.29 Breakthrough curve for salt tracer, 12 m downgradient from a salt induction well (MW-A). Data points represent vertical averaging of electrical conductivity values in the monitoring well. Areas under the electrical conductivity curve were calculated relative to the baseline level (0.20 mS, indicated by the dotted horizontal line) for determination of center of mass.

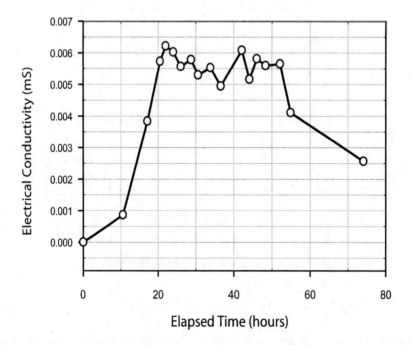

FIGURE 12.30 Breakthrough curve for salt tracer, 88 m downgradient from the salt induction well (MW-C). Data points represent deviations from baseline electrical conductivity measurements, averaged across the screened interval (22 to 39 m below ground surface). The observed electrical conductivity signals were very small, relative to the baseline values (0.257 to 0259 mS).

formation. If remedial action is still needed following source removal, a fast-acting process, such as in site chemical oxidation, would likely be very effective.

CASE STUDY 5

Recently, field studies were initiated in an alluvial fan aquifer in the Western United States and the results illustrate some of the problems encountered in high-flow, anisotropic aquifers. The aquifer is approximately 2.4-m thick in this location, comprising high-energy deposits of coarse gravels and sands, with a strong groundwater surface elevation gradient (0.01 cm/cm). This study utilized a high-volume injection (154 m³), calculated to deliver strong tracer detections at dose response wells located 4 and 9 meters from the injection well, even if the mobile pore fraction was as high as 20 percent. Tracer observation wells were located 15 and 25 m from the injection well, oriented approximately along the expected groundwater flow path, in the layout shown in **Figure 12.31**. The two dose response wells were oriented perpendicular to the expected groundwater flow path, also shown in **Figure 12.31**. The tracer response was expected to arrive first at the 4-m location and pumping was planned to continue until the tracer arrival was registered at the 9-m location. That was expected to generate a broad tracer cloud, maximizing the likelihood of one or both observation wells intercepting the passing tracer cloud.

The aquifer has high total dissolved solids, generating a strong background electrical conductivity (> 1,000 uS), which ruled out the use of electrical conductivity for tracer monitoring. Bromide concentrations in the aquifer are low (0.3 mg/L), so bromide at 230 mg/L was selected as the tracer solution and a bromide-ion-specific probe was selected for tracer monitoring, with periodic grab sampling for laboratory confirmation of the continuous probe measurements. Lab tests were conducted prior to field deployment which determined that the probes over-estimated concentrations in the aquifer background range (0.3 mg/L Br-) and that only probe readings of 4

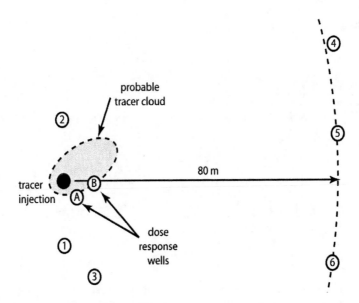

FIGURE 12.31 Layout for Case Study 5, high-volume tracer injection in an alluvial fan aquifer, using an existing monitoring well network. Wells A and B, located 4 and 9 meters, respectively, from the injection well, were used as dose response well. Wells 1 through 5 were monitored for the appearance of bromide tracer, which arrived at tracer observation well 5.

mg/L or greater were indicative of elevated bromide ion concentrations. These results make it clear that specific ion probe behaviors must be carefully investigated, prior to field deployment. Our experience also indicates that probe-to-probe differences may be significant.

The bromide ion probes were placed at the mid-point of the monitoring well screened intervals, logging bromide ion concentrations at 1-minute intervals throughout the study. The probes were used to profile each monitoring well daily, using the lift-drop-replace method depicted in **Figure 12.10**, above. This allows the probe to capture concentration values at the top and bottom of the screened interval and to generate vertical mixing within the well.

The tracer breakthrough curve shown in **Figure 12.32** was observed at the dose response well located 9 m from the injection well, MW-B. Tracer arrived rapidly, reaching 50 percent of its maximum value for this location after 24.7 m³ of tracer injection. Tracer arrival at the 9-m location significantly preceded arrival at the 4-m dose response point. Based on an aquifer thickness of 2.4 meters, Equation (12.2) yields an estimate of 4 percent for the mobile porosity, θ_m. This represents a lower limit on the mobile porosity, because the tracer cloud appears to be elliptical and Equation (12.2) is based on a circular tracer distribution.

The dose monitoring wells shown in **Figure 12.31** were profiled at least daily during the entire breakthrough curve development. The first profile at MW-B, marked as item B on **Figure 12.32**,

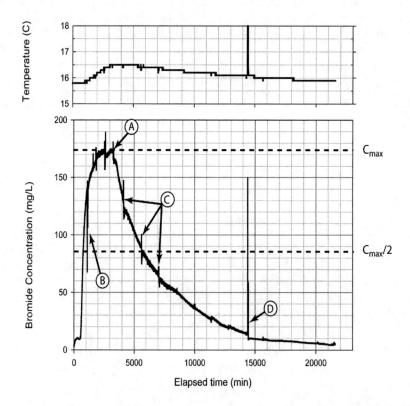

FIGURE 12.32 High-volume tracer test in an alluvial fan aquifer — results for dose response well B, located 9 m from the injection. 154 m³ of 230-mg/L bromide tracer was injected in just over 3,000 minutes (A). Concentrations reached 50 percent of the maximum value at 720 minutes, after 24.7 m³ of tracer had been injected. The first in-well profile (B) showed that the highest bromide concentration was passing the depth of probe placement. Later profile events (C) indicated the passage of tracer peaks at other depths in the screened interval. The probe was recalibrated after 10 days, (D), resulting in a small adjustment.

showed that the fixed probe depth was aligned with the highest bromide concentration. Examination of the raw data indicates a single measurement at 69 mg/L, roughly one-half of the mid-screen concentration (details of the profile can be recorded separately to analyze details of tracer passage). There was not a significant mixing effect from the tracer profiling, indicating that concentrations were relatively uniform across the screened interval.

After the injection process was completed, marked as item A on **Figure 12.32**, tracer concentrations began a rapid decline and background bromide concentrations were restored after 20,000 minutes. Vertical profiles collected during tracer washout, indicated as items C in the figure, showed that the fixed probe position was representative of the passing tracer cloud. The rapid washout of tracer at this location suggested that this dose response well was located near the edge of the tracer cloud, in the upgradient half of the injected tracer cloud.

The injected tracer was slightly warmer than the aquifer, approximately 17 °C, versus 15.7 °C. At the 9-m dose response well, the temperature peak arrived later than the chemical tracer peak, as expected — thermal diffusivity exceeds chemical diffusivities — and the peak temperature was less than the injected temperature, due to diffusive losses.

The bromide probes were re-calibrated after 10 days. Item D in **Figure 12.32** shows the results of a recalibration event at a dose response well and the probe at this location required only minor correction. At other monitoring locations, the probes were unstable and the continuous logging data was unusable. This highlights the importance of grab sample collection for confirmation of probe data and to provide a record of tracer movement, in the event of a probe failure. The probes had all passed a brief logging test period before deployment, but an extended bench-scale trial may be required to identify probes that are unstable over periods of several days. Small recalibration adjustments, of the magnitude shown in **Figure 12.32**, are acceptable, but larger calibration adjustments render the data unusable. Probes requiring daily adjustments greater than a few percent should not be deployed.

A second dose response well, MW-A, located only 4 meters from the injection well, reached 50 percent of its peak volume only after 38.5 m^3 of tracer injection, suggesting that the tracer cloud distribution was strongly asymmetrical. **Figure 12.33** provides the tracer breakthrough curve for the 4-m dose response well, with data from the specific ion probe (solid line) as well as laboratory confirmation data, shown as open triangles. The peak concentration was much closer to the injected concentration at this location than at the 9-m dose response location, consistent with lesser diffusive losses over the shorter path length. The peak temperature was also higher at the 4-m dose response location, also consistent with the shorter path. The later peak arrival suggests a lower hydraulic conductivity along the path from the injection well to the 4-m dose response well, when compared to the rate of tracer cloud propagation toward the 9-m dose response well. The laboratory data suggest that tracer arrival may have occurred slightly earlier than was recorded on the bromide probe, which could be caused by the probe placement in the screened interval missing the first arrival.

Observation well at locations 1 and 2 in **Figure 12.31** did not intersect the tracer, contrary to the expected groundwater flow. The rapid tracer washout at the two dose response wells was also inconsistent with the conceptual site model developed by the project team. **Figure 12.34** shows the injection well concentrations, which were examined to check the washout rate in the center of the tracer cloud (it wasn't possible to deploy a recording probe in the injection well, so only laboratory data are available at this location). The data show a normal tracer washout, with tracer concentrations remaining at more than 5 percent of the injected concentration, 30 days following completion of the injections. That rules out an excessively quick washout of the entire tracer cloud and forces a re-examination of the conceptual site model to account for rapid washout of the dose response wells and failure of tracer to appear in wells 1 and 2.

The likely shape of the tracer cloud was inferred from observations in the dose response wells and from the lack of significant tracer appearance in observation wells 1 and 2 and is shown on **Figure 12.31**. The project team began to check wells along the apparent flow path, in tracer

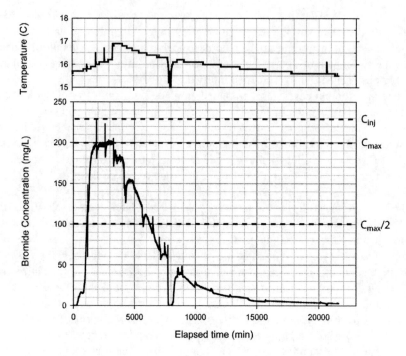

FIGURE 12.33 High-volume tracer test in an alluvial fan aquifer – Part 2. Results for dose response well A, located 4 m from an injection well, with combined continuous probe and intermittent profiling. Solid lines represent probe data and open triangles are laboratory analysis results.

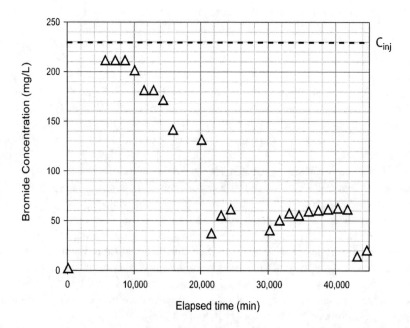

FIGURE 12.34 High-volume tracer test in an alluvial fan aquifer – Part 3. Laboratory analysis results for tracer injection well. No samples were collected during tracer injection.

observation wells 4, 5 and 6. Because the team didn't expect groundwater flow in this direction, bromide probes had not been deployed in this sector and samples weren't collected for laboratory analysis until the tracer peak had already passed the area. **Figure 12.35** shows results for the tracer observation well at location 5, which registered tracer approximately 20-fold higher than background, in the first sampling which wasn't conducted until 22 days after the tracer injection was started. The data shown in **Figure 12.35** captured the tracer tail, but missed the peak arrival.

The combined behaviors of the dose response and tracer observation wells indicate a highly anisotropic aquifer formation, with extremely high groundwater velocities along some of the flow paths. Because the project team did not respond to data that contradicted their conceptual model, an opportunity was missed to quantify tracer peak and groundwater flow velocities from the first tracer episode. It is clear, however, that groundwater velocities in the mobile pore space appear to exceed 6 m/day along some pathways. This suggests that the regions of the contaminated aquifer downgradient from the source zone are likely to respond very quickly to contaminant source mass reductions. The site conceptual model has been adjusted accordingly and a portion of the remedial efforts initially allocated to downgradient sectors of the plume have been re-directed to the source zone.

SUMMARY OF KEY TRACER STUDY FINDINGS

When we decided to require tracer studies as a step in the design process for all our injection-based remedial technologies, a flood of new information began to flow to us on aquifer structure, groundwater flow velocities and solute transport patterns in real systems. The tracer study results forced major revisions to many of the conceptual site models (few conceptual site models emerge from tracer studies without some revision) and, in several cases, remedial strategies were revised or abandoned as a result.

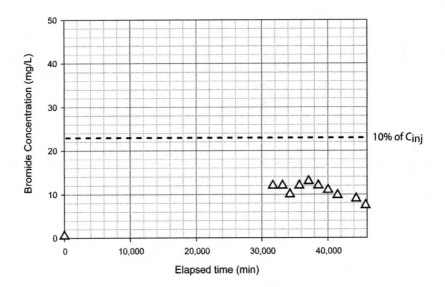

FIGURE 12.35 High-volume tracer test in an alluvial fan aquifer – Part 4. Laboratory analysis results for tracer observation well 5. Post-injection sample collections were not initiated until 20 days after injections were completed, missing the tracer peak arrival. The dashed line indicates 10 percent of the injected tracer concentration, for reference.

Aquifer heterogeneities — The tracer studies repeatedly showed that solute transport is occurring in multiple parallel pathways over narrow vertical intervals. This result is consistent with studies that examine detailed aquifer structure, but it was somewhat surprising to observe the large number of separable peaks passing 1.5-m screened intervals (5 or more in Case Study 3, for example). Tracer observations ratify the view of solute migration expressed in Figure 5.2.

What a monitoring well "sees" — The observation of multiple separable peaks passing screened intervals, the sharp stratification of concentrations within wells and the impact of disturbance of the water column on measured concentrations combine to direct us toward a re-evaluation of what monitoring wells are actually "seeing" and how sampling techniques impact our observations.

Groundwater velocities — All these studies showed groundwater traveling at velocities far exceeding the average and it is clear that our expectations for contaminant and reagent transport velocities must be revised to account for these observations. Average velocities remain of interest, but it is important to understand that tracer studies are telling us the mobile fraction is small and actual groundwater velocities are going to be 3 or 4-fold higher than the average velocity in most cases.

Solute transport and storage — All the tracer breakthrough curves we have examined are strongly lognormal, consistent with the expectation that diffusive interactions between mobile and immobile porosities will control the rate of solute mass transport in natural aquifers. Comparison of high and low-transfer tracers (chemical versus thermal, as seen in Case Studies 3 and 5) supports the conclusion that diffusive mass transfer plays a significant role in longitudinal dispersion of the tracer cloud. This impacts the concentration distribution for a reagent pulse and provides the mechanism for contaminant mass storage in mature plumes.

Cost-effectiveness — The frequency of conceptual site model revisions and modification of remedial strategies that are occurring after tracer study implementation makes it clear that tracer studies are one of the most cost-effective tools we have at our disposal.

12.8 USING TRACERS TO DETECT DISPLACEMENT

The potential for injected reagents to displace contaminated groundwater is one of the persistent challenging problems for remedial designers and regulatory officials. When monitoring well concentrations decrease following reagent injections, was the decrease a result of contaminant destruction or was it simply a displacement of contaminated water by the reagent mixture? Tracers can be used to detect the dilution of contamination that results from displacement without treatment, as shown in the following hypothetical example:

A nutrient solution has been injected to enhance aerobic biodegradation of benzene, that has been consistently observed at 50 ug/L in groundwater. Aerobic biodegradation of benzene can be rapid and the carbon dioxide produced from the decomposition of 50 ug/L would not be detectable against the aquifer background geochemistry. If the benzene concentration declines, how can we tell if the decrease is due to decomposition or simple dilution by the injected nutrient solution? Superimposing a conservative tracer on the injected nutrient solution provides a means of determining whether the dilution and displacement that necessarily occurs with reagent injection was the dominant mechanism causing contaminant concentration declines.

Figure 12.36 shows two hypothetical scenarios. In the upper panel, the benzene concentration drops by 50 percent, and the concentration of a bromide tracer added to the nutrient solution is observed at 50 percent of its injected concentration. In this instance, we can conclude that the nutrient solution displaced half the groundwater and no treatment has occurred. In the lower panel, the bromide is observed at less than 10 percent of the injected concentration and the benzene has been reduced by more than 5-fold. Here, we can conclude that the reagent injection has the desired effect and the benzene concentration reduction was due to decomposition.

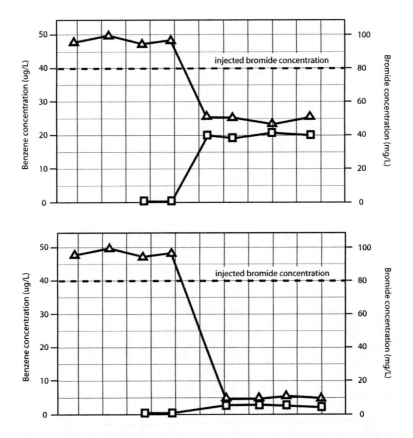

FIGURE 12.36 Use of tracers to detect dilution effects of reagent injection — a hypothetical example. In the upper graph, a bromide tracer injected with the treatment reagent is diluted to 50 percent of the injected concentration, while the benzene concentration decreases by the same proportion, indicating dilution, rather than treatment. In the lower graph, the benzene concentration decrease far exceeds the tracer increase, indicating that the treatment has decreased benzene, with only a slight dilution due to injected reagent.

13 Injection Based Reactive Zone Design

Injection and distribution of treatment reagents and re-injection of treated groundwater are two of the most challenging tasks undertaken by remediation professionals. When we began to recognize the limits on transverse dispersivity in aquifers, the prescribed reagent injection volumes increased dramatically, eliciting questions of feasibility from practitioners and concerns regarding lateral displacement of contaminated groundwater, by site owners, regulators and practitioners, alike. **Figure 13.1** shows the volume-radius relationship for injected fluids, based on values for the mobile porosity, θ_m. For the range of typical mobile porosities, the volume required to reach a prescribed radius of coverage is quite large. The technical challenge can be distilled to the question: *How does an aquifer accommodate injected fluid?* The simple answer is: *The aquifer volume expands.* A more complete answer is developed for Section 13.1.

The mobile porosity is our avenue of access for placing reagents in contact with contaminants and the relative proportions of mobile and immobile porosity and the structural detail of the porosities (i.e., mass transfer geometry introduced in Chapter 7) become very important determinants of our ability to effectively distribute reagents. **Figure 13.2** shows a hypothetical distribution of trichloroethene among mobile and immobile, dissolved and sorbed fractions, for typical aquifer conditions. We assumed that the organic carbon fraction would be greater in the less permeable aquifer soils and that the mobile fraction is 29 percent of the total porosity. For aquifers that match these assumptions, only 28 percent of the trichloroethene mass resides in the mobile pore fraction; more than two-thirds of the mass would be in the immobile porosity. Contaminant accessibility to injected reagent is certainly a challenge and the ability to map contaminant locations and effectively deliver reagents is likely to be a significant factor underlying successful aquifer restorations.

The difficulty of aquifer restoration is compounded by the fact that fluid injection is much more difficult than extraction. The conventional wisdom among large-scale water supply and injection specialists is that, at best, we can only inject at one-quarter to one-third of the flow we could extract from any given well — and that's when we get to choose the location for an injection program. When the injection location is dictated by the requirements of a remedial design, the prospects for subsurface are often much more troublesome. Fluid accommodation is the starting point for our discussion of injection-based remedial designs.

In the sections that follow, we describe the mechanisms that limit fluid injections into the subsurface, arriving at the concept of the vertical accommodation rate, which explains the asymmetry between aquifer resistances to fluid injection and extraction. Then, we present methods for calculating an aquifer's tolerance for pressurized injection, which gives system designers a basis for setting strict pressure limits on reagent injections. Finally, we examine strategies for reagent distribution and provide a brief look at equipment requirements for reagent injection.

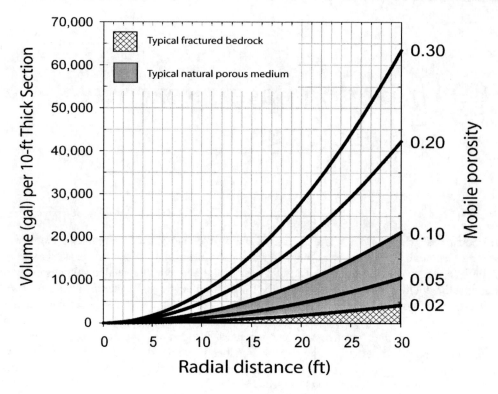

FIGURE 13.1 Relationship between the injection volume required to reach a planned radial distance and the mobile porosity of the aquifer. Typical values for mobile porosity are 0.02 to 0.10 (2 to 10%) for natural porous media (shaded area) and less than 2% for fractured bedrock media (hatched area). Values for porous media higher than 0.10 are atypical among natural porous media.

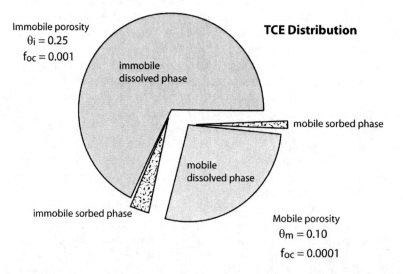

FIGURE 13.2 Distribution of trichloroethene (TCE) in an aquifer with 25% immobile porosity and 10% mobile porosity. The total organic carbon in the immobile aquifer portion is assumed to be 1,000 mg/kg (foc = 0.001) and the organic carbon in the mobile fraction is 100 mg/kg (foc = 0.0001). The organic carbon partition coefficient for trichloroethene is 94 L/kg. The aquifer bulk density was assumed to be 1,800 kg/m^3 for this calculation.

13.1 HOW AN AQUIFER ACCOMMODATES INJECTED FLUID

Inertia, Friction and Water Mounding

For all the scenarios we encounter in remediation hydrogeology, water is effectively an incompressible fluid. When we apply a force to the groundwater in an aquifer, that force can be (theoretically) transmitted to the entire connected water mass, which may extend for several kilometers. If we intend to move water through the impulse of an applied force, the mass that we have to set in motion may be quite large. If we can overcome the inertia and get the groundwater moving, it meets resistance to flow through the porous matrix, due to the frictional losses of viscous flow (Chapters 1 and 3). The effects of resistance are additive along the flow path, so the energy that must be expended to overcome frictional resistance can be immense.

Figure 13.3 shows a conceptual force balance that we encounter when we try to add fluid to an aquifer from a radial well embedded in the formation. The inertial masses below and laterally, as well as frictional resistance to flow over large lateral distances, cause the net accommodation of injected fluid to be through vertical expansion of the aquifer volume. **Figure 13.4** shows a conceptual fluid injection, with the near-well fluid motion occurring along the most permeable strata. These flows are offset by vertical displacement of the groundwater and invading fluid volumes in the overlying strata, as suggested in the figure. The fingering flows shown conceptually in the figure are consistent with the multi-peak tracer breakthrough curves we consistently observe (Chapter 12), as well as the results of high-resolution permeability observations reported in the literature (e.g., Julian, et al., 2001 and Rivett, et al., 2001).

If the aquifer is relatively homogeneous, the net upward displacement forms a water mound overlying the injected reagent volume. **Figure 13.5** speculates on the dissipation of the water mound and its lateral migration along the water table surface. After the mound forms, it encounters the same inertial and frictional forces that the original injection encountered, and the only force acting

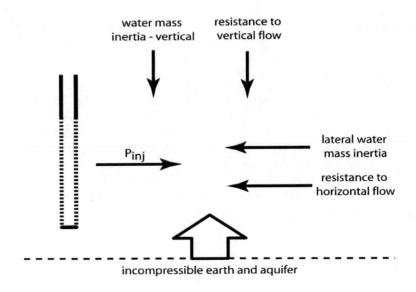

FIGURE 13.3 Conceptual diagram showing forces acting on the injection of incompressible fluid into an aquifer.

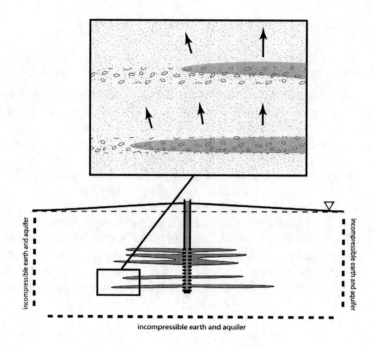

FIGURE 13.4 Injection of incompressible fluid into an aquifer. The injected fluid volume must be offset by an expansion of the aquifer volume. The fluid injection propagates along permeable pathways and the system volume increase is accommodated by upward displacement of resident groundwater and reagents flowing along permeable strata at higher elevations in the formation. To avoid structural failure of the aquifer matrix, the injection rate must be within the vertical fluid accommodation potential.

to move the water downward in an unconfined system is gravity. We expect gravity to flatten the mound and to generate displacement of underlying groundwater, but we don't expect the mound to simply drain back into the aquifer, as if the fluid is immediately re-inserted below the aquifer surface.

Radial Flow Example

To demonstrate the importance of resistance to radial flow as a driver in the net vertical movement of water, we created a simple thought experiment on the injection of water into a very simple, confined system. **Figure 13.6** shows the system: a vertical well, fully screening a conductive stratum that is bounded by low-permeability strata, above and below. The permeable stratum discharges to a zero-resistance boundary at a distance of 10 meters from the injection well and the flow is uniform and radial. We can calculate the pressure drop from the injection well to the zero-resistance boundary, if we set values for hydraulic conductivity, unit thickness and pumping rate. Recall Equation 3.45 for radial flow in a uniform matrix:

$$\Delta P_{a \rightarrow b} = \frac{Q}{K \cdot 2 \cdot \pi \cdot h} \cdot \ln\left(\frac{b}{a}\right) \tag{13.1}$$

where b is the outer radius of the injection zone and a is the inner radius (well radius) of the radial flow. If we wish to inject reagent at a continuous rate of 500 ml/s (7.5 gpm), into a relatively permeable

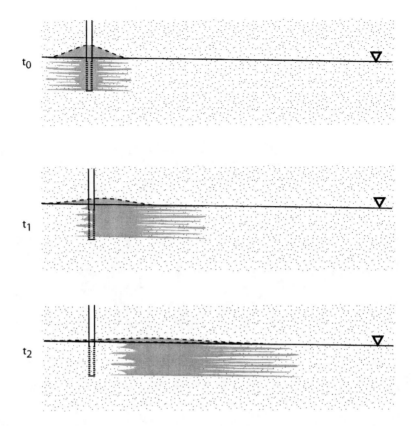

FIGURE 13.5 Conceptual diagram showing the development and dissipation of a water mound, overlying the location of a reagent injection into an aquifer. The mound forms as the aquifer volume expands vertically to accommodate injected fluid. Gravity flattens the mound against the original groundwater surface and the groundwater surface elevation gradient drives migration in the same direction as the underlying reagent pulse. The mound may migrate more slowly than the underlying reagent mass, due to reduced hydraulic conductivities in the partially saturated capillary fringe.

stratum (assume K = 1 x 10⁻³ cm/s), 150 cm thick, the pressure drop that must be established from the outer radius of the well, to the outer boundary of the porous matrix, is given by:

$$\Delta P_{a \to b} = \frac{500}{1 \times 10^{-3} \cdot 2 \cdot \pi \cdot 150} \cdot \ln\left(\frac{1,000}{5}\right) \quad \frac{cm^3}{s} \cdot \frac{s}{cm} \cdot \frac{1}{cm} \cdot \frac{cm}{cm} \tag{13.2}$$

$$\Delta P_{a \to b} = 2,810 \quad cm \ H_2O$$

If we use a 5-cm radius well (4-in diameter) the pressure drop across 10 m of porous matrix required to sustain a 0.5 L/s flow is 28 m water column, or approximately 42 psi (2.75 x 10⁶ dynes/cm²). Two questions arise: 1) How much work is being exerted per unit time (power) in this example (i.e., can we realistically pump this fast)? and 2) How does the pressure compare to the effective stress of the system shown in **Figure 13.6**?

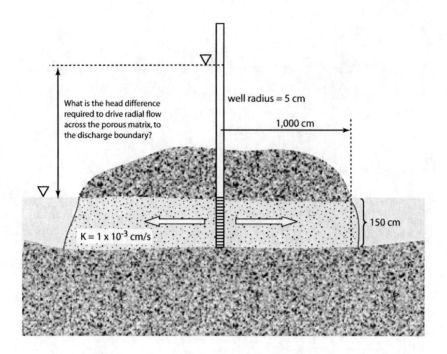

FIGURE 13.6 Estimating pressure required to drive radial flow across a porous medium, to a zero-resistance discharge boundary located 10 m from the injection well. The permeable matrix is bounded above and below by impermeable materials.

Power — The work expended, per unit time, to push water across the 10-m radial distance is calculated from the 500 cm³/s of water, falling through the 2,810 cm of head loss that is required to drive the flow. If the density of the water is 1.0, and the force is gravity, acting over the 2,810-cm distance, the work is:

$$500\,\frac{g}{s}\times 980\,\frac{cm}{s^2}\times 2,810\,cm = 1.38\times 10^{10}\,\frac{erg}{s} = 1.76\,HP$$

The power required is well within typical pump capacities.

Pressure – If we compare the hydrostatic pressure at the well screen to the effective stress of the surrounding formation, we can see that a significant overburden down-force is required to prevent structure failure of the confining matrix.

Effective stress is calculated from Equations (2.21) to (2.23) and we can use them to calculate the overburden thickness required to just balance the 2,810 cm H₂O pressure. From Equations (2.20) and (2.21):

$$\sigma_w = \rho_w\times g\times h_w = 1\,\frac{g}{cm^3}\times 981\,\frac{cm}{s^2}\times 2,810\,cm = 2.76\times 10^6\,\frac{dynes}{cm^2} \qquad (13.3)$$

From Equations (2.18) and (2.19):

$$\sigma_s = \rho_s \times g \times h_s = 1.8 \frac{g}{cm^3} \times 980 \frac{cm}{s^2} \times h \ cm = 1,764 \times h \quad \frac{dynes}{cm^2} \tag{13.4}$$

We can use Equations (13.3) and (13.4) to solve for the overburden thickness that matches the pressure at the top of the well screen, assuming the bulk density for unsaturated soil is 1.8 g/cm³. This is the condition at which $\sigma_w = \sigma_s$ and the effective stress, σ_e, equals 0.

$$h_s = \frac{\rho_w \cdot h_w}{\rho_s} \quad cm \tag{13.5}$$

To just offset the 2,810-cm water head ($\sigma_e = 0$), we need 1,561 cm of overburden soil. This tells us that if the overburden is less than 15 meters in thickness, the pressure required to inject 0.5 L/s into the permeable matrix will cause structural failure of the overburden, instead of delivering water at the prescribed rate to the 10-m boundary.

The simple thought experiment of **Figure 13.6** helps to place the injection process into a useable perspective: First, it is clear that the force required to push fluids over modest radial distances can be quite large and, second, the forces commonly applied in fluid injections can easily exceed the effective stress. This leads us to suggest that injection pressures should be limited, to prevent formation failure. The following "rule of thumb" equation combines the contributions of vadose zone thickness (h_{dry}) and saturated matrix thickness above the injection point (h_{sat}) to develop a maximum injection pressure (a 60-percent safety factor is applied):

$$P_{max,injection} = 0.6 \times \sigma_e = 0.6 \times \left[\left(\rho_{dry} \cdot g \cdot h_{dry} + \rho_{sat} \cdot g \cdot h_{sat} \right) - \rho_w \cdot g \cdot h_{sat} \right] \quad \frac{dynes}{cm^2} \tag{13.6}$$

This limit is often exceeded in reagent injections and, in these cases, it is likely that the injected reagents are distributed quite unevenly in the formation, with a significant portion arriving in the vadose zone. Section 13.2 develops the subject of soil structural capacities in much greater detail and introduces more extensive calculation tools to determine injection system pressure limits.

The pressure actually applied at an injection well screen reflects the pressure applied to the line (wellhead gauge pressure), plus the elevation head from the ground surface to the injection screen (this is the part that is often overlooked by field crews). **Figure 13.7** outlines the variables used to calculate fluid delivery pressure at the well screen. If the pressure measurement point is on a feed line offset from the well head, pipe friction pressure losses further complicate the estimate of pressures at the well screen.

Figure 13.8 shows a possible result of excess pressures applied to reagent injection. If the effective stress falls to zero or goes negative, the aquifer matrix is fractured and the overburden deforms, either becoming compacted or deforming at the ground surface[1]. The field crew can discern the point at which structure failure occurs, through their monitoring of flow rates and pressures, although at that point, irreversible damage to the formation is likely to have occurred. **Figure 13.9** shows the pressure-flow graph for a site where structural aquifer matrix failure occurred during a tracer injection test. The aquifer soils were compressible silts and clays, interbedded with thin layers of more permeable materials. As the tracer solution was injected, nearby wells immediately responded with static level increases and the nearest well actually became flowing artesian. The point of structure failure can be seen on the graph when the flow rate increases without added

---

[1] Surface deformation is routinely measured to track the propagation of hydrofracturing processes.

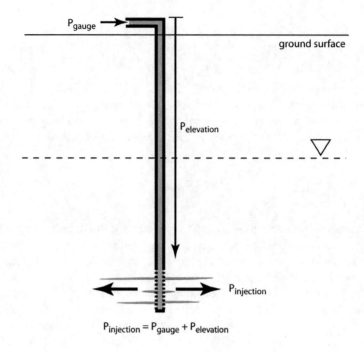

$$P_{injection} = P_{gauge} + P_{elevation}$$

FIGURE 13.7 Estimation of injection pressure applied to the formation.

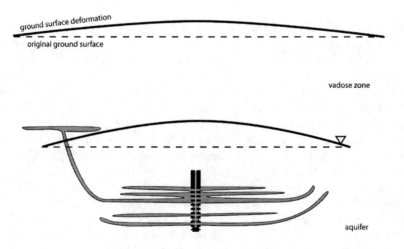

FIGURE 13.8 Injection of incompressible fluid into an aquifer, accompanied by structural failure of the aquifer matrix. Structural failure occurs when vertical fluid displacement is too slow to accommodate the fluid injection rate. Deformation can often be quantified at the ground surface with sensitive tilt meters.

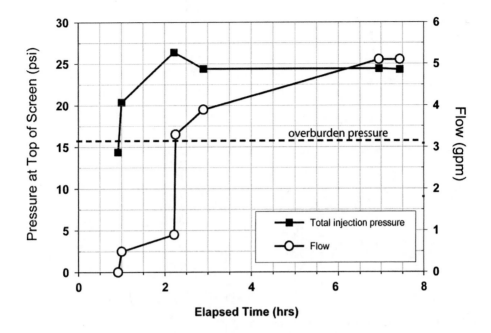

FIGURE 13.9 Well head pressures and volumetric flow associated with an over-pressurization failure of an aquifer matrix at an injection well.

pressure (in fact, the pressure dropped slightly in this case, clearly marking the point of failure). The failure occurred after the applied pressure exceeded the overburden pressure (the effective stress fell below 0). Despite water elevation fluctuations, no tracer was observed in the observation wells, the nearest of which was located 1.5 m from the injection well, screened at the same depth. The injected fluid caused structural failure of the porous matrix, compressing the silts and clays, forcing water out of the matrix. **Figure 13.10** shows a conceptual model of the site structure that can account for the observations.

IMPACTS OF DRILLING METHODS AND LINE LOSSES

The ability to monitor pressures applied to a formation during fluid injection are confounded by sources of resistance to flow, other than from the formation. Friction losses in the feed lines, well screen losses, compaction of the formation during drilling and poor well development practices all contribute to gauge pressures observed at the ground surface, which may obscure the pressure being applied to the formation at any time.

Line loss — Frictional losses can be calculated for feed lines, using Equations (1.12) through (1.15). These losses can be subtracted from pressures at the screen; however, it is important to remember that if the flow drops to zero, line loss falls to zero and the full gauge pressure is being applied to the formation.

Well screen losses — The pressure drop through well screens should be minimized through the use of sufficient diameters and high-open-area screen materials. Wire-wound stainless

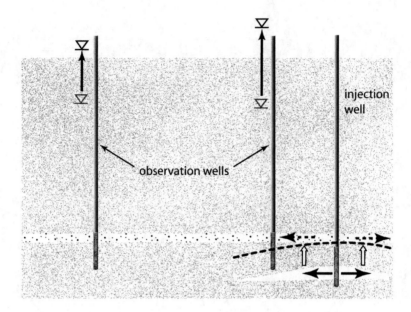

FIGURE 13.10 Conceptual model of compression and displacement effects of aquifer matrix failure, due to over-pressurization of a tracer injection well. Hydraulic responses were observed immediately in neighboring wells, but no tracer arrived.

steel screens are preferred. Refer to Chapter 10 for further details on screen and well pack designs.

Well development — Proper well development minimizes pressure losses between the well screen and the formation. Use of the wire-wound screen simplifies the well development process.

Formation compaction — Drilling practices can compact the formation, significantly increasing the pressure required to push fluid past the bore hole wall. **Figure 13.11** shows a diagrammatic representation of a direct-push point compacting a compressible soil formation. Auger drilling can also cause formation compaction. Care should be taken to avoid compaction especially in silty or clay soils, which are more compressible than well-sorted sands and gravels. It may be necessary to push core tools ahead of injection probes when direct-push rigs are used for fluid injection.

VERTICAL ACCOMMODATION RATE

The net groundwater displacement is vertical in most systems and the rate at which vertical water movement can be sustained is of critical importance. We have termed an aquifer's capacity to sustain fluid mounding as the *vertical fluid accommodation rate*. **Figure 13.12** shows a cross-sectional view of an aquifer receiving fluid injection, with strata representing a range of horizontal hydraulic conductivities. The figure provides a framework for discussing fluid accommodation.

The vertical fluid accommodation rate is *the maximum vertical flow, per unit area, that can be sustained without exceeding the effective stress at the injection point*. It is a function of the vertical hydraulic conductivity and the effective stress at the injection depth. Vertical flow is likely

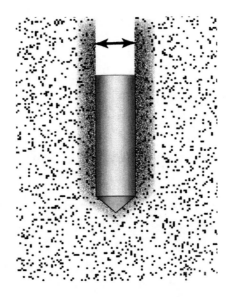

FIGURE 13.11 Compaction can occur when solid probes are pushed through compressible soils.

to encounter multiple strata and the vertical pathline flow resistance is the sum of the series of resistances generated by each stratum. Most importantly, as the mound rises, the downward force of the water column counteracts the force of the injection pumping and a point is reached where there is no further vertical flow at the injection pressure limit. The mound can only accommodate flow by lateral drainage.

For an injection system like that shown in **Figure 13.12**, we can calculate the vertical accommodation rate at the injection pressure estimated by Equation (13.6). We are interested in the pressure at the top of the screened interval, assuming that our injection target is the conductive stratum at the base of the formation (fully screened interval). First, we calculate the injection pressure limit, based on the maximum acceptable effective stress. From the figure, we obtain the key parameters:

$$h_{dry} = 660 \ cm \quad and \quad h_{sat} = 140 \ cm$$

We can assume that:

$$\rho_{dry} = 1.7 \ \frac{g}{cm^3} \quad , \quad \rho_{sat} = 2.0 \ \frac{g}{cm^3} \quad and \quad \rho_w = 1.0 \ \frac{g}{cm^3}$$

From Equation (13.6),

$$P_{max,injection} = 6.87 \times 10^5 \ \frac{dynes}{cm^2} \tag{13.7}$$

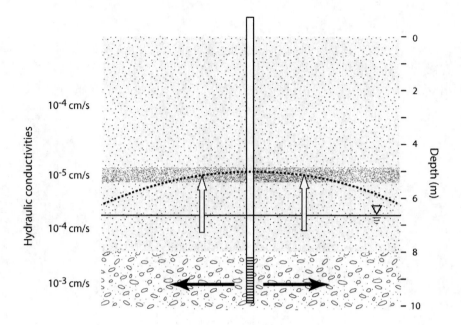

FIGURE 13.12 Vertical fluid accommodation. The rate at which an aquifer can accommodate injected fluid without structural failure is most often determined by the rate of vertical fluid accommodation.

That is the pressure limit at the well screen, to avoid formation failure. It converts to 82 kPa , 9.92 psi or 6.96 m H_2O. Note that the injection pressure limit allows the buildup of a water column that could reach the ground surface, given the assumptions that were made for the calculation.

There are a number of strategies to calculate the vertical accommodation rate. A simple approach is to calculate an effective hydraulic conductivity for the interval from the top of the screen to the ground surface, to use in subsequent calculations, as described earlier in Equation (3.35):

$$K_z = \frac{d}{\sum \frac{d_i}{K_i}} \qquad (13.8)$$

For the aquifer in **Figure 13.12**, we assumed that the vertical hydraulic conductivities were 10 percent of the corresponding horizontal value.

$$K_z = \frac{800}{\frac{270}{1\times10^{-5}} + \frac{50}{1\times10^{-6}} + \frac{480}{1\times10^{-5}}} = 6.4\times10^{-6}\ \frac{cm}{s} \qquad (13.9)$$

The vertical flux is a function of the available injection pressure and the gravitational downforce exerted by the water mound as it builds.

$$\frac{Q}{A} = K_{eff} \frac{P_{inj} - \rho_w \cdot g \cdot h}{h} \tag{13.10}$$

When the mound height, h, reaches the pressure limit, P_{inj}, the vertical flow rate decreases to 0. The vertical flux is undefined (infinite) at h = 0. **Figure 13.13** provides a family of curves solving for vertical flux density, as a function of mound height for several values of K_{eff}. In the example aquifer from **Figure 13.12**, the initial mound height is 140 cm and the starting vertical accommodation rate is decreased significantly from the value derived for a nil initial mound.

It is important to understand the role of the vertical accommodation rate — there is a limit to any aquifer's ability to accommodate injected fluid and we can calculate limits for that rate — but it is not feasible to directly calculate the rate of water mound propagation and lateral drainage for any realistic system. We can use analytical approaches such as those outlined in **Figure 13.13** to anticipate patterns of system behavior, but it is unlikely that we can match real site behavior with these calculations. Commercial groundwater modeling software also has limited capacity to represent injected fluid flow patterns, especially when flow rates induce structural changes in the matrix. This is a case for which it is necessary to conduct field trials to support injection designs, especially when injections are intended for extended periods.

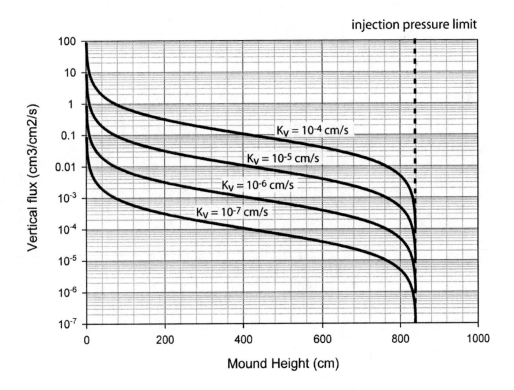

FIGURE 13.13 Vertical accommodation rates for the pressure limit described in the scenario developed for Figure 13.12 (824,000 dynes/cm², or 11.9 psi). Each curve represents the flux for an average hydraulic conductivity along the vertical flow path.

LARGE-SCALE FIELD TRIAL

The challenges associated with designing large-scale, extended injection operations can be illustrated with the results of a recent field trial in the Ogallala Aquifer in Texas. The aquifer in the study area lies in alluvial fan deposits, comprising interbedded sands, silts and gravels, with a saturated thickness of approximately 50 feet. The water table surface is approximately 100 feet below the ground surface in this area, with interbedding extending 50 feet into the vadose zone. Injection trials were conducted to support the design of a large-scale reactive-zone treatment system for the reductive dechlorination of chlorinated solvents. In the system, groundwater will be extracted, amended and re-injected into a network of 50 injection wells, each required to sustain 5 gpm minimum delivery for two years.

A 6-inch-diameter test well was constructed with a stainless steel wire-wound screen, spanning the saturated thickness of the aquifer (50 feet). The well was test pumped for 20 hours at groundwater extraction rates from 10 to 80 gpm. The drawdown response is shown in **Figure 13.14**, along with a curve fit using the Theis function in AQTESOLV®. Assuming a vertical-to-horizontal hydraulic conductivity ratio of 1:10, analysis of the response indicates the average hydraulic conductivity in this area is 2.5 x 10^{-2} cm/s (70 ft/day).

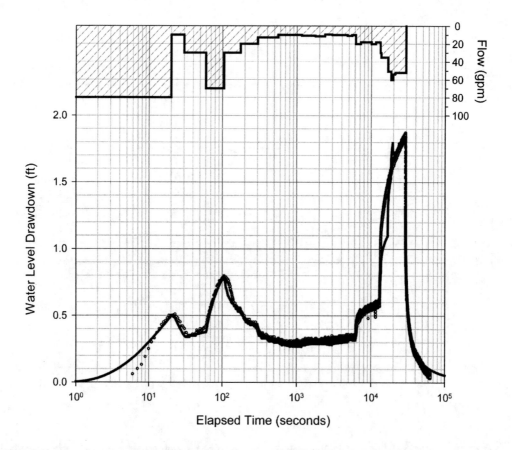

FIGURE 13.14 Results of groundwater extraction testing in the Ogalalla aquifer in Northwestern Texas. The well was 6-inch diameter with wire-wound stainless screen covering the uppermost 50 feet of the aquifer.

A 28-day injection test was then conducted at a 10 gpm continuous flow and the results are shown on **Figure 13.15**. The Theis equation was used to generate an expected water level rise at the 10 gpm pumping rate, based on the aquifer parameters gathered from the extraction test. The predicted injection head is indicated by the solid line on the figure. The initial injection head was approximately 7 ft, approximately 7-fold greater than the expected drawdown at a comparable extraction rate. Then, after two days of injection, the water rise began to diverge even more strongly from the extraction-based expectations. Even though the average hydraulic conductivity was very high in this formation, the injected fluid could not be accommodated at the observed extraction head (for the specified flow) at any time. The observed head became prohibitive after 20 days, due to the risk of increasing pressures to a point that could induce formation damage or spillage at the ground surface.

Clearly, the injection and extraction processes are not symmetrical at this site and we probably shouldn't expect injection-extraction symmetry, in any case: 1) Because the net fluid movement is vertical, it is the vertical component of hydraulic conductivity that controls resistance to injected

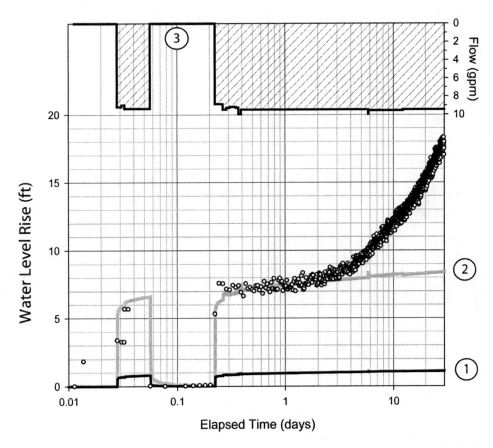

FIGURE 13.15 Comparison of predicted (solid line) and observed (dots) water elevations at an injection well constructed in the Ogalalla Aquifer in Texas. The Theis-predicted heads are shown by the solid line (1), that was based on aquifer parameters obtained from an extraction test (Figure 13.14). The early-time injection behaved as if the hydraulic conductivity was 7.4-fold lower, as indicated by the Theis-fitted gray line (2). After three days of injection, the water level began to deviate significantly from the early rate of rise, indicating that the water mound had encountered a discontinuity in the formation. Flow rates are indicated by the shaded area in the upper panel of the graph. Note that the injection pump was inadvertently shut down during the first day (3).

fluid, rather than horizontal conductivity — that, alone, could generate a 10-fold increase in this case, based on site-specific estimates. 2) The injected fluid is forcing invasion of partially-saturated aquifer matrix. As noted in the Brooks-Corey relationship, the hydraulic conductivity of a porous matrix is reduced dramatically under partial saturation conditions (refer to **Figure 4.15**). These two factors, alone, can account for the initial 7-fold difference between observed extraction and injection heads at 10 gpm flows.

When the injection flow was halted, the observed head quickly dropped to the early time head (approximately 7 feet), indicating that the pressures greater than the 7 feet were generated by pumping radially into a confined formation. **Figure 13.16** provides a cross-sectional interpretation of the injection test results. The static condition is indicated in the figure as "t_0" and snapshots of the mounding process are presented at even time steps from t_1 through t_4. The vertical accommodation rate is satisfactory during the first two days of pumping (**Figure 13.15**) as indicated by the water mound development at t_1 and t_2 in **Figure 13.16**. Then, the rising water appears to have contacted a less permeable stratum overlying the developing water mound, forcing the fluid laterally through the formation. Between t_2 and t_3, there was a transition from a net vertical flow to a net horizontal (radial) flow and the resistance to flow increased as the radial path length increased.

The project team adapted by running the injection process intermittently and was able to inject sufficient reagent to achieve project objectives. However, it is very important to notice that the limitations of injecting water into a formation cannot be assessed by a short-duration delivery of fluid. Fluid must be injected at a rate and for a length of time consistent with operational planning, to assess aquifer fluid accommodation behavior and to unmask heterogeneities that could restrict the fluid injection capacities. This is especially important with systems involving large-volume, long-duration injections.

DISPLACEMENT CONCEPTS

The large volumes associated with reactive zone injections raise concern among stakeholders that contaminants could be displaced by the injected reagent, possibly spreading contamination outside the treatment zone. Displacement certainly occurs and the concerns are legitimate, although the magnitude of displacement should be small in most sites. If we assume that an aquifer is radially symmetrical surrounding an injection location, then **Figure 13.17** provides a synopsis of the displacement that occurs with an injection episode. The injected fluid pushes its way out to the injection radius, r_{inj}. That displaces the fluid originally occupying a portion of the aquifer thickness (the mobile porosity, θ_m), from radius 0 to r_{inj} outward, to then occupy a band from r_{inj} to r_2, as indicated on the figure. The area covered by the injection and the area it displaces are equal, hence:

$$\pi \cdot r_2^2 = 2 \cdot A_{inj} \tag{13.11}$$

Expanding the injection area term, A_{inj},

$$\pi \cdot r_2^2 = 2 \cdot \pi \cdot r_{inj}^2 \tag{13.12}$$

Solving for r_2,

$$r_2 = \sqrt{2} \cdot r_{inj} \tag{13.13}$$

Extending the analysis,

$$r_3 = \sqrt{3} \cdot r_{inj} \tag{13.14}$$

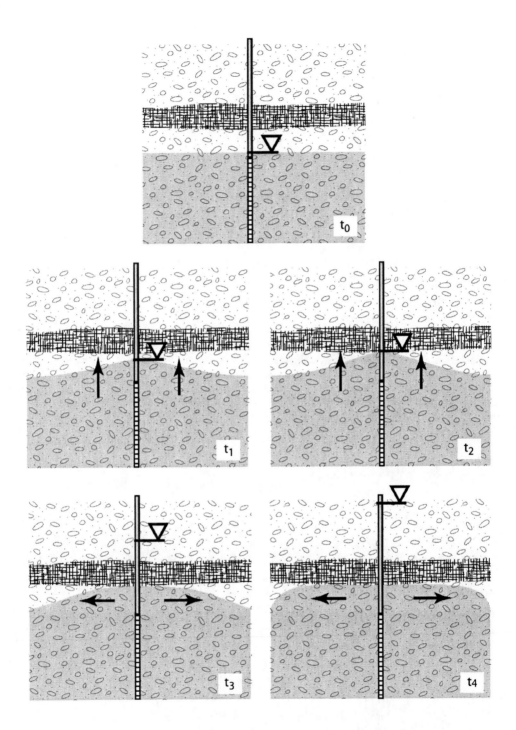

FIGURE 13.16 Cross-sectional interpretation of changes in head required to drive a 10 gpm injection flow in the Ogallalla formation (Figure 13.15). Prior to injection (t_0), there was an unsaturated, high-permeability stratum. During the first day of pumping (t_1 and t_2), vertical fluid accommodation was sufficient to accept the injected fluid with a net upward movement of water. At t_2, the water mound contacted a confining layer and subsequent injection flows became net horizontal (radial), as indicated for t_3 and t_4. The resistance to radial flow is large and the head required to drive the 10 gpm injection flow began an exponential increase.

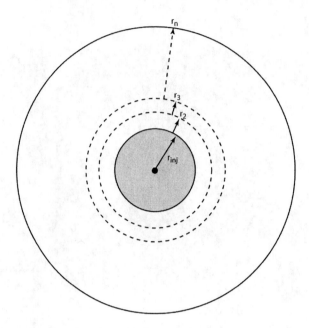

FIGURE 13.17 Conceptual diagram of successive radial displacement volumes. Injection of a fluid to the radius r_{inj} causes displacement of groundwater originally occupying the zone of injection, pushing groundwater out to radius r_2. That displacement, in turn, causes displacement to r_3 and beyond, as suggested by r_n. The radial distances are successively smaller and the displacement that could occur in a hypothetical frictionless, massless aquifer is calculated by Equations (13.16) and (13.18). Due to inertia and frictional resistance to flow, the primary net displacement is actually vertical.

And generalizing,

$$r_n = \sqrt{n} \cdot r_{inj} \tag{13.15}$$

If water is massless (no inertia to overcome in the injection) and travels with no friction in the aquifer (neither of which are realistic assumptions) the displacement process could proceed, *ad infinitum*. To project what the displacement might be at a specified distance, x, from the injection point, we can solve for n as follows:

$$n_{x,r_{inj}} = \left(\frac{x}{r_{inj}} \right)^2 \tag{13.16}$$

The theoretical maximum displacement for any location is the difference between two successive radii at that radial distance. For example,

$$\Delta_{r_2 \to r_3} = \sqrt{3} \cdot r_{inj} - \sqrt{2} \cdot r_{inj} \tag{13.17}$$

and simplifying,

$$\Delta_{r_2 \to r_3} = r_{inj} \cdot (\sqrt{3} - \sqrt{2}) \qquad (13.18)$$

Generalizing for the nth band of displacement,

$$\Delta_{r_n \to r_{n+1}} = r_{inj} \cdot (\sqrt{n+1} - \sqrt{n}) \qquad (13.19)$$

Sample calculation –

$$injected\ radius = 20\ ft$$

$$solve\ for\ displacement\ at\ x = 50\ ft$$

$$n_{x, r_{inj}} = \left(\frac{x}{r_{inj}} \right)^2$$

$$n_{50, 20} = \left(\frac{50}{20} \right)^2 = 6.25$$

$$\Delta_x = r_{inj} \left(\sqrt{n_{x, r_{inj}} + 1} - \sqrt{n_{x, r_{inj}}} \right)$$

$$\Delta_x = 20 \left(\sqrt{7.25} - \sqrt{6.25} \right) = 3.85\ ft$$

If the fluid injection reaches a radius of 20 feet, it is possible to generate a 3.85-foot radial displacement, 50 feet from the injection location. If the injection is not radially symmetrical, the displacement pattern would also be asymmetrical – some areas receiving more and some receiving lesser displacement. **Figure 13.18** shows the theoretical displacement that could be expected from injections to 10, 30 and 50-foot radii. It is important to remember that pressure effects travel farther than actual displacement (refer to Chapter 11 and Case Study 1 in Chapter 12).

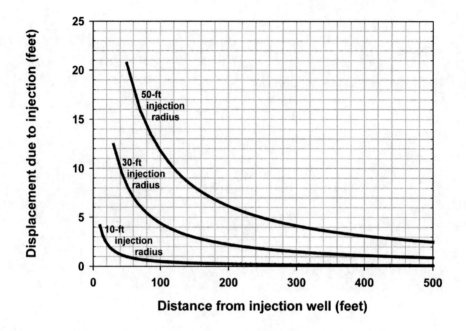

FIGURE 13.18 Theoretical lateral displacement of groundwater that would occur if there were no inertia of the water mass to overcome and no resistance to flow in the porous matrix.

13.2 PRESSURE LIMITS AND FORMATION FAILURE

As discussed above, injection wells are less efficient than extraction wells because of the work required to displace groundwater from the surrounding formation. Under pumping conditions, the cone of depression assists in the groundwater recovery system through gravitational potential, whereas the cone of pressure associated with injection is counterproductive to the injection process. The water table "mound" that is created during the injection process results in a negative feedback loop that counteracts the injection process, requiring a greater magnitude of positive head under injection than the negative head (drawdown) that accompanies groundwater extraction under comparable extraction flow rates. The result is that injection wells typically yield 25 to 33-percent of the flow capacity that an extraction well would in the same aquifer formation. More importantly, our experience shows that the hydraulic equations used for pumping do not accurately characterize the system behavior under injection conditions. This is due to the fact that the common hydraulic analyses do not account for compressibility of the aquifer or inertia forces that are significantly more important under conditions of injection. We recommend injection well capacity testing to evaluate the flow versus head requirements and develop a realistic design expectation.

Because injection wells require a higher injection head/pressure to achieve the same flow rate as extraction wells, injection wells require special design and construction techniques to accommodate the operating pressures necessary to avoid well-seal failures (this topic was discussed in Chapter 10). In addition, relatively minor gauge injection pressures (less than 5 psi [34.5 kPa]) can result in hydraulic fracturing and formation damage. The water table mound or pore-pressure increase that is created during injection reduces the effective stress in the formation, potentially leading to hydraulic fracturing and fluidization of the formation, if the pressure increase is beyond the formation's strength, or when effective stress is reduced to zero. Hydraulic fracturing can lead to development of permanent flow conduits that disrupt the natural flow paths in the formation or formation damage that reduces the injection capacity of the well and limits the ability to effectively distribute reagents throughout the contaminated profile within the aquifer. The purpose of this section is to provide practitioners with injection well design strategies and performance testing procedures to ensure proper injection well performance.

It is important to note that injection system failure can occur before fracturing or fluidization is induced in the formation. Under shallow water table conditions, fluid injection can also result in short-circuiting through improperly designed monitoring wells or improperly abandoned borings. Modest gauge pressure at the injection well can provide adequate hydraulic head to build a water-table mound above the ground surface. Borings that are backfilled with native materials or plugged using poor-quality bentonite or cement-bentonite mixtures can act as vertical conduits for short-circuiting because of the poor-quality of seal with the formation and relatively high-permeability of the materials. This can lead to unpredictable distribution patterns in the subsurface. In cases where the aquifer system consists of multiple water-bearing units, the reagent can short-circuit to shallower units that are unaffected and do not require treatment. At many sites, short-circuiting is expressed at the ground surface, creating an obvious problem; however, when the vadose zone is stratified, such as the case where permeable fill material overlies the native aquifer materials, it is possible to have short-circuiting occur below grade such that the issue goes unnoticed. In this case, the system failure might not be detected for a considerable period of operation.

We recommend developing injection design strategies by considering the potential failure mechanisms introduced above. The initial design should consider the following failure modes:

- *Short-circuiting* — Short-circuiting refers to travel along man-made pathways in the subsurface. Examples include utility trenches, improperly abandoned geotechnical borings (a very common occurrence), nearby monitoring well borings and the injection well boring, itself. Gravity injection (e.g., zero gauge pressure) is the simplest design option to prevent short-circuiting fluid from reaching the ground surface, but it does not completely prevent injected fluid travel along these pathways. Operating pressures in the injection well should not exceed the equivalent water-column height, if potential short-circuiting is an issue. In cases where subsurface short-circuiting is likely, the designer should evaluate the equivalent water-column pressure based on the height to the vertical barrier, or consider preemptively abandoning improperly design wells and borings to mitigate this issue. Poorly abandoned borings and improperly grouted wells can generate partial losses of injected fluid from the intended injection zone, at virtually any workable injection pressure.

- *Hydraulic fracturing* — Uncontrolled hydraulic fracturing can occur if the effective stress is reduced to the point where the shear strength of the aquifer matrix is exceeded. Failure in the initial stage is controlled by the ratio of the major and minor principal effective stresses and the soil's internal friction angle. As the magnitude of the applied pressure is increased beyond vertical effective stresses in the formation, it is possible to initiate and propagate horizontal fractures. In homogeneous systems, fracturing is initiated through shearing along planes of weakness that are determined by the principal stresses in the system and the soil's resistance to shearing. In heterogeneous systems, flow is concentrated in fine-scale, high-permeability depositional features, commonly inducing fracturing along the bedding structures within the aquifer. While hydraulic fracturing methods can be used to enhance distribution of reagents under certain circumstances, our treatment here is designed to specifically avoid this process.

- *Fluidization or liquefaction* — Liquefaction of unconsolidated formations occurs when the vertical and horizontal effective stresses in the aquifer are reduced to zero through the application of injection pressures that exceed the overburden pressure in the formation. When the effective stress is reduced to zero, the grains comprising the aquifer matrix lose all contact and the formation strength is reduced to zero. This results in a quick sand condition, where the formation is fluidized and the structure of the formation can be "homogenized." This approach is used as an advanced design approach to initiate fractures, particularly with directional nozzles that concentrate flow and enable creation of fractures at prescribed depth intervals; however, when fluidization occurs under uncontrolled conditions, the results are unreliable and counterproductive to *in situ* injection design.

Design considerations to avoid formation damage and hydraulic fracturing in this book were developed using conventional soil strength analysis. We recommend the following methods to estimate safe injection pressures when developing remedial designs and performance objectives. However, it is important to note that the methods were developed for static groundwater systems that do not consider fluid flow, particularly the potentially concentrated flow, associated with injection systems. For this reason, we always recommend performing injection capacity tests (within safe pressure ranges) before full-scale implementation to ensure that the design calculations match the actual site conditions.

SOIL STRENGTH

The strength of unconsolidated, cohesionless soils comprising aquifer matrices is a function of the intergranular stress (effective stress) and the soils' internal resistance to shearing, or **friction angle**, ϕ. Recall from Chapter 2 that the effective stress is the difference between the total stress and pore-pressure for aquifer systems. Soil strength can be described using Mohr-Coulomb theory which states that failure will occur along the plane for which the ratio of the shear stress to the normal stress reaches a critical limiting value as shown in Equation (13.20) after Perloff (1976):

$$\tau_{ff} = \sigma'_{ff} \tan\phi + C \tag{13.20}$$

where τ_{ff} is the shear stress on the failure plane, σ'_{ff} is the normal effective stress to the failure plane, ϕ is the soil friction angle and C is the cohesive strength of the formation. For cohesionless sand and gravel soils that commonly comprise aquifers, the cohesive strength is minor to negligible. For our purposes, we assume that the cohesive strength is zero unless it is measured in the geotechnical laboratory.

Friction angles for cohesionless sands and gravels typically range between 28 and 45 degrees. Generally, soil friction angles increase with grain size; however, the friction angle also depends on the degree of compaction, or relative density, the sorting of the aquifer matrix and the angularity of the grains. Friction angles are typically measured using direct shear tests or tri-axial compression tests in the geotechnical laboratory, but reasonable field estimates can be obtained using the standard penetration test or CPT testing. **Table 13.1** presents friction angles for the broad classes of soils comprising aquifer matrices. **Table 13.2** illustrates the influence of sorting and grain shape on friction angles for a USCS-classified medium-grained sand.

Mohr-Coulomb theory provides useful relationships to define soil strength under limiting (incipient failure) conditions. For simplicity, we present the analysis for the two dimensional case for normally consolidated soils, where σ'_v is the **vertical effective stress** and σ'_h is the **horizontal effective stress**:

$$\sigma'_{ff} = \sigma'_{h_f} \left(1 + \sin\phi\right) \tag{13.21}$$

$$\sigma'_{ff} = \sigma'_{v_f} \left(1 - \sin\phi\right) \tag{13.22}$$

TABLE 13.1
Approximate values of internal friction angle, ϕ, for cohesionless soils as a function of grain size and compaction after Perloff (1976) and California DOT (1990).

USCS Soil Classification	Compaction or Density	Friction Angle, ϕ (degrees)
Gravel, Gravel-Sand Mixture, Coarse Sand	Dense	41-45
	Moderate	37-41
	Loose	29-34
Medium Sand	Dense	35-38
	Moderate	32-35
	Loose	27-30
Fine Sand	Dense	30-32
	Moderate	26-28
	Loose	25-27
Silty-Fine Sand, Sandy Silt	Dense	28-30
	Moderate	26-28
	Loose	25-27
Silt	Dense	26-27
	Moderate	24-25
	Loose	23-24

TABLE 13.2
Influence of grain shape and sorting on internal friction angle for USCS classified medium sand. Adapted from Perloff (1976).

	Friction Angle, ϕ (degrees)	
Compaction or Density	**Rounded Grains, Well-Sorted**	**Angular Grains, Poorly-Sorted**
Loose-very loose	28-30	32-34
Moderately dense	32-34	36-40
Very dense	35-38	44-46

The definitions in Equations (13.21) and (13.22) assume that the vertical effective stress is the major principal stress, σ'_1, and the horizontal effective stress is the minor principal stress, σ'_2.

The shear strength of a normally consolidated soil, τ_{ff}, is a function of the vertical and horizontal effective stresses and the soil's friction angle as shown in Equation (13.23):

$$\tau_{ff} = \left(\frac{\sigma'_{v_f} - \sigma'_{h_f}}{2}\right) \cdot \cos\phi \tag{13.23}$$

From Equations (13.21) through (13.23), Mohr circle failure analysis indicates that two potential failure planes will develop, oriented +/- (45° + ϕ/2) to the principal plane.

$$\frac{\sigma'_{v_f}}{\sigma'_{h_f}} = \frac{(1+\sin\phi)}{(1-\sin\phi)} = \tan^2\left(45° + \frac{\phi}{2}\right) \tag{13.24}$$

$$\frac{\sigma'_{h_f}}{\sigma'_{v_f}} = \frac{(1-\sin\phi)}{(1+\sin\phi)} = \tan^2\left(45° - \frac{\phi}{2}\right) \tag{13.25}$$

In normally consolidated material soils, the vertical effective stress is typically much greater than the horizontal effective stress. Equations (13.21) through (13.25) are written for normally consolidated soils, where the major principal stress is the vertical effective stress and the minor principal stress is the horizontal effective stress.

For earth systems at equilibrium, the ratio of the vertical effective stress to horizontal stresses can be evaluated using the **coefficient of earth pressure at rest, K_o,** after Perloff (1976):

$$K_o = \frac{\sigma'_h}{\sigma'_v} = 1 - \sin\phi \tag{13.26}$$

Equation (13.26) is an empirical relationship that is widely used in geotechnical engineering to estimate horizontal stresses in **normally consolidated soils** — that is soils where the current vertical effective stress condition represents the maximum stress in the soil's history. In simple terms, soils are considered normally consolidated when the vertical effective stress is due to the weight of the current overburden. Substituting the range of friction angles for cohesionless sands and gravel soils (ϕ between 28 and 45 degrees) into Equation (13.26) yields values for K_o between 0.29 and 0.53, which implies that the horizontal effective stresses are approximately 30 to 50-percent of the vertical stresses. During injection, normally consolidated soils typically fail along vertical to sub-vertical planes because the horizontal effective stress is the limiting condition — as the horizontal stress is progressively reduced the soil loses its strength along the horizontal direction allowing vertical fractures and failure planes to form.

In **over-consolidated cohesionless soils**, where historical maximum vertical stresses exceed present conditions, K_o can range between 0.5 and more than 10 (Perloff, 1976). Soils with values of K_o between 0.5 and 3 are termed **nominally over-consolidated** because the horizontal effective stress is higher than Equation (13.26) would predict; however, the behavior is consistent with normally consolidated soils.

The **over-consolidation ratio, OCR**, is defined as the ratio of maximum historical vertical effective stress to present vertical effective stress, due to the weight of the soils:

$$OCR = \frac{\sigma'_{v_{maximum}}}{\sigma'_{v_{present}}} \qquad (13.27)$$

It is common that natural soils may exhibit some degree of over-consolidation due to past loading associated with glaciation in northern latitudes, significant erosion of overlying sediments in the past or compaction associated with construction techniques in fill materials. Although the historical loads have been removed, the horizontal effective stress dissipates more slowly than the vertical effective stress in soils, resulting in a state where the horizontal effective stress exceeds the vertical effective stress. The implication is that the horizontal effective stress is the major principal stress and the vertical effective stress is the minor principal stress: $\sigma'_h = \sigma'_1$ and $\sigma'_v = \sigma'_2$.

For **over-consolidated** soils, Equations (13.12) through (13.25) for Mohr-Coulomb failure analysis are expressed as follows:

$$\sigma'_{ff} = \sigma'_{v_f}\left(1+\sin\phi\right) \qquad (13.28)$$

$$\sigma'_{ff} = \sigma'_{h_f}\left(1-\sin\phi\right) \qquad (13.29)$$

$$\tau_{ff} = \left(\frac{\sigma'_{h_f}-\sigma'_{v_f}}{2}\right)\cdot\cos\phi \qquad (13.30)$$

$$\frac{\sigma'_{h_f}}{\sigma'_{v_f}} = \frac{\left(1+\sin\phi\right)}{\left(1-\sin\phi\right)} = \tan^2\left(45° + \frac{\phi}{2}\right) \qquad (13.31)$$

$$\frac{\sigma'_{v_f}}{\sigma'_{h_f}} = \frac{\left(1-\sin\phi\right)}{\left(1+\sin\phi\right)} = \tan^2\left(45° - \frac{\phi}{2}\right) \qquad (13.32)$$

In over-consolidated soils, failure commonly occurs on the horizontal to sub-horizontal plane because the vertical effective stress normal to the plane is the limiting condition. The earth pressure at rest for over-consolidated soils, $K_{o\,OC}$ is defined using the over-consolidation ratio and Equation (13.26) (Perloff, 1976):

$$K_{o\,OC} = \frac{\sigma'_h}{\sigma'_v} = (1-\sin\phi)\cdot(OCR)^{\sin\phi} \qquad (13.33)$$

The over-consolidation ratio is determined using triaxial compression tests in the lab, or using a bore-hole dilatometer test to measure the horizontal effective stress. The over-consolidation ratio can also be estimated using CPT methods to compare the measured vertical effective stress to the calculated value based on soil density. The reader is encouraged to consult a geotechnical engineer to evaluate appropriate testing methods to determine the state of over-consolidation at each site.

Unless geotechnical testing data are available, we advise that design calculations should be completed assuming normally consolidated conditions using Equations (13.21) through (13.26).

MATRIX FAILURE AND HYDRAULIC FRACTURING

When injection is used to distribute reagents, the applied injection pressure increases the pore pressure in the formation and reduces the effective stress. Because the horizontal effective stress is the least principal stress in normally consolidated formations, it is the limiting constraint with respect to formation failure and hydraulic fracturing. Matrix failure is induced when the horizontal stress is reduced so that the ratio of the vertical to horizontal effective stress is increased to a critical limiting value that is determined by the soil's internal friction angle. Hydraulic fracturing is initiated in porous media when the injection pressure reduces the horizontal effective stress to zero.

Aquifer failure can be evaluated using the coefficient of earth pressure at rest, K_o, to estimate the horizontal effective stress condition and Mohr-Coulomb failure theory to estimate the maximum pore pressure increase that can be tolerated based on the soil's internal fraction angle. When injection pressure is applied to an aquifer, the vertical and horizontal effective stresses are reduced proportional to the increased pore pressure, $\Delta\sigma_u$:

$$\sigma'_v = \sigma'_{v_i} - \Delta\sigma_u \tag{13.34}$$

$$\sigma'_h = \sigma'_{hi} - \Delta\sigma_u \tag{13.35}$$

where the subscript i on σ'_v and σ'_h in Equations (13.34) and (13.35) denotes the initial vertical and horizontal effective stress before increasing the pore pressure through injection. Combining Equation (13.26) with Equation (13.35) yields the horizontal effective stress in terms of the initial vertical effective stress for normally consolidated soils:

$$\sigma'_h = \sigma'_{h_i} - \Delta\sigma_u = \left(\sigma'_{v_i} \cdot (1-\sin\phi)\right) - \Delta\sigma_u \tag{13.36}$$

Consider a normally consolidated, angular, moderately-dense, medium-grained sand aquifer with an initial vertical effective stress of 4.0 x10⁶ dynes/cm² and a friction angle of 38 degrees. What is the initial vertical and horizontal effective stress in the system, assuming normally consolidated conditions in the formation? What injection pressure can be applied without inducing formation failure? What pressure would induce hydraulic fracturing? The solution is derived using the formulas provided and the Mohr's circle to graphically illustrate the process, as shown on **Figure 13.19**.

Substituting the initial vertical effective stress of 4.0 x 10⁶ dynes/cm² and friction angle of 38 degrees into Equation (13.36) yields:

$$\sigma'_{hi} = 4.0 \times 10^6 \ \frac{dynes}{cm^2} \cdot (1 - \sin(38)) = 1.54 \times 10^6 \ \frac{dynes}{cm^2}$$

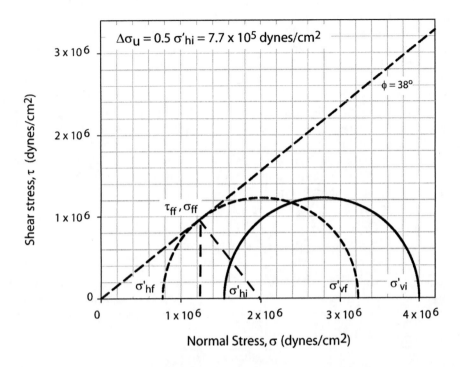

FIGURE 13.19 Mohr's circle graphical analysis to determine the allowable pore pressure increase prior to structural failure. The solid line circle is the initial stress condition, which is defined by the initial vertical and horizontal effective stresses, $\sigma'_{vi} = 4 \times 10^6$ dynes/cm² and $\sigma'_{hi} = 1.54 \times 10^6$ dynes/cm². The soil is "normally" consolidated, so the vertical stress is the major principal stress. Under injection pressure, the pore pressure increase reduces the vertical and horizontal effective stresses proportionately, shifting the Mohr's circle leftward until it is tangent to the failure envelope defined by ϕ. The long-dashed lines highlight the similar triangles that are used to estimate the shear stress at failure, τ_{ff} and the normal stress on the failure plane, σ_{ff}.

The initial stress condition can be plotted on the Mohr's circle, recognizing that the radius of the circle

$$r = \frac{(\sigma'_{vi} - \sigma'_{hi})}{2}$$

$$r = 1.23 \times 10^6 \ \frac{dynes}{cm^2}$$

and the mid-point is

$$MP = \sigma'_h + r$$

$$MP = 2.77 \times 10^6 \ \frac{dynes}{cm^2}$$

The failure envelope is defined by the internal friction angle, ϕ=38 degrees.

Equations (13.34) and (13.35) indicate that increasing the pore pressure in the system will shift the Mohr's circle to the left by an amount equal to $\Delta\sigma_u$. To solve for the change in pore pressure at failure, $\Delta\sigma_{uf}$, we need to determine the change in pore pressure that shifts the Mohr's circle so that it is tangent to the failure envelope. Recognizing that the radius of the Mohr's circle is constant with the change in pore pressure enables use of geometry to solve for the change in the pore pressure for the failure condition. Note also that the center of the Mohr's circle shifts left by $\Delta\sigma_u$. Therefore, we can solve for the change in pore pressure by solving for the mid-point at failure, recognizing that there are two similar triangles in the construction. The sum of the normal stress components from these similar triangles yields the mid-point of the failure condition. The first component, the normal stress at failure, σ_{ff}, is related to the shear stress at failure, τ_{ff}, using Equation (13.23) and recognizing that the radius of the Mohr's circle is invariant:

$$\sigma'_{ff} = \frac{\tau_{ff}}{\tan\phi} = \frac{r\cos\phi}{\tan\phi} \tag{13.37}$$

The second normal stress component is the contribution from the similar triangle formed by the radius of the failure circle on the normal stress axis and τ_{ff} with the complementary angle of ϕ, $(90-\phi)$. The normal stress contribution for this segment is then given by:

$$\sigma' = \frac{\tau_{ff}}{\tan(90-\phi)} = \frac{r\cos\phi}{\tan(90-\phi)} \tag{13.38}$$

Combining Equations (13.37) and (13.38) yields the equation for the **failure circle mid-point**, **MP$_{ff}$**:

$$MP_{ff} = \frac{r\cos\phi}{\tan\phi} + \frac{r\cos\phi}{\tan(90-\phi)} \tag{13.39}$$

Finally, we can solve for the pore pressure that induces failure, $\Delta\sigma_{uf}$, by subtracting the mid-point at failure from the initial Mohr's circle midpoint, **MP$_i$**:

$$\Delta\sigma_{uf} = MP_i - MP_{ff} \tag{13.40}$$

$$\Delta\sigma_{uf} = \left[\left(\sigma'_{vi} - \sigma'_{hi}\right) + \sigma'_{hi}\right] - \left[\frac{r\cos\phi}{\tan\phi} + \frac{r\cos\phi}{\tan(90-\phi)}\right] \tag{13.41}$$

Substituting r = 1.23×10^6 dynes/cm², MPi=2.77×10^6 dynes/cm², and ϕ=38 degrees into Equation (13.40) yields:

$$\Delta\sigma_{uf} = 2.77x10^6 - \left[\frac{1.23\times10^6 \cdot \cos(38)}{\tan(38)} + \frac{1.23\times10^6 \cdot \cos(38)}{\tan(52)}\right] \frac{dynes}{cm^2}$$

$$\Delta\sigma_{uf} = 7.70x10^5 \quad \frac{dynes}{cm^2}$$

With $\Delta\sigma_{uf}$ from the calculation above, we can find the vertical and horizontal effective stresses at failure by substituting the initial stresses into Equations (13.34) and (13.35):

$$\sigma'_{vf} = \sigma'_{vi} - \Delta\sigma_u = 4.0\times10^6 - 7.70\times10^5 \quad \frac{dynes}{cm^2}$$

$$\sigma'_{vf} = 3.23\times10^6 \quad \frac{dynes}{cm^2}$$

$$\sigma'_{hf} = \sigma'_{hi} - \Delta\sigma_u = 1.54\times10^6 - 7.70\times10^5 \quad \frac{dynes}{cm^2}$$

$$\sigma'_{hf} = 7.70\times10^5 \quad \frac{dynes}{cm^2}$$

The shear stress at failure, τ_{ff}, and normal stress at failure, σ_{ff}, were embedded in the prior calculations, but can be calculated explicitly using Equations (13.23) through (13.25):

$$\tau_{ff} = \left(\frac{\sigma'_{vf} - \sigma'_{hf}}{2}\right)\cos\phi$$

$$\tau_{ff} = \left(\frac{3.23\times10^6 - 7.7x10^5}{2}\right)\cos(38) \quad \frac{dynes}{cm^2}$$

$$\tau_{ff} = 9.69\times10^5 \quad \frac{dynes}{cm^2}$$

$$\sigma'_{ff} = \sigma'_{hf}(1+\sin\phi) = 7.70\times10^5(1+\sin(38)) \quad \frac{dynes}{cm^2}$$

$$\sigma'_{ff} = \sigma'_{vf}(1-\sin\phi) = 3.23\times10^6(1-\sin(38)) \quad \frac{dynes}{cm^2}$$

$$\sigma'_{ff} = 1.24 \times 10^6 \ \frac{dynes}{cm^2}$$

Substituting the values for τ_{ff} and σ_{ff} into Equation (3.20) yields the friction angle of 38 degrees:

$$\tan^{-1}\phi = \frac{\tau_{ff}}{\sigma'_{ff}} = \frac{9.69 \times 10^5 \ \frac{dynes}{cm^2}}{1.24 \times 10^6 \ \frac{dynes}{cm^2}} = 38^o$$

The failure planes will be oriented +/- (45° + ϕ/2) to the major principal plane, +/- 24 degrees to the vertical or +/- 64 degrees from the horizontal. In normally consolidated soils, the major principal plane is the vertical stress and failure planes tend toward the vertical, rather than horizontal. For over-consolidated soils with K_o greater than one, fracture planes tend toward the horizontal, since the major principal plane is horizontal.

It is not a coincidence that the increase in pore pressure at failure equals ½ of the initial horizontal effective stress in normally consolidated soils:

$$\Delta\sigma_{uf} = \sigma'_{hi}/2 \tag{13.42}$$

This relationship stems from the ratio of vertical and horizontal stresses according to the earth pressure at rest coefficient, and holds for all internal friction angles in **normally consolidated soils**. Therefore, the limiting injection pressure that induces failure can be expressed in terms of the initial vertical effective stress and soil fraction angle as shown in Equation (13.43) below:

$$\Delta\sigma_{uf} = \sigma'_{vi}(1-\sin\phi)/2 \tag{13.43}$$

For friction angles between 28 and 45 degrees, the quantity (1-sinϕ), or K_o, ranges between 0.53 and 0.29. Equation (13.43) implies that for equivalent vertical effective stress conditions, soils with the least friction angle will tolerate the highest increase in pore pressure prior to failure.

For **over-consolidated soils** with K_o greater than or equal to one, the vertical effective stress becomes the limiting condition. Failure is initiated when the change in the pore pressure reduces the vertical effective stress to the limiting value on the Mohr's circle. In this case, it is necessary to use Equations (13.39) and (13.40) to solve for limiting pore pressure at failure, since the simple relationship does not always hold.

Injecting at pressures above the limiting pore pressure can result in hydraulic fracture initiation in normally consolidated soils when the horizontal effective stress is reduced to zero. When the horizontal effective stress is reduced to zero in normally consolidated soils, there is no strength normal to the horizontal plane and vertical fractures develop. **Figure 13.20** shows the Mohr circle for the stress condition where the horizontal effective stresses are reduced to zero:

$$\Delta\sigma_u = \sigma'_{hi}$$

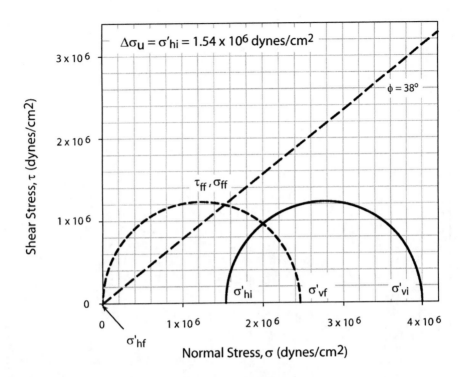

FIGURE 13.20 Mohr's circle analysis, showing the influence of injection pressure equal to the initial horizontal effective stress. The initial stress condition is shown using a solid line circle and the failure condition is shown with a dashed line. The final stress condition results in zero horizontal effective stress, which moves the Mohr's circle leftward by 1.54×10^6 dynes/cm^2. Under these stress conditions, the principal effective stress normal to the horizontal plane is zero, resulting in the potential to create vertical fractures in the formation.

For over-consolidated soils, the horizontal effective stress is the major principal plane. Reducing the vertical effective stress to zero eliminates the strength normal to the vertical plane and horizontal fractures develop. Hydraulic fracturing methods exploit these relationships by focusing flow using nozzles, which induce fluidization in discrete intervals of the formation and results in propagation of horizontal fractures.

UNCONFINED AQUIFERS

The total head, h_t, at the injection well is the combination of the head due to the water column in the injection well, h_{wc}, and the applied gauge-pressure or head measured at the well head, h_{wh}. Injecting in unconfined aquifers at pressures above the water column head can create short-circuiting through the ground surface, if potential conduits such as borings or improperly constructed wells exist to allow unimpeded vertical flow. Application of positive gauge-pressure heads leads to mounding that exceeds the intial "free-board" of the vadose zone (depth to water from ground surface), which can ultimately lead to excursions of reagents through the ground surface. Because most aquifers exhibit hydraulic conductivity anisotropy due to stratification, this mode of failure typically occurs only in shallow systems with depth to water less than a few meters; however, our experience indicates that this mode of failure can occur at depths greater than 20 meters (65 feet). Injection well testing is required to verify that this failure mode is not the limiting case.

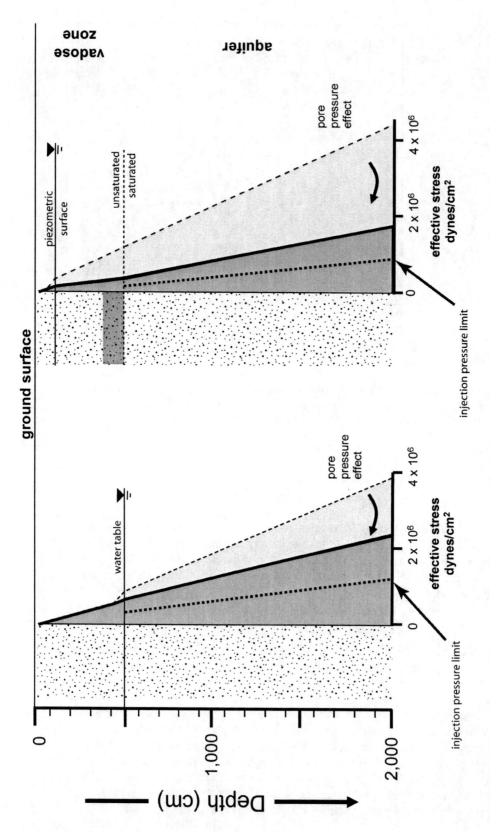

FIGURE 13.21 Comparison of effective stress in confined and unconfined conditions. The formation on the left is in water table condition (unconfined), as introduced earlier in Figure 2.7. On the right, a similar aquifer matrix is confined by a low-permeability stratum and the piezometric surface rises 400 cm above the top of the aquifer. The safe injection pressure in the confined aquifer is approximately two-thirds of the allowable value for the unconfined system.

The safe injection pressure for an unconfined aquifer can be determined using the relationships discussed above. We focus our analysis at the top of the injection screen because the incremental change in effective stress will be greatest at this location in the formation. The design concept is illustrated using the example for effective stress from Chapter 2 and summarized in **Figure 13.21**. For this case, assume that a 6-meter injection well (20-foot) is planned to be installed between 14 and 20 meters below grade. The current water table is approximately 5 meters below grade, leading to the initial total- and effective-stress distribution shown on the left side of **Figure 13.21**. Assume that the soil is normally consolidated with a friction angle of 30 degrees. The dry weight of the soil, γ_d, is 1,686 dynes/cm² and the saturated weight of the soil, γ_s, is 2,029 dynes/cm². The unit weight of water, γ_w, is 980 dynes/cm². From Equation (2.22), the total vertical stress at 14 meters depth is:

$$\sigma_v = \gamma_d h_d + \gamma_s h_s$$

$$\sigma_v = 1,686 \ \frac{dynes}{cm^3} \cdot 500 \ cm + 2,029 \ \frac{dynes}{cm^3} \cdot 900 \ cm$$

$$\sigma_v = 2.67 \times 10^6 \ \frac{dynes}{cm^2} \quad or, \quad 2.67 \times 10^3 \ kPa$$

The pore pressure at 14 meters depth, which is initially 9 meters below the water table, is given by Equation (2.21):

$$\sigma_w = \gamma_w h_w$$

$$\sigma_w = 980 \ \frac{dynes}{cm^3} \cdot 900 \ cm$$

$$\sigma_w = 8.82 \times 10^5 \ \frac{dynes}{cm^2} \quad or, \quad 8.82 \times 10^2 \ kPa$$

Substituting the results from above into Equation (2.23) yields the initial vertical effective stress condition:

$$\sigma_v' = \sigma_v - \sigma_w$$

$$\sigma_v' = 2.67 \times 10^6 - 8.82 \times 10^5 \ \frac{dynes}{cm^2}$$

$$\sigma'_v = 1.79 \times 10^6 \; \frac{dynes}{cm^2} \quad or, \quad 1.79 \times 10^3 \; kPa$$

The initial horizontal effective stress for the aquifer is determined using Equation (13.26):

$$\sigma'_{hi} = \sigma'_{vi} \left(1 - \sin(30)\right) = 1.79 \times 10^6 \; \frac{dynes}{cm^2} \cdot (0.5)$$

$$\sigma'_{hi} = 8.95 \times 10^5 \; \frac{dynes}{cm^2}$$

Applying the results in Equation (13.43) yields a maximum injection pressure of 4.48×10^5 dynes/cm^2. The Mohr circle for the example is presented in **Figure 13.22.**

For the site conditions, this implies that applying an injection pressure of (increasing the pore pressure by) 4.48×10^2 kPa could result in formation damage. Dividing the fracture pressure

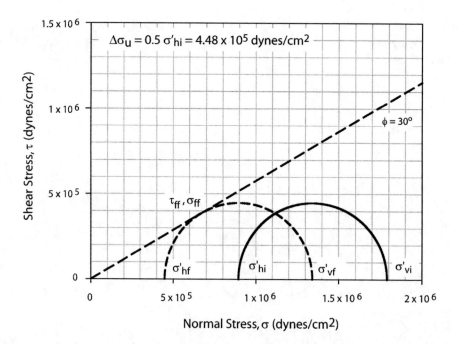

FIGURE 13.22 Mohr's circle for the unconfined injection scenario. The solid line shows the initial vertical and horizontal effective stresses, $\sigma'_{vi} = 1.79 \times 10^6$ dynes/cm^2 and $\sigma'_{hi} = 8.95 \times 10^5$ dynes/cm^2, based on a friction angle of 30 degrees. The Mohr's circle shifts left by $\Delta\sigma_u = 0.5 \times \sigma'_{hi}$,to the failure envelope, when the limiting pressure is applied. Because $K_0 = 0.5$ for $\phi = 30$ degrees, increasing the injection pressure to match the initial horizontal effective stress would shift the Mohr's circle further leftward, such that $\sigma'_h = 0$, causing vertical fracturing with limited additional pressure.

in dynes/cm² through the unit weight of water yields the equivalent water column pressure for formation failure of 457 cm, or approximately 4.6 meters. Since the depth to water is approximately 5 meters, this implies that a gauge-pressure would induce failure in the system.

Vertical fracturing would be induced when the initial horizontal effective stress was matched by the injection pressure, or a total injection pressure of 8.95 x 10² kPa (913 cm water column). As a result, injecting at a gauge-pressure head of 412 cm water column (4.04 x 10² kPa) would induce vertical fracturing. Fluidization of the formation will occur when the vertical effective stress in the formation is reduced to zero, which would be a total injection pressure of 1.79 x 10³ kPa, or a gauge-pressure head of 1,324 cm water (1.3 x 10³ kPa).

CONFINED AQUIFERS

Injection into confined aquifers requires higher operating pressures, because the vertical accommodation is near zero (hence, the basis of confinement). The initial confining pressure leads to a condition of lower effective stress in the aquifer from the outset, which can ultimately constrain the allowable injection rates and pressures in confined systems. The formation failure and fracture initiation criteria remain unchanged; however, the calculations need to be completed carefully to ensure that the initial stress conditions are evaluated properly.

The design approach is illustrated by extending the unconfined example as shown on the right side of **Figure 13.21**. For this example, the confined aquifer is bounded by a 1-meter thick clay unit at a depth of 4 meters with a dry unit weight of 1,500 dynes/cm³ and a saturated unit weight of 1,650 dynes/cm³. The unit weights for the aquifer matrix are the same as the unconfined example. For this example, the potentiometric surface of the confined aquifer is situated 1 meter below grade, but the sands above the clay confining unit are unsaturated. Assume normally consolidated conditions with a friction angle of 30 degrees. What injection pressure would be allowed in using a well screened between 14 and 20 meters below grade in the confined system to avoid formation failure and fracturing?

The total vertical stress is calculated using the weight of the overlying sediment, accounting for contributions from the unsaturated sand above the clay, the clay and the saturated sand:

$$\sigma_v = \gamma_d h_d \ (sand) + \gamma_s h_s \ (clay) + \gamma_s h_s \ (sand)$$

$$\sigma_v = 1,686 \ \frac{dynes}{cm^3} \cdot 400 \ cm + 1,650 \ \frac{dynes}{cm^3} \cdot 100 + 2,029 \ \frac{dynes}{cm^3} \cdot 900 \ cm$$

$$\sigma_v = 2.67 \times 10^6 \ \frac{dynes}{cm^2} \quad or, \quad 2.67 \times 10^3 \ kPa$$

Note that the minor difference in clay soil density had a negligible impact on the total vertical stress in the confined system as compared to the unconfined example above. The pore pressure is calculated using the height of the potentiometric surface above the top of screen, which is situated at 14 meters below grade:

$$\sigma_w = \gamma_w h_w = 980 \ \frac{dynes}{cm^3} \cdot 1,300 \ cm$$

$$\sigma_w = 1.27 \times 10^6 \; \frac{dynes}{cm^2} \quad or, \quad 1.27 \times 10^3 \; kPa$$

Applying Equations (2.23) and (13.26) yields an initial vertical effective stress in the confined system of 1.40×10^6 dynes/cm² (1.40×10^3 kPa) and horizontal effective stress of 7.0×10^5 dynes/cm² (7.0×10^2 kPa). The Mohr's circle for the confined aquifer is shown on **Figure 13.23**. As shown, increasing the pore pressure to 0.5 times the initial horizontal effective stress moves the stress circle to the failure envelope. For the confined condition the limiting injection pressure is 3.5×10^5 dynes/cm² (3.5×10^2 kPa).

The confining pressure in this example reduced the initial vertical effective stress by 3.9×10^5 dynes/cm² compared to the unconfined case. Note that the horizontal effective stress was reduced by 1.95×10^5 dynes/cm² compared to the unconfined case, due to multiplier of $K_o = 0.5$. As a result, the limiting injection pressure is only reduced by 50- percent of the confining pressure. Similarly, vertical fractures would be induced at an injection pressure equal to the initial horizontal effective stress, or 7.0×10^5 dynes/cm², which is reduced by 50-percent of the confining pressure when compared to the unconfined case.

OVER-CONSOLIDATED SOILS

Consider the unconfined example above, but assume that geotechnical testing results indicate that the aquifer is over-consolidated with a calculated OCR equal to 10. What is the limiting

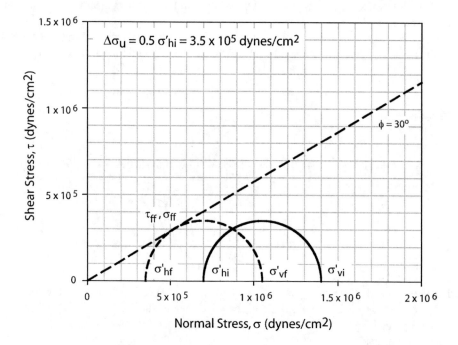

FIGURE 13.23 Mohr's circle for the confined aquifer injection scenario. The initial vertical effective stress, $\sigma'_{vi} = 1.79 \times 10^6$ dynes/cm², the initial horizontal stress, $\sigma'_{hi} = 2.83 \times 10^6$ dynes/cm², and the soil friction angle is 30 degrees. The injection pressure is reduced by 50 percent of the confining pressure.

injection pressure that would induce Mohr-Coulomb failure? What injection pressure would induce horizontal fractures?

From the initial example, the vertical effective stress is 1.79 x10^6 dynes/cm^2, which is derived from the weight of the soil and initial pore pressure. Using Equation (13.33) from above to estimate the horizontal effective stress with a friction angle, ϕ=38 degrees, and OCR=10 yields:

$$\sigma_h' = \sigma_v'(1-\sin\phi)\cdot(OCR)^{\sin\phi}$$

$$\sigma'_h = 1.79x10^6 \cdot(0.5)\cdot(10)^{0.5} \quad \frac{dynes}{cm^2}$$

$$\sigma_h' = 2.83x10^6 \quad \frac{dynes}{cm^2}$$

The resulting initial stress condition is shown on the Mohr's circle on **Figure 13.24**. The increase in pore pressure that causes failure is obtained using Equation (13.39) to solve for the midpoint of the failure circle. Substituting values for the vertical and horizontal effective stress from above yields:

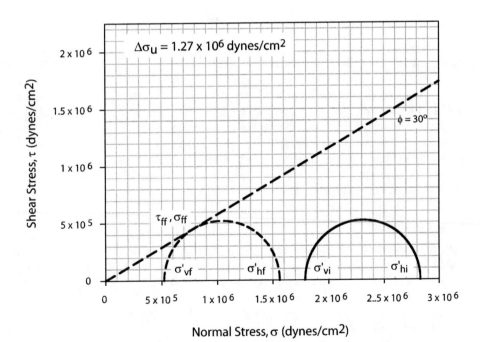

FIGURE 13.24 Mohr's circle for the over-consolidated, unconfined injection scenario. The solid line shows the initial vertical and horizontal effective stresses, σ'_{vi} = 1.79 x 10^6 dynes/cm^2 and σ'_{hi} = 2.83 x 10^6 dynes/cm^2. The change in pore pressure that induces failure, $\Delta\sigma_{uf}$ = 1.27 x 10^6 dynes/cm^2, which is greater than 50 percent of the minor principal stress, is due to the additional shear resistance that arises from over-consolidation.

$$\Delta\sigma_u = MP_i - MP_{ff}$$

$$MP_i = \frac{(\sigma_h' - \sigma_v')}{2} + \sigma_v' = \frac{2.83x10^6 - 1.79x10^6}{2} + 1.79x10^6 \ \frac{dynes}{cm^2}$$

$$MP_i = 2.31x10^6 \ \frac{dynes}{cm^2}$$

$$MP_{ff} = \frac{r\cos\phi}{\tan\phi} + \frac{r\cos\phi}{\tan(90-\phi)}$$

$$MP_{ff} = \frac{5.2x10^5(0.866)}{0.577} + \frac{5.2x10^5(0.866)}{1.732} \ \frac{dynes}{cm^2}$$

$$MP_{ff} = 1.04x10^6 \ \frac{dynes}{cm^2}$$

Substituting the midpoints into Equation (13.40) yields $\Delta\sigma_{uf}$=1.27 x10^6 dynes/cm² (1.27 x10^3 kPa). Note that allowable injection pressure is not equal to 50-percent of the initial minor principal stress (vertical effective stress), as the over-consolidation state leads to greater resistance to shear failure. Dividing the allowable pore pressure increase by the unit weight of water leads to an allowable injection pressure equal to 1,294 cm water column, or 794 cm water gauge pressure for an initial depth to water of 500 cm. This result is 381 cm greater than the normally consolidated case for the unconfined aquifer example. Based on the Mohr's circle shown on **Figure 13.24**, it is clear that injection at a total pressure of 1.79 x10^6 dynes/cm² (1,825 cm water pressure) would cause horizontal fractures, as the vertical effective stress would be reduced to zero.

ALLOWABLE INJECTION PRESSURES

Based on the discussions above, it is clear that increasing the pore pressure through injection can induce failure in aquifer matrices. The examples showed that soil strength under injection conditions depends on the soil's internal friction angle, the overburden pressure due to the weight of the soil column and the stress condition based on past loading history. Our goal in operating injection wells is to avoid formation damage and failure. To determine the safe injection pressure for remedial design, we solve for the increase in pore pressure that results in failure, $\Delta\sigma_f$, and reduce it by an appropriate factor of safety. Because the soil strength relationships do not account for fluid flow and represent the average soil behavior rather than the limiting case, we recommend applying a factor of safety of 20 percent to ensure that applied injection pressures do not result in formation damage. Therefore, the **design injection pressure**, P_{design}, is related to the maximum allowable pressure, P_{max}, as shown below:

$$P_{design} = \frac{P_{max}}{1.2} = \frac{\Delta\sigma_{uf}}{1.2} \tag{13.44}$$

We recommend using the strength relationships for normally consolidated soils to solve for the increase in pore pressure, $\Delta\sigma_{uf}$, which results in failure as shown in Equation (13.41), unless geotechnical testing data show that overconsolidated conditions exist. Thus, for normally consolidated frictionless soils, the maximum allowable injection pressure is given by combining Equation (13.41) with a 20 percent factor of safety:

$$P_{design} = \frac{\Delta\sigma_{uf}}{1.2} = \frac{1}{1.2}\left[\left(\sigma'_{vi} - \sigma'_{hi}\right) + \sigma'_{hi}\right] - \left[\frac{r\cos\phi}{\tan\phi} + \frac{r\cos\phi}{\tan(90-\phi)}\right] \tag{13.45}$$

For **overconsolidated soils**, the horizontal effective stress is greater than the vertical effective stress, which changes the order of the horizontal and vertical effective stresses in Equation (13.45) above:

$$P_{design} = \frac{\Delta\sigma_{uf}}{1.2} = \frac{1}{1.2}\left[\left(\sigma'_{hi} - \sigma'_{vi}\right) + \sigma'_{vi}\right] - \left[\frac{r\cos\phi}{\tan\phi} + \frac{r\cos\phi}{\tan(90-\phi)}\right] \tag{13.46}$$

INJECTION WELL CAPACITY TESTING

We recommend injection well capacity testing to verify the flow versus pressure head relationship prior to upscaling system designs. The primary reason is that the geotechnical approach developed above does not consider the potential adverse impacts of flowing water in the aquifer on the soil strength. In addition, the relationships were applied assuming that the formation was homogeneous, such that the measured or estimated friction angles and over-consolidation ratios represent the limiting condition. In reality, heterogeneities in the aquifer will tend to concentrate flow in the more permeable facies, potentially leading to failure well before the geotechnical calculations would indicate.

Injection well capacity tests should be performed using the estimated range of allowable pore pressure increase as an upper limit for operating pressure. The test should be conducted much like a step-drawdown test, where the injection pressure and flow rate are monitored through several "steps" in the injection flow. The field team should log the flow and water level or pressure data frequently, or preferably use pressure transducers to automate the process. Each flow-rate step should be conducted until the pressure/water level in the well stabilizes, or the change in pressure divided by the change in time approaches zero, or the slope of the pressure curve becomes horizontal. The test should be repeated for 2 or 3 additional flow rates so that a pressure versus flow rate curve can be developed.

The initial flow rate can be estimated assuming the injection well capacity is 25- to 33-percent of the well under extraction conditions. If the well was properly developed, the field team should have a reasonable estimate of specific capacity to use as a starting point. The goal is to test the well at 3 or 4 different flow rates and we advise starting the test at ¼ the specific capacity under extraction conditions, or 10 to 20-percent of the design injection flow rate. We advise using a 20-percent factor of safety for the design injection pressure, or dividing the maximum allowable pressure by a factor 1.2, to avoid formation damage:

$$P_{design} = \frac{P_{max}}{1.2} \qquad (13.47)$$

Note that the pressure in this case refers to the total pressure at the top of the screen, which includes the pressure from the water column, plus any gauge pressure measured at the well head. By convention, we use heads for these calculations, as it simplifies the assessment in the field and allows direct comparison with the water column head, h_{wc}, and gauge pressure at the well-head, h_{wh}. The total of the water column head and the gauge pressure head is the **pressure head** for the injection well.

The flow rate for each subsequent step can be estimated by comparing the allowable head increase to the incremental steady-state pressure head increase resulting from the flow rate steps. The steady-state pressure head increase is estimated by projecting the pressure head trend (slope of pressure head versus time curve) to the duration of injection. Based on the assumption that the flow versus pressure head relationship is linear, one would expect that doubling the flow rate would double the incremental pressure head at the injection well. This is a good first approximation for the flow versus head relationship; however, the results of several steps provide a more reliable estimate for the well capacity because the head change is often non-linear. If the flow versus pressure relationship appears linear, then the linear approach can be used to estimate the subsequent flow rates. If the pressure increases at a rate higher than the flow rate, subsequent flow rates should be incremented cautiously to avoid overshooting the allowable pressure.

The design well capacity, or design flow rate is estimated using the step flow rate that results in a projected pressure head that corresponds to the allowable pressure head at the injection well. If the step test duration is the same as the design duration, the flow rate from the step test can be used to directly estimate the actual design flow rate. If the duration of the step test is much shorter than the duration of injection for the final design, we recommend completing a longer-term injection test to verify the pressure head for the actual design injection duration. The method is illustrated using the following example.

For the over-consolidated unconfined aquifer example that would fail at 12.5 meters pressure head, the design pressure head for the test should be set at 10.4 meters, to maintain a 20-percent factor of safety. Assume for this case the well development indicated a specific capacity for extraction of 20 L/minute per meter of drawdown. Assuming we can inject at 25 percent of the extraction rate (for any displacement), we select an initial flow rate of 5 L/minute for the step test, which is expected to yield approximately 1 m of displacement, or 10 percent of the safe injection pressure. **Figure 13.25** illustrates the graphical method of analysis for evaluating the step test in this example.

As shown in **Figure 13.25**, the initial injection rate, Q1 = 5 L/minute, results in a projected pressure head of just less than 2 meters compared to a design pressure head of 10.4 meters. This implies that an initial estimate of total injection well capacity would be approximately 5 times 5 L/minute, or 25 L/minute. As this pressure head is approximately 20-percent of the design, we double the flow rate for the second step, so that Q2 = 10 L/minute. The projected pressure head for the second step is approximately 4.3 meters, indicating that the well capacity should be approximately 2.4 times Q2, or 20 L/minute. We increase the flow rate for the third step to evaluate the pressure near 75% of capacity. In this case, Q3 was increased to 15 L/minute resulting in a projected pressure head of 9.2 meters. At this point, the results are close to the design pressure head, but the increase is becoming non-linear. A fourth step is conducted using 17 L/minute to verify this trend and the result is a project pressure head that approaches the failure threshold for the well based on strength. In this case, we would select a flow rate near 15 L/minute for the design pressure head to avoid failure. Because a slight increase in flow beyond 15 L/minute resulted in

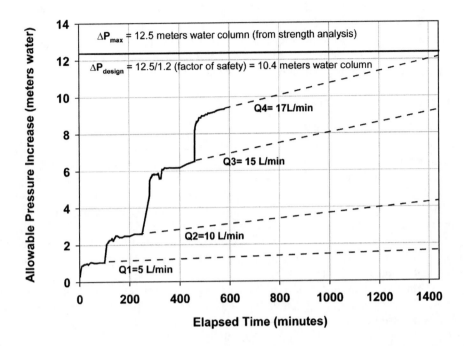

FIGURE 13.25 Example step injection test, illustrating the approach used to evaluate injection well capacities. The maximum injection pressure ΔP_{max} = 12.5 meters (water column pressure + gauge pressure) is determined from soil strength analysis. The design pressure, ΔP_{design}, is calculated by dividing ΔP_{max} by 1.2 to provide a 20-percent margin of safety. The first flow step was established by increasing flow until the injection pressure reached 10 percent of ΔP_{max}. That rate, Q1, was 5 L/min. After the injection pressure line for Q1 stabilized, the line was projected to 1440 minutes, to estimate the pressure after one day of injection (dashed line). After projecting the pressure over an extended injection period, the flow was increased and the projection process was repeated, until the projected pressure reached ΔP_{design}. If the injection program is intended to run longer than 1 day, an extended test is required to determine whether the projected pressures are valid over an extended interval (refer to Figure 13.15 and the related discussion).

too high a pressure head, we would perform a long-term injection test at this location if the design duration were much greater than 1 day.

The results of the injection well capacity testing should be compared to the preliminary design estimates. If the injection well capacity is less than planned, it will be necessary to revise the injection strategy using lower flow rates at multiple injection wells simultaneously, or using longer injection periods to inject the prescribed volumes. If the results indicate that flow rates higher than originally planned can be safely maintained, the result will be greater flexibility in implementation, potentially reducing operating costs over the life cycle of the project.

13.3 CREATING REACTIVE ZONES THROUGH REAGENT INJECTION

The most common *in situ* remedial technologies currently in use rely on the distribution of reagents in the subsurface, to establish the reaction conditions necessary for biological treatment, destructive chemical oxidation or reduction, or precipitation. The earlier discussions of flow concentration in aquifers (Chapter 3), limits on transverse dispersivity (Chapter 5) and mass transport patterns in aquifers (Chapter 7) all laid important groundwork for understanding the challenges of reagent distribution in aquifers. The most important realization from those discussions is that, as a result of flow concentration and near-zero transverse dispersivity, it is necessary to achieve complete lateral coverage of the aquifer cross-section, to be certain that reagents are contacting all of a contamination plume. Transverse dispersivity cannot generate a wide swath of reagent coverage from a small initial distribution. There are two main reagent strategies that can achieve the necessary coverage – natural-gradient distribution and recirculation.

NATURAL GRADIENT REAGENT DISTRIBUTION

The most common injection based technologies superimpose reagents on a natural-gradient groundwater flow. The injections may occur once or twice in short episodes, as is often the case for oxidant applications, or may be repeated at regular intervals, as for biological systems. The objective in each case is to establish an *in situ* reactive zone, as originally described by Suthan Suthersan (2002) and elaborated by Suthersan and Payne (2005). The basic reactive zone strategies are summarized in **Figure 13.26**. In each case, reagents are injected to create and sustain a chemical or biological reaction habitat, over an aquifer segment that is the reactive zone, and managed for the destruction of contaminants. These systems may be used to destroy source mass, to eliminate contamination from a well developed plume or to set up a long-term barrier to contaminant migration in flowing groundwater.

In the simplest reactive zone strategies, injected reagents are superimposed on natural gradient groundwater flows. **Figure 13.27** tracks an injected reagent pulse, migrating with the groundwater flow. The transverse coverage is established during the injection process, when reagent forcibly displaces and mixes with groundwater in the injection zone (**Figures 13.4** and **13.5**). The reagent-groundwater mixture travels along the aquifer transport pathways and diffusive exchange and reaction losses decrease reagent concentrations over time. Because reactive consumption and diffusive loss of reagent are time-dependent (not distance-dependent) processes, it is often helpful to think of points along the flow path by the time in transit from the injection location, as suggested by **Figure 13.27**, rather than the Cartesian distance.

Figure 13.28 provides a graphical representation of a biological reactive zone system in the Midwestern United States. The system has been operated as an enhanced reductive dechlorination barrier for 8 years, at the current time, and its operations have been reported in Payne, et al. (2001) and Suthersan and Payne (2005). The average groundwater velocity through the treatment zone is 1 ft/day and the injections are planned to reach a 30-foot diameter, or one month of natural gradient flow. The injections are repeated at 30 to 60-day intervals, which is sufficient to maintain a reactive zone to a distance of 100 feet (100 days) downgradient from the line of injection wells. The operational results for the first 1,500 days are shown in **Figure 13.29**. The pre-treatment influent and effluent chlorinated ethenes totaled 10 uM. During the first 400 days of operation, the process mined large amounts of contaminant from the immobile and non-aqueous phases and the effluent reached 68 uM ethene and ethane (the final dechlorination products) at the peak of this phase. Since that initial surge of removal, the reactive zone has behaved as an *in situ* bioreactor, with 10 uM combined perchloroethene, trichloroethene and cis-dichloroethene influents, and 10 uM of ethene and ethane effluent.

The success of natural-gradient systems depends on achieving sufficient lateral coverage at the time of injection — there's no help from transverse dispersivity — and a dose repeat frequency that is sufficient to sustain reagent concentrations between injection events. Sites with high-mass-transfer aquifer geometries are likely to perform the best and sites with low mass-transfer geometries may be very difficult to treat. Tracer studies provide an important tool in testing aquifer responses to reagent injection before a commitment is made to full-scale system installation.

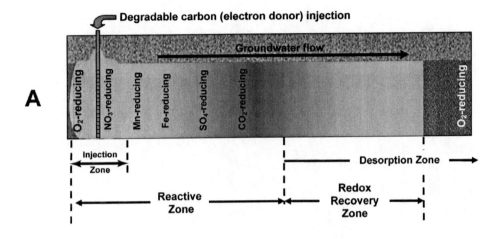

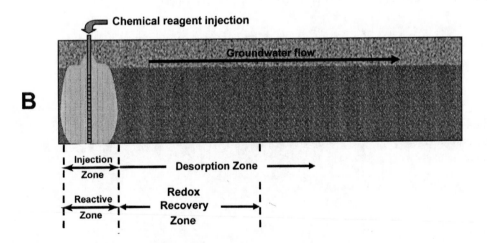

FIGURE 13.26 Conceptual diagram of *in situ* biological (A) and chemical (B) reactive zones. The redox effects of both technologies are reversed in the redox recovery zones. Recontamination may occur from mass stored in the immobile porosity in areas downgradient from the reactive zone.

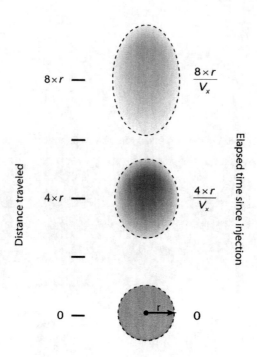

FIGURE 13.27 Reagent distribution by repeated injection and natural groundwater flow. The center of solute mass travels at the average groundwater velocity and the solute mass becomes elongated along the groundwater flow axis, due to diffusion interaction between high- and low-conductivity strata in the porous medium. Field studies indicate that transverse spreading is very limited.

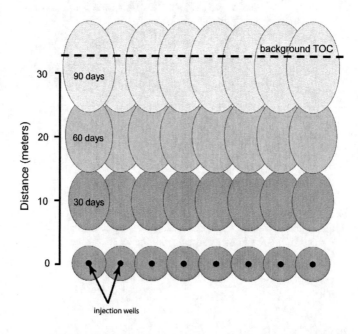

FIGURE 13.28 Repeated injections to establish an *in situ* reactive zone in a natural gradient aquifer.

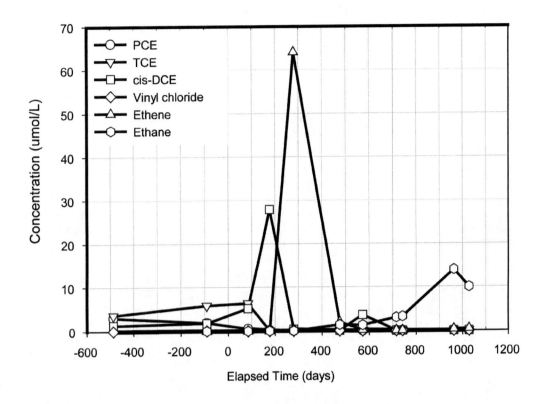

FIGURE 13.29 Results of an *in situ* reactive zone acting as a migration barrier at a site in the Midwestern United States (data extended from Payne, et al., 2001).

RECIRCULATION APPROACHES

For small to moderate-scale reactive zone applications, natural gradient systems are the most cost-effective approach. The system shown in **Figure 13.28**, for example, covering an area of 35 x 50 m, with a 3 to 5-m aquifer thickness, is well within the limits of a system well-suited to natural-gradient reagent distribution. There are three circumstances in which recirculation is needed to support reagent injection:

1. *Low-velocity aquifers* — In very low velocity aquifers, the reagent distribution may be inadequate with natural gradient flow and reagents can be fully consumed near the injection point. Recirculation can be used to establish coverage over large areas in these cases.
2. *Large-volume treatments* — When the treated volume becomes large (50,000 m², 15 m aquifer thickness, for example), recirculation is often the most cost-effective means of reagent distribution.
3. *Confined aquifers* — Injection pressure limitations usually dictate that injections be balanced by a comparable volume of fluid extraction.

We use the term recirculation to denote injection balanced by extraction. In most cases, it is appropriate to treat (if needed) extracted groundwater, amend it with reagents and re-inject it into the treated volume — hence the term recirculation. In some cases, water chemistry or regulatory constraints dictate that extracted water be pumped to a surface water discharge point. In these cases it is necessary to use public water supplies or other uncontaminated sources to support reagent injection.

DIPOLE SYSTEM

The simplest recirculation approach is a dipole, pairing a single extraction and single injection well. The general layout and expected streamflow lines are shown in **Figure 13.30**. If there were no groundwater flow through the treatment volume, the streamflow lines would be symmetrical around an axis between the injection and extraction points. Groundwater flow distorts the flow pattern, causing compaction of streamlines on the upgradient side and stretching of the downgradient streamlines. The shorter pathlines are expected to dominate transport between the injection to extraction locations. The efficiency (recapture rate) of the dipole will be a function of the pumping-induced pore volume exchange rate, relative to the groundwater flow-induced exchange rate. If the groundwater flow is significant, the recapture rate will be low, as indicated on the right half of **Figure 13.30**.

Reagent breakthrough time, t, for a dipole system can be approximated from Equation (13.48), developed by Craig Divine of ARCADIS, from formulas given by Charbeneau (2000) (Craig Divine, personal communication).

$$t = \frac{4 \cdot \pi \cdot \theta_m \cdot b \cdot L^2}{Q} \left(\frac{1 - \pi \cdot \dfrac{C_{extr}}{C_{inj}} \cdot \cot\left(\pi \cdot \dfrac{C_{exr}}{C_{inj}} \right)}{\sin\left(\pi \cdot \dfrac{C_{extr}}{C_{inj}} \right)} \right) \tag{13.48}$$

where:

    L = distance between injection and extraction wells
    θ_m = mobile porosity
    b = saturated thickness of the aquifer
    Q = injection and extraction flow rate
    C_{extr} = Reagent concentration observed at the extraction well
    C_{inj} = Injected reagent concentration

The equation can be iterated over values of C_{extr}/C_{inj} to obtain corresponding estimates of t. **Figure 13.31** shows a family of curves developed from Equation (13.48), for selected values of θ_m ranging from 0.02 to 0.2, in a 3-m thick formation (b = 3). The wells were spaced at 10 m (L = 10) and the injection and extraction flows (Q) were both set at 1 m³/hr (4.33 gpm).

From **Figure 13.31**, we can see that the initial breakthrough occurs fairly quickly at low mobile porosities, but breakthrough occurs much more slowly at mobile porosities greater than 0.05. Because this analysis assumes stagnant groundwater, the plots of C_{extr}/C_{inj}, derived from Equation (13.48), approach 1.0 at very large values of t. If the equation accounted for groundwater flow, the asymptote would be at a lower portion of the injected concentration. It is clear that a dipole design can achieve reagent distribution (although at a fraction of the injected concentration), but it will be necessary to inject at much higher concentrations than the minimum required to achieve remedial objectives.

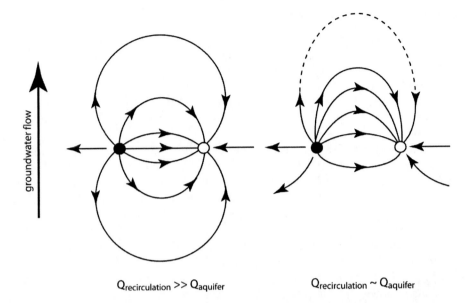

$Q_{recirculation} >> Q_{aquifer}$ $Q_{recirculation} \sim Q_{aquifer}$

FIGURE 13.30 Expected streamflow lines for a dipole groundwater recirculation system. If the recirculation rate is large, compared to groundwater flow, the recapture efficiency is high and the streamflow lines are symmetrical around the axis between the injection and extraction wells. When the groundwater flow is in the same range as the recirculation rate, the streamflow lines are distorted and the recapture rate decreases significantly.

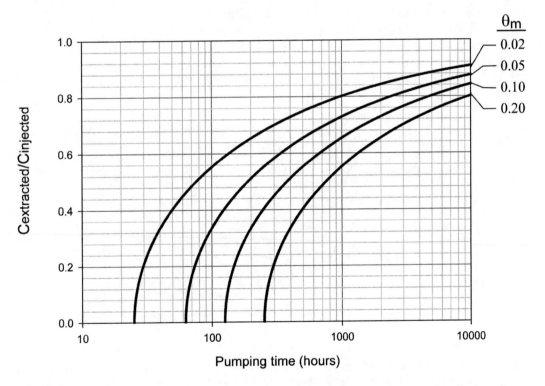

FIGURE 13.31 Estimated breakthrough at the extraction well in an injection-extraction dipole well set, for selected values of mobile porosity, θ_m. The injection and extraction wells are spaced 10m apart and the saturated thickness is 3m. The injection/extraction flow was set at 1 m³/hr (approximately 4.3 gpm). Each curve was generated from Equation (13.48).

FLOW NETWORKS IN CONFINED SYSTEMS

The limitations of the dipole system can be overcome by the construction of grid networks of injection and extraction wells at regular spacing. **Figure 13.32** shows a set of injection and extraction wells, excepted from a larger network or regularly spaced wells. No-flow boundaries occur between wells of equal polarity (i.e., between neighboring pairs of extraction wells or injection wells). These no-flow boundaries form a series of intersecting vertical planes in the porous medium, along the dashed lines shown in the figure. At the intersection of no-flow boundary planes, nodes of very little flow occur, shown by the shaded area in the figure.

The no-flow boundaries in a grid network can be re-oriented by changing the polarity of every second well in the network (Payne and Kilmer, 1994), as shown in **Figure 13.33**. When this occurs, the location of the no-flow node shifts to an area that was covered by strong flows in the alternate network flow allocation. This simple construct allows higher reagent distribution efficiency than the dipole system and something like this would be required to establish reagent injections in many confined aquifer systems.

Unconfined aquifers provide greater design flexibility for large-scale networks. We recommend developing initial designs with hydraulic and mass transport modeling, then flow-testing the design for extended periods in the field. The testing program should run over an interval sufficient to expose hydraulic limitations that will control full-scale remedy implementation. Tracer testing should also be conducted to determine the reagent transport properties of the formation — in many cases, significant design adjustments will be needed to bring reagents into contact with their target contaminants.

INJECTION EQUIPMENT EXAMPLE

From the preceding discussion of fluid distribution patterns and soil structural capacities in the subsurface, it is clear that the reagent injection process must be engineered with strict limits and control on the flow rates and pressures generated by an injection process. **Figure 13.34** offers a schematic of the basic elements of a subsurface reagent injection system. In most remedial systems, the reagent is purchased in concentrated form and diluted with water. Pumps feed the reagent and

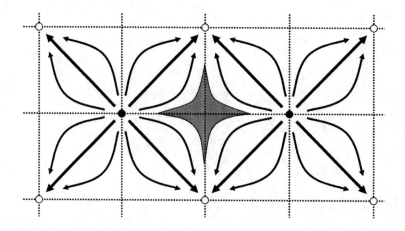

FIGURE 13.32 No-flow boundaries and flow streams that can be expected for a balanced grid of injection and extraction wells in a confined aquifer. Dotted lines indicate no-flow boundaries and the shaded area indicates a zone of very little flow that develops at the intersections of no-flow boundaries not occupied by wells (open circles indicate extraction and solid circles indicate injection). Adapted from Payne and Kilmer (1994).

FIGURE 13.33 Well field rotation — alternative flow patterns can be established in a gridded network of balanced injection and extraction wells, to provide flow coverage to all areas of the formation. The no-flow zones (indicated by shaded areas in each panel) develop at the intersection of no-flow boundaries (indicated by dotted lines) between wells of equal polarity. By shifting the polarity (injection or extraction) of every other well, the flow network is rotated and the position of no-flow zones shifts, creating flow in zones that were previously "no-flow." Adapted from Payne and Kilmer (1994).

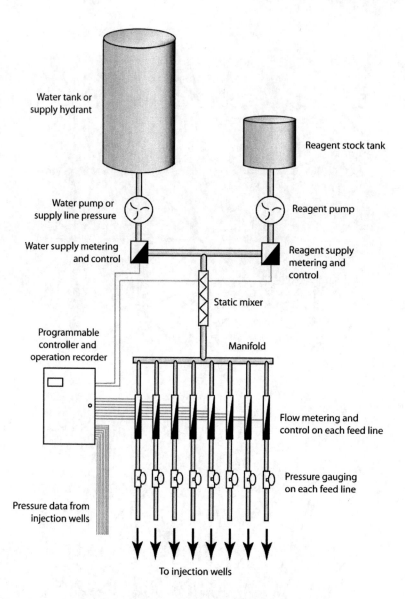

FIGURE 13.34 Schematic layout of a subsurface reagent injection system.

water through mass flow controllers that manage the reagent dilution process to maintain design reagent concentrations. A static mixer is normally required to assure that the reagents are well-mixed with dilution water before reaching the distribution manifold.

Flow metering and control is required for each injection line, as the back-pressure from piping and the aquifer matrix will be different for each line. Independent pressure gauges are also required, as shown. In systems with long piping runs to the well head, and at sites with long vertical drops to the groundwater surface, it may be necessary to install pressure monitoring in the well (*in the screened interval*), to be certain that injection pressures do not exceed allowable levels. Even with a programmable controller, it is very difficult to account for pressure loss due to pipe friction and the pressure readings at the manifold are an unreliable representation of pressure applied to the formation.

Figure 13.35 shows a mobile reagent injection system, being used for the injection of carbohydrate solution in support of enhanced reductive dechlorination. The truck-mounted tank carries undiluted carbohydrate feedstock and the water supply is drawn from a nearby hydrant. The mixing, control and distribution gear are mounted on the trailer, to the rear.

The trailer-mounted distribution and control equipment is shown on **Figure 13.36**. Each injection line is separately metered to track total fluid injection and the injection flow rate. Flexible lines connect the manifold to the individual well feed lines, accessed in the sub-grade vault. Combination pressure-vacuum gauges are used at the connection, because in a properly constructed injection well the vertical head difference between the screen elevation and the ground surface normally exerts a suction at the ground surface (this must be recognized in the design of solenoid controls for automated systems — the suction can pull open controllers and fluid can be injected inadvertently).

Systems such as that shown in **Figures 13.35** and **13.36** provide a cost-effective method for conducting repetitive reagent injections that are normally required in natural-gradient reagent distribution. For very large scale operations, permanent central mixing stations connected by subsurface piping to distribution nodes may be the most cost-effective approach.

In every case, it is essential that pre-design testing be conducted to fully understand the aquifer structure and its sensitivity to injection pressures. Pre-design testing must include tracers to assure that the effects of aquifer heterogeneities have been recognized. After a design is selected, pilot injection testing should be conducted, at design flows and durations, to be certain that the aquifer can accommodate the volumes and pressures required to achieve remedial objectives. We have seen many cases in which significant modifications to the remedy were required as a result of pre-design or pilot-scale testing.

FIGURE 13.35 Reagent injection setup. Concentrated reagent solution is carried in a stock tank on the truck bed. Dilution water is drawn from a nearby hydrant and reagent mixing and pumping occurs on the attached trailer.

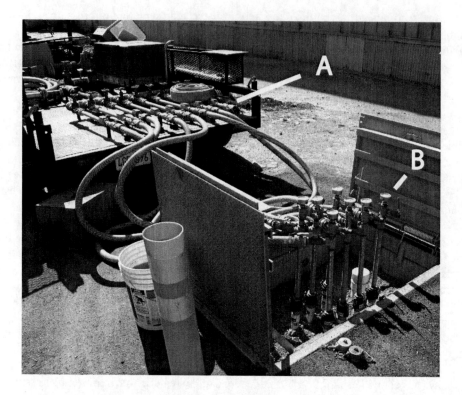

FIGURE 13.36 Injection manifold, showing flow rate and cumulative volume metering (A) and combination (pressure-vacuum) gauges (B) attached to each injection well feed line.

14 Flow Controlling Reactive Zone Designs

In many low-permeability aquifers, it is very difficult to construct well systems that can reliably inject reagents — the concentration of flow in silt and clay aquifers is often extreme, with thin lenses of conductive matrix embedded within much larger bodies of compressible material. It is often difficult to map the contaminant transporting strata and drilling tends to smear the lower-permeability soils across the borehole, blocking the flow conduits. Attempts to clear the smearing during well development are often unsuccessful, yielding reagent injection wells that are disconnected from the contaminant transport pathways. Flow-controlling reactive zone designs offer an alternative strategy to overcome the limitations on reagent distribution in low-permeability (silt and clay dominated) aquifers. They can also provide a cost-effective alternative to groundwater pumping, when flow containment is the primary objective.

The flow-controlling reactive zone strategies can be divided into two basic classes:

Flow concentrating — Groundwater flow is captured from the aquifer and directed through a reagent containing treatment volume. We will examine three flow concentrating remedial strategies:
- Funnel and gate systems
- Permeable reactive barriers
- Hydraulic fracturing

Flow blocking — *In situ* stabilization is a long standing approach that secures, but does not destroy, contaminants in the subsurface. A recent alteration of that strategy merges the stabilization process with contaminant destruction:
- Clay — Zero-valent-iron treatment

For each of these strategies, we provide a brief description of the method and case study observations from bench and field trials.

14.1 FUNNEL AND GATE

Funnel and gate systems comprise a groundwater flow barrier (the funnel) constructed perpendicular to the axis of flow, and a system of water collection from the upgradient side of the barrier that funnels groundwater through a treatment or sorption chamber (the gate) and directs the treated water to the downgradient side of the barrier. Starr and Cherry (1994) provided a general description of the funnel and gate concept, showing low-permeability barriers in several configurations. Because the impermeable funnel section blocks flow, groundwater builds up on the upgradient side of these systems. The diagrams in Starr and Cherry (1994) represented relatively simple systems. In practice, we prefer to prevent the buildup of groundwater on the upgradient side of the systems, opting instead to provide collection trenches upgradient of the barrier, channeling the flow through the barrier to distribution trenches on the downgradient side of the system.

Figure 14.1 shows a schematic of a funnel and gate strategy that was developed for a site in the Western United States. A shallow silt aquifer at the site (3 to 6 m below ground surface) was contaminated by chlorophenols and the silty formation was underlain by clayey siltstone, beginning at the 6-m depth. The contamination extended across a seepage front of approximately 120 m. The funnel and gate strategy developed for the site incorporated a bentonite grout slurry wall to block groundwater flow, with peastone collector trenches on the upgradient side and peastone filled distributor trenches on the downgradient side of the system. The collector trenches fed into sumps filled with granular activated carbon and the discharge side of each sump was piped to a corresponding peastone distribution trench. **Figure 14.2** shows a general overview of the construction process.

The slurry wall barrier was constructed to 190 m in length, designed to extend beyond the width of the contaminated aquifer. The groundwater flow across the barrier was estimated at 76 L/min (20 gpm) seasonal maximum and a system of four collector-distributor systems, each responsible for one-quarter of the flow, was arrayed along the length of the system. The slurry wall extended vertically from the ground surface to the siltstone.

The gate sumps were built from corrugated metal pipe, 1.25 m in diameter, as seen in **Figure 14.3**. Feed lines from the peastone collector trenches, made from filter-sock covered polyethylene drain tile, were connected to the gate sump and the assembly was placed below-grade, as shown

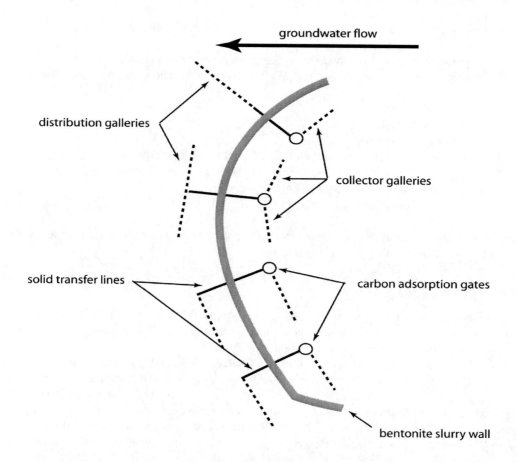

FIGURE 14.1 Site layout for a funnel and gate system in the Western United States.

FIGURE 14.2 Overhead view of funnel and gate system construction.

FIGURE 14.3 Funnel and gate construction, showing setup of activated carbon gate.

in **Figure 14.4** and filled with granular activated carbon (**Figure 14.5**). Groundwater monitoring wells were placed upgradient and downgradient from the gate sumps, to track the in-feed and discharge of chlorophenols at the gate sump (primarily pentachlorophenol).

Collector and distributor lines were constructed by trenching to near the bedrock surface and partially backfilled with peastone. The collector and distributor drain tile lines were then spooled out into the trenches, as shown in **Figure 14.6**, to place the tiles below the pre-construction groundwater surface, then peastone backfilling was completed.

FIGURE 14.4 Funnel and gate system, showing final connection of collector lines to gate tank.

FIGURE 14.5 Filling gate tank with granular activated carbon.

After all piping systems had been completed, the funnel trench was back-filled with a bentonite slurry, shown in **Figure 14.7**. Steel plates were placed atop each of the four gate sumps and the site surface was restored. After 4 years of system operation, monitoring well data indicated that carbon breakthrough was occurring and the sumps were opened for carbon replacement. The spent carbon was extracted by vacuum, into a temporary storage vessel and dewatered. New carbon was placed in the gate sumps and the system was re-covered to continue operation.

FIGURE 14.6 Collector line installation for funnel and gate system.

FIGURE 14.7 Constructing the slurry wall component of a funnel and gate system.

14.2 PERMEABLE REACTIVE BARRIERS

In the funnel and gate system, the treatment process is concentrated in a small segment of the system and it is necessary to force groundwater flow into the small gate area. Because the funnel blocks groundwater flow, there is always a requirement for head buildup on the upgradient side of the system to drive flow into the gate and groundwater may flow around or under the constructed barrier, escaping treatment. An alternative approach is to provide a filled trench that is more permeable than the surrounding aquifer matrix, effectively attracting groundwater flow to the treatment system. In this case, the treatment process must be distributed along the entire permeable collector trench. For aerobically-degradable compounds, the trench can be backfilled with stone and air lines built into the lower reaches, to provide an aerobic treatment system. More commonly, the target compounds are dissolved heavy metals, such as chromium, or chlorinated solvents such as perchloroethene or trichloroethene. The most common treatment processes are anaerobic, driven by reaction products from the corrosion of zero-valent iron metal embedded in the permeable wall.

Hexavalent chromium reduction can be reduced, then precipitated, by the zero-valent iron system reactions shown in Equations (14.1) and (14.2). There is a net consumption of protons in the reaction sequence and pH can be expected to rise in, and downgradient from, a zero-valent iron reactive barrier.

$$CrO_4^{2-}{}_{(aq)} + Fe^0{}_{(s)} + 8H^+{}_{(aq)} \rightarrow Fe^{3+}{}_{(aq)} + Cr^{3+}{}_{(aq)} + 4H_2O_{(l)} \tag{14.1}$$

$$(1-x)Fe^{3+}{}_{(aq)} + (x)Cr^{3+}{}_{(aq)} + 2H_2O_{(l)} \rightarrow Fe_{(1-x)}Cr_{(x)}OOH_{(s)} + 3H^+{}_{(aq)} \tag{14.2}$$

There are also many reaction pathways for zero-valent iron that compete with the target reactions by unproductively oxidizing the iron, including the oxidation reactions in Equations (14.3) and (14.4)

$$2Fe^0{}_{(s)} + O_{2(g)} + 2H_2O_{(l)} \rightarrow 2Fe^{2+}{}_{(aq)} + 4OH^-{}_{(aq)} \tag{14.3}$$

$$Fe^0{}_{(s)} + 2H_2O_{(l)} \rightarrow Fe^{2+}{}_{(aq)} + H_{2(g)} + 2OH^-{}_{(aq)} \tag{14.4}$$

Each of these reactions generates hydroxyl anion, which adds to the pH increase associated with the productive reactions of Equations (14.1) and (14.2). Molecular hydrogen is also produced, which could serve as an electron donor for various microbial species, including dechlorinating bacteria. A more extended discussion of *in situ* oxidation, reduction and precipitation reaction mechanisms is provided by Suthersan and Payne (2005).

The reductive dechlorination of alkenes (e.g., perchloroethene and trichloroethene) has been observed and Roberts, et al. (1996) proposed a generalized reduction process that is summarized by Equation (14.5).

$$PCE \xrightarrow{2e^- + H^+} TCE \xrightarrow{2e^- + H^+} DCE \xrightarrow{2e^- + H^+} VC \xrightarrow{OH^-} acetylene \tag{14.5}$$

Several permeable reactive barriers have been studied intensively and are well reported in the literature. One such example is a zero-valent iron permeable reactive barrier, constructed at the U.S. Coast Guard facility in Elizabeth City, North Carolina in 1996. The system was constructed in 1996 and is deployed as a "hanging wall," intercepting the upper portion of the site aquifer, immediately upgradient of its discharge to the Pasquotank River. The wall is 46 m long, extending to an 8-m depth and its operation is described in detail by Puls, et al. (1999). Blowes, et al. (2000) provide a summary of operational results for Elizabeth City system and several other zero-valent iron and related reactive wall systems.

Permeable reactive barriers are typically installed by excavating a trench in the aquifer matrix and backfilling with a mixture of inert, highly porous material (e.g., coarse sand) and reagent (e.g., zero-valent iron). For sites at which the target zone is below the reach of excavating equipment, reagent injections may be used to create a reactive barrier, an approach that falls within the hydraulic fracturing methods, described in Section 14.3. The permeable reactive barrier strategy has proven successful at many sites and the installation process has evolved since the earliest installations. The construction process must avoid excessive mixing of aquifer matrix material (usually low-permeability soils) with the porous fill, so that maximum hydraulic conductivity can be maintained in the wall. If this separation is not maintained, it is possible to construct a relatively low-permeability barrier that blocks groundwater flow. It is also important to maintain a consistent placement of fill, because variations in permeability will channel flows to some areas and bypass others, shortening the useful reagent life span.

Figures 14.8 – 14.10 show some of the construction strategies used in a recent permeable reactive wall installation in the Eastern United States. The site lies adjacent to a small stream and the aquifer soils are primarily silts and sands, with trichloroethene contamination traveling to the stream along the sandy conduits. This site was introduced earlier, in **Figures 9.12** and **9.13**. The reactive barrier extends along the entire contaminated face, to a depth of approximately 35 feet.

FIGURE 14.8 Vacuum well point dewatering for groundwater suppression, prior to zero-valent iron reactive wall construction.

FIGURE 14.9 One-pass trencher preparing to install permeable zero-valent iron reactive wall.

FIGURE 14.10 One-pass trencher installing zero-valent iron and sand slurry to a 35-foot depth.

The site was prepared by depressing the groundwater surface with a vacuum dewatering system, shown in **Figure 14.8**. The reactive barrier was constructed using a "one-pass" trencher, shown in **Figures 14.9** and **14.10**. The trench is excavated by the chain cutter system seen in **Figure 14.9**, while the reactive mixture is fed to the unit on a conveyor and backfilled through the hopper, both visible in **Figure 14.10**. The trench is held open for backfilling by the side shields seen in **Figure 14.9**.

The system shown in **Figures 14.8-14.10** was completed shortly before this writing and the first quarterly sampling episode indicated that pH inside the wall exceeded 9, as expected, and chlorinated ethenes inside the wall were below detection, also as expected. Decreases in chlorinated ethene concentrations downgradient from the barrier are expected to develop over the succeeding 12 months. Most importantly, groundwater elevation observations indicate that the wall is not blocking groundwater flow.

There is an important point of caution on the construction of permeable reactive barriers – The sites where permeable reactive barriers are utilized often comprise low-compressive-strength soils. When heavy equipment is run on the site surface, especially equipment that vibrates during operation, the underlying soils can be compacted, decreasing the hydraulic conductivity of the formation adjacent to the permeable wall. A properly constructed wall should increase net hydraulic conductivity on the flow path from upgradient to downgradient of the wall. We have observed at least one site where there was a significant decrease in net hydraulic conductivity that likely resulted from soil compaction during construction. In that case, the compaction was severe[1] and groundwater elevations rose several feet on the upgradient side of the wall, following construction. Contaminated groundwater reached the surface and drained away laterally, untreated.

14.3 HYDRAULIC FRACTURING

In bedrock and in some silt or clay dominated aquifers, contaminant mass transport may occur in fractures or in very thin conductive strata embedded within a low-permeability matrix. It can be very difficult to inject reagents into these formations and the transport pathways may be quite irregular. Hydraulic fracturing provides a means of 1) increasing fluid (and reagent) injectability, while 2) concentrating flow in the newly formed conductive stratum. The concentration of flow can decrease the rate of mass exchange between mobile and immobile porosities, as described earlier in Chapter 7 (see **Figure 7.10**, for example). However, the reduction of contact is more than offset by the improved ability to distribute reagents.

Hydraulic fracturing techniques were adapted from the oil industry for application in shallow formations. Much of the early research was conducted with support of U.S. EPA and documentation was provided in EPA (1994) and Murdoch (1995). Many examples of hydrofracturing developed in the succeeding years, but the technique is not widely practiced, because controlled fractures are feasible in relatively few formations. The typical process has two components, a carrier fluid and a proppant. The proppant is a coarse, inert material such as a coarse sand and the carrier fluid is typically water, with a guar gum thickener. An enzyme is added to the mixture to decompose the guar.

Figure 14.11 shows a hydraulic fracturing setup. An uncased borehole is cut into the formation and notches are cut into the side walls to serve as fracture initiation points. The deflated fracturing assembly is then inserted into the target zone, as shown on the left side of the system. The packers are then inflated to isolate and pressure seal the target zone and the carrier fluid – proppant mixture is pumped into the formation, as indicated on the right side of the figure. A support equipment rig is

---

[1] Observers noted that groundwater accumulated at the site surface at locations where heavy equipment was operating, even though static water elevations were several feet below the ground surface.

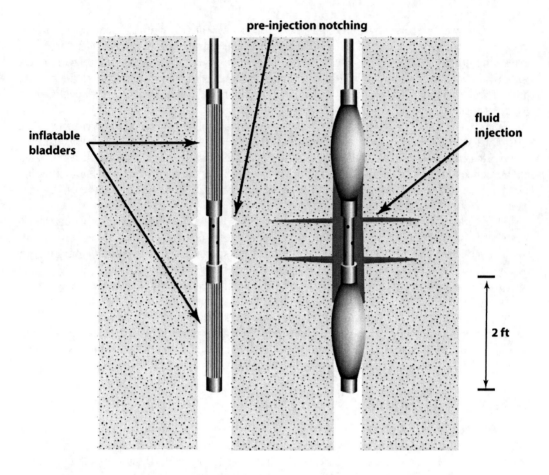

FIGURE 14.11 Packer assembly for development of hydraulic fracturing of porous media. The borehole is prepared by notching the formation at the levels where fractures are desired, and the deflated packer assembly is lowered into the borehole (left side of the diagram). The packing bladders are then inflated to build a seal against the borehole wall and fluid can then be injected, under pressure, into the formation. The intent is to propagate fractures from the initiation notches.

shown on **Figure 14.12**, carrying tanks for proppant and carrier fluid mixing, as well as the pumps and controls needed to support the process.

To create a useable system, fractures must propagate in the horizontal plane. As we described in Chapter 13, horizontal fracturing occurs in over-consolidated soils, where the horizontal compaction exceeds the compaction that can be generated by existing stresses. Over-consolidated soils are encountered where the loads that compacted soils to their current levels have been removed. This occurs as a result of post-depositional erosion and in glaciated regions, where the weight of continental ice sheets compacted soils and then was removed as the ice melted. The vertisols of the Texas Gulf Coast are also often over-consolidated, as a result of cyclical wetting and drying, while near-surface fractured bedrock formations may be over-consolidated, as a result of erosional losses at the surface (EPA, 1994).

FIGURE 14.12 Hydrofracturing trailer setup, showing tanks for proppant and thickening agent (guar gum).

The hydrofracturing system depicted in **Figures 14.11** and **14.12** was used to create a set of horizontal fractures at the site of perchloroethene contamination in a shallow sandstone formation near Denver, Colorado. Fracture propagation was monitored at the ground surface by a network of tilt meters, linked to a monitoring computer that generated a mapping of the fracture zone in real time. It is important to note that each fracture lifts the overlying material by approximately 1 to 1.25 cm and two fractures can be propagated at once with the setup shown in the figures. Sensitive tilt meters can easily register the development of the fractures in the underlying bedrock.

After the guar thickener was solubilized by the enzyme additive, the fracture wells were used as injection wells, to distribute a carbohydrate electron donor solution, supporting the development of enhanced reductive dechlorination to treat the solvent. **Figure 14.13** provides a summary of the enhanced reductive dechlorination process, as observed at a monitoring well within 50 feet of the nearest reagent injection well. Electron donor injections, which began at 0 days elapsed time on the figure, increased soluble organic carbon (TOC) to concentrations exceeding background ($<$ 5 mg/L) levels. The microbial community responded quickly, achieving complete conversion of perchloroethene (PCE) to cis-dichloroethene (cis-DCE) in less than 6 months. In the succeeding 2 years, cis-DCE and vinyl chloride (VC) were dechlorinated to ethene and ethane and all constituents were below drinking water criteria after 1,200 days.

In evaluating the outcome of a fractured-support remedy, it is important that contaminant concentrations reached the remedial goals, but it is also important to be certain that the improvement was not simply a result of dilution due to displacement of contaminated groundwater by injected reagent. This can occur in any injection-based remedy, but fracture systems, due to their very small mobile porosity, are the most susceptible to reagent displacement. In the case shown in **Figure 14.13**, we can see that the results are not displacement dominated, because the total alkene molarity increased more than 50 percent as the dechlorination became established.

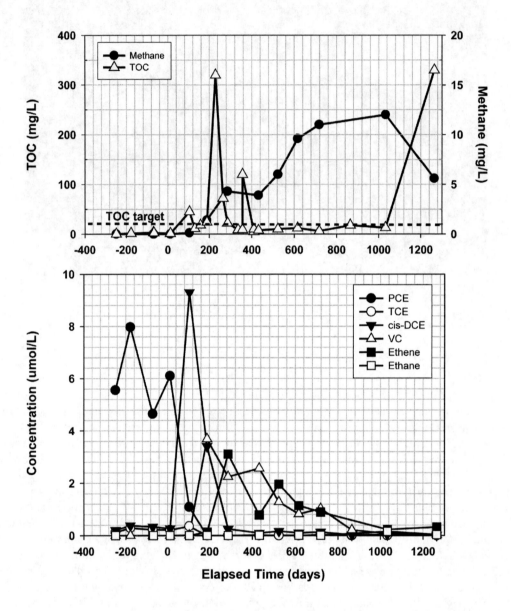

FIGURE 14.13 Results of enhanced reductive dechlorination in a fracture-enhanced bedrock formation in the Western United States. Total alkene molarity increased by 50 percent after the first electron donor injection (t = 0), then declined gradually over the following 3 year period, during which all chlorinated alkenes fell to less than drinking water protection criteria. Each electron donor injection event was accompanied by a spike in the total organic carbon (TOC) concentration.

14.3 CLAY – ZERO VALENT IRON SYSTEMS

In some sites, especially DNAPL source zones, there is a significant risk of displacing contaminant when reagents are injected or even through the compression caused by excavation equipment. An alternative approach has been developed to combine zero-valent iron (ZVI) with bentonite clay and mix the two into DNAPL source zones with large auger rigs. The approach was developed by DuPont and the resulting patent was transferred to Colorado State University, which now manages licensing. The clay-ZVI system blocks groundwater migration through the source zone and leaves reactive iron in the resulting monolith to dechlorinated solvent that enters the aqueous phase. The iron usage is much more efficient in this case than in permeable reactive walls, because there is not a continuous flow of corrosive material (oxygen or nitrate, for example) through the matrix. The iron–solvent ratios can much more closely approach stoichiometric levels in these systems. Shackelford, et. al., (2005) provide an introduction to the clay-ZVI technology and show results of early treatments.

The clay-ZVI method requires bench-scale testing to determine the optimal iron and clay mix ratios that are needed to achieve the desired hydraulic conductivity decreases and solvent dechlorination rates. **Figure 14.14** shows a laboratory setup for bench testing the technology. The process is run at the meso-scale, with samples collected from the field placed in large columns

FIGURE 14.14 Zero-valent iron-clay technology bench study setup. Clockwise from left: bench mixing apparatus; bench-scale mixing auger; finished column with Tedlar® bag for gas collection. Photos courtesy Dr. Tom Sale, Colorado State University.

and mixed in the apparatus shown on the left side of the figure. The mixing blade, shown on the upper right, simulates the action of the full-scale field rig. After mixing, treated test columns are incubated at site aquifer temperatures, to track the progress of dechlorination reactions (lower right in the figure). A Tedlar® bag is shown, attached to the column, to measure hydrogen gas

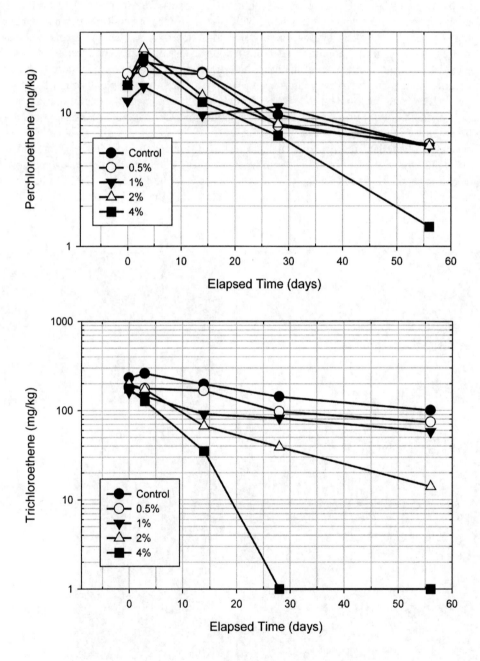

FIGURE 14.15 Results of bench-scale testing of ZVI-clay mixing technology for DNAPL source zone treatment. Bentonite clay (1% w/w) was mixed into DNAPL source zone soils from a site in the Midwestern United States, where chlorinated solvents were disposed with oils and paint solvent wastes. Zero-valent iron addition ranged from 0.5 to 4 percent (w/w) and a control column was mixed with 1% clay, only.

production associated with the zero-valent iron chemistry [note the hydrogen produced in Equation (14.4)] and to track any associate gas-phase solvent losses.

The results of a bench trial for a mixed solvent-oil DNAPL are shown in **Figure 14.15**. Field samples were collected from a site in the Midwestern United States, where paint and degreasing solvents, waste oil and other hydrocarbons were disposed in open pits, then backfilled. The fluid properties of this material were described earlier, in **Table 1.3**. The high non-chlorinated content of this material distinguishes this treatment attempt from previous work reported by Shackelford, et al. (2005)

The results of the bench trial are shown in **Figure 14.15**, comparing the performance of 1-percent bentonite clay (control) and mixtures of 1-percent bentonite with zero-valent iron ranging from 0.5 to 4 percent, w/w. Control concentrations decreased more than expected, which is not yet fully understood. Both the 2 and 4-percent zero-valent iron mixtures generated satisfactory rates of dechlorination and the 2-percent mix rate was selected for field-scale implementation.

FIGURE 14.16 Auger injector-mixer unit for installation of the zero-valent iron and clay source mass treatment system.

The field equipment used for the clay-ZVI application is shown in **Figure 14.16**. To achieve a mixing depth of 30 feet or more, a long hollow Kelly bar is used, supported by a mobile crane. The bentonite-ZVI slurry is injected into the top of the Kelly bar, through a swivel coupling, shown in the top panel of the figure. The fluid is pushed out into the formation through injector-mixer blades, shown in the lower panel of the figure. A vapor collection shroud is suspended over the mixing location and air is continuously drawn through the shroud with a vacuum pump, to extract any solvent vapors that are driven off by the mixing process. The vapors are collected in beds of granular activated carbon for later treatment.

15 Summary of New Developments and Their Implications

This book outlined a series of changes in the interface between hydrogeology and remediation engineering that can have a significant impact on our approaches to site investigation and cleanup. The interface was defined, up to this time, by the advection-dispersion equation, which dominated thinking on how contaminant plumes migrated and how injected reagents would be distributed in the subsurface. *Remediation Hydraulics* was conceived to create a more realistic and adaptive interface between the richly detailed science of hydrogeology and the increasingly demanding requirements of the remediation engineering discipline. We conclude here with a brief summary of some of the implications of this adaptation in thinking and observations we have made in the field as we developed this understanding.

15.1 A SHIFTING MODEL OF THE SUB-SURFACE

One of the important factors prompting us to undertake the studies that underlie Remediation Hydraulics was the need to improve the design and operation of reagent injection networks supporting *in situ* reactive zone remedies. The earliest theme we developed related to the mechanisms underlying solute spreading in aquifers and the growing pool of observations suggesting solute plumes (injected or spilled) did not spread, as expected under the standard advection-dispersion model. The two main points of this, developed in Chapter 5, were:

Transverse (flow-dependent) dispersivity is very small — Observed transverse dispersivities are quite small and generate only minor spreading of solutes in groundwater transport.

Formation-dependent and time-dependent (diffusion) dispersivities can be large — The transverse spreading that may be observed is likely due to anastomosing or braided channel forms and temporal variations in groundwater gradient. The longitudinal spreading that is observed is probably a result of diffusive transfer of solute mass between the mobile and immobile porosities.

The implications of these observations are certainly very significant to all stakeholders with an interest in groundwater restoration and, at the time we began developing *Remediation Hydraulics*, it was a major focus of our deliberations. However, the more significant changes in our understanding were driven by the results of tracer studies we began to require[1] from all *in situ* remedy sites. As tracer observations accumulated from Europe and the Americas, the divergence

---

[1] ARCADIS has established tracer studies as a standard practice for all sites where reagent injections are planned as part of a remedy. Tracer studies are encouraged for all sites at which contaminant mass transport may be important, regardless of remedial plans.

between average groundwater velocities and the actual transport velocities unmasked by tracer studies was stunning. Ten-fold, or greater, differences between the average (superficial) velocity and the observed velocities are common.

Aquifers are heterogeneous and anisotropic, as a rule — Depositional processes that form aquifer matrices typically deposit materials in patterns containing strong heterogeneities that are aligned in regular anisotropic patterns. Hydrostratigraphic analysis provides a method for assessing the depositional processes, but some amount of formation mapping is always required to determine the aquifer structure. Hydrostratigraphic analysis provides an important tool to inform the mapping process.

Groundwater elevation gradients may not be a reliable indicator of groundwater flow direction — The groundwater elevation gradient for an aquifer expresses the gravitational driving force supporting groundwater flow, but it is not the only determinant of groundwater flow direction. The hydraulic conductivity structures of heterogeneous, anisotropic aquifers are often not aligned with the fall line of the groundwater elevation gradient. In these cases, relatively small hydraulic conductivity contrasts can direct groundwater flow and contaminants in directions significantly off the elevation fall line.

Groundwater flow always takes the path of least cumulative resistance — Significant concentrations of groundwater flow can accumulate in high-conductivity channels, embedded in natural aquifers during their deposition. Although these flow concentrations are theoretically reflected in groundwater elevation mappings, they are unlikely to be detected by any realistic monitoring well network. In the model aquifer presented in **Chapter 3**, extreme concentrations of flow were barely discernable in a water surface elevation mapping constructed from observations made on 1-foot centers across the entire 300 x 400-foot grid (120,000 observation points).

Injection and extraction of groundwater are asymmetrical — The accommodation of fluid injected for any formation does not occur at the same rate as fluid extraction - extraction flows are primarily horizontal, while injection flows are predominantly vertical. The difference between vertical and horizontal hydraulic conductivities in most formations causes a significant asymmetry between the injection and extraction processes and extraction well yields cannot be used to predict sustainable fluid injection rates.

Average groundwater velocities significantly understate actual transport velocities in aquifers — Groundwater tracer studies repeatedly showed that groundwater velocities in the mobile pore space dramatically exceeded the average values obtained for the respective formations. Moreover, the tracer studies showed that the mobile porosity for aquifers often comprises multiple parallel pathways, each carrying groundwater at a distinct velocity. In one case study, at least five distinct peaks were observed passing a 1.5-m long well screen. Groundwater tracer studies also showed strongly lognormal breakthrough curves, a pattern predicted to occur when aqueous-phase diffusion drives solute migration from mobile to immobile pore spaces.

Every monitoring well sampling method yields strongly biased results — Every available monitoring well sampling method carries significant biases that must be understood to allow proper interpretation of results. The interval sampled and the volume extracted determine the results. High-volume pumping best represents potential exposure, while discrete interval sampling best represents transport. Interval-averaging techniques (e.g., passive diffusion bags) are most likely to detect the presence of contamination.

Long-term aquifer pumping tests understate average hydraulic conductivity — Pumping tests tend to underestimate the contributions of highly conductive elements in anisotropic formations, as described in Chapter 3. This is acceptable and conservative for water supply development, but severely limits the value of pumping tests, particularly long-term tests, for remediation hydraulic analysis.

Aqueous-phase diffusion controls transport behavior in aquifers — Diffusive mass transport from mobile to immobile pore spaces causes the immobile pore space to act as a persistent contaminant storage location, along the entire length of any contaminant plume. If the contaminant source is sustained over long periods, large amounts of mass can reach immobile storage and back-diffusion will be dramatically slower than inward diffusion, due to lesser concentration gradients.

The immobile pore space is the primary storage location for contaminants in aquifers — Remedial system designs, as well as expectations for the achievable rate and extent of cleanup that can be achieved for any site, must account for the contaminant mass that is stored in immobile pore spaces. Advective flows do not directly contact most of this mass and successful remedies must either perfuse reagents into the immobile pore spaces or maintain treatment in the mobile pore spaces long enough to outlast back-diffusion of contaminants.

Evolving conceptual site models — Remedy implementation is the most rigorous test applied to any conceptual site model and no site model is immune to modification after the remedy begins. The conceptual site model must continue to evolve throughout the entire remedial process and adaptive design provides the flexibility required to respond as new information on site structure, contaminant distribution and reagent actions become available. Every data collection event presents an opportunity to validate the conceptual site model.

All of these shifts in our understanding of groundwater contamination and remedy behavior are rooted in the recognition that heterogeneities in aquifer structure control what we observe and what we can accomplish. Essentially, we have concluded that it isn't possible to effectively model all of these behaviors — it is fundamentally necessary to map them.

15.2 MAPPING HETEROGENEITIES – TOWARD A QUANTITATIVE CONCEPTUAL SITE MODEL

As discussed in **Chapter 7**, it is the continuity of the mobile pathways and interconnectedness between the mobile and immobile segments of the aquifer that control the transport behavior of contaminants and reagents in aquifers. We recognize that our ability to improve the reliability of *in situ* remedies is determined by our capacity to accurately map the hydrostratigraphy and plume at the remediation hydraulics scale. The hydrostratigraphy concepts introduced in Chapter 9 highlight the importance of mapping the architecture of the mobile and immobile facies in aquifer systems. Our approach advocates integrating high-resolution stratigraphy and plume mapping with small-scale, transient methods of conductivity testing to enable quantitative correlations between the spatial distribution of the plume and the permeability of the mobile pathways and immobile reservoirs. The challenge continues to be how to accurately characterize the heterogeneities at intermediate scales, which are often below our radar screen.

There is a critical role for remediation hydrogeologists. It is critical that properly trained geologists and hydrogeologists make soil classification interpretations that extend beyond the USCS and ASTM requirements. Hydrostratigraphy interpretations should be framed in the context of the depositional environment, with the goal of describing the facies trends and depositional elements that comprise the mobile and immobile segments of the system. Practitioners should strive to develop facies models for regional aquifer systems, much as the reservoir engineers do in the oil patch or as research geologists do in the glaciated northern latitudes. We agree with de Marsily et al. (2005) that developing facies libraries and databases will dramatically improve our understanding of aquifer systems in particular localities, but also allow the peer review and information exchange necessary to improve the process at the remediation hydrogeology scale.

High-resolution hydrostratigraphy mapping is required to develop 3-dimensional interpretations of aquifer architecture. Continuous sampling or logging methods are required. When continuous soil boring programs are used, we recommend stratigraphic logging techniques to facilitate facies trend analysis between boreholes. The visual classification of the soil should be augmented with focused sieve testing and grain-size analysis to make interpretations more quantitative and reliable. We recommend using CPT (cone-penetrometer) soundings when possible because the approach provides a consistent tool for discriminating between the mobile and immobile facies. However, CPT soundings should always be calibrated versus continuous borings to ensure that facies interpretations are accurate.

High-resolution plume mapping is an integral component of hydrostratigraphy analysis, as it is the correlation between the plume distribution and facies in the aquifer that control the transport behavior. Vertical aquifer profiling is preferred over indirect, field-screening methods such as MIPs (membrane interface probes) for quantitative distribution mapping; however, the utility of MIPs can be significant when calibrated against site-specific vertical aquifer profile sampling and analysis. The key to success is to initiate the program with frequent sampling in the vertical profile to capture the correlation between facies and plume distribution. Once the correlation is established, biased sampling based on the facies can be a very effective and efficient means of mapping the plume in three-dimensions. It is important to note that this relationship can change over the length of the plume, not just because of facies variability, but also because of the plume maturation process as described in Chapter 7.

High-resolution hydraulic conductivity mapping quantitatively distinguishes mobile from immobile porosities. Small-scale, transient methods should be employed to develop a correlation between hydraulic conductivity and hydrofacies trends. We strongly recommend slug-testing and mini-pumping test methods over conventional water-supply hydraulics tests. The objective is to evaluate the range of variability that characterizes the system, not to determine the average conductivity for the system. This approach enables the hydrogeologist to make more quantitative interpretations regarding the mobile and immobile porosity in the system and how the interchange of contaminant and reagent mass will affect the transport behavior and ultimately the effectiveness of the remedy.

There's no truth like tracer truth. Tracer testing methods provide the most reliable means of evaluating contaminant transport and reagent distribution behavior in aquifer systems. Due to the site-specific nature of the mass transfer between the mobile and immobile segments of the aquifer, we recommend using a transport test to directly observe the behavior of injected reagents, rather than to estimate it indirectly using hydraulic methods. The advantages of well-planned and executed tracer tests are many:

- *Injection well design parameters* — The designer can obtain direct measures of injection well capacity, flow versus pressure rating curves; radius versus volume relationships and injected fluid distributions.
- *Mobile and immobile porosity estimates* — Tracers provide a direct measure of the mobile and immobile porosities. With this information, we can better characterize transport behaviors for contaminants, as well as injected reagents. Tracers are especially useful for estimating the injected reagent concentrations required to achieve working strength reagent coverage at prescribed distances from an injection location.
- *Groundwater flow* — By monitoring the advance of the tracer cloud, the designer can determine groundwater flow direction and velocity.

   • *Injected reagent reaction rates* — Finally, extending the tracer test to include an
     inert tracer and the reagent of choice allows the designer to estimate reaction rates and
     system reagent demands that are essential in developing a final *in situ* remedy design.
     None of these critical design parameters can be reliably estimated using hydraulics
     and conventional modeling approaches.

   The application of currently available quantitative hydrogeology methods yields an ample return
on investment, when these methods are used to support full-scale remedial system design. Each of
these current methods is applied at relatively small scales and numerous observations are required
to develop a quantitative conceptual site model. New, intermediate-scale, sample collection and
analysis techniques are under development and promise to be even more cost effective than current
methods.

15.3 DEVELOPING METHODS IN QUANTITATIVE HYDROGEOLOGY

Geophysical methods, including seismic, resistivity, and hydraulic interference testing, have
potential application for mapping the geometry of facies bodies at intermediate scales that are
inaccessible through conventional exploration programs. Similar to other indirect methods such
as CPT and MIPs, the reliability and utility of the methods can be improved through a systematic
calibration of the method to the site conditions. The continued evolution of data acquisition systems
and PC workstation power makes it easier and more cost-effective than ever to incorporate these
emerging techniques into our characterization strategy.

   Resistivity tomography — Resistivity tomography has been successfully applied using sur-
     face and cross-hole methods to discriminate between sands and clays in aquifers, improv-
     ing the hydrostratigraphic interpretations between boreholes. In the past, the practical
     limitation has been the 2-dimensional nature of the data acquisition, or ambiguous results
     when proper calibration efforts are not completed. The value of this method is its capacity
     for three-dimensional and even four-dimensional data acquisition and interpretation. For
     example, 3-dimensional soundings of background electrical resistivity (to obtain broad
     aquifer matrix classifications) can be followed by the injection of electrically conductive
     tracer to highlight transport pathways. With continued advances in instrumentation and
     analysis tools, it may be possible to map the spreading of tracer in near real-time. This
     could be the ultimate application of tracer testing techniques.
   Hydraulic tomography methods may provide three-dimensional mapping of depositional
     elements and show great promise for mapping the distribution of hydraulic conductivity,
     directly. As we discussed in **Chapter 11**, the method is tailored to collecting and ana-
     lyzing transient responses of the aquifer to synchronized pressure pulses. It is possible
     that hydraulic tomography and resisitivity tomography could be used in a complementary
     fashion, since these techniques measure two related material properties – hydraulic con-
     ductivity and electrical resisitivity, both of which depend on grain-size distribution.

   Numerical modeling methods provide opportunities to test hypotheses developed from
quantitative conceptual models, providing feedback for further model development. While we
are still limited by computational demands associated with modeling of flow and transport at the
remediation hydrogeology scale, continued advances in PC workstation capacity, and development
of parallel processing codes that can distribute the work among many computers, are rapidly making
the problem more tractable. Although we still lack the capability to explicitly map heterogeneities
at appropriate scales, models can be developed at more realistic scales, using the hydrostratigraphy
principles and emerging techniques described above. When combined with tracer testing and
inverse modeling methods, we envision being able to better describe the behavior of systems,

without relying on contrived approximations such as dispersion or "up-scaled" parameters to estimate dual-domain mass transfer processes.

15.4 ADAPTIVE DESIGN AND OTHER CLOSING POINTS FOR CONSIDERATION

Stakeholders are accustomed to hearing remediation engineers and marketers making bold claims that some particular technology is well-suited to a site, prior to any testing or detailed site investigation. The lessons of *Remediation Hydraulics* show that brave declarations on the applicability of any technology are inappropriate and that it is critical that remediation scientists and engineers remain open to changes in the conceptual site model and, ultimately, the selected remedy for any site. Adaptive design is an approach to remedy development that allows for modification of the reagent delivery (i.e., reagent delivery schedule, concentrations, volumes) or the remedial mechanisms, themselves (i.e., biological, chemical, physical), as the implementation proceeds.

Adaptive design requires very effective operational decision making and that decision making is founded on three fundamentals:

- *Knowledge of process mechanisms* — The remedy design should be based on proven cause-effect relationships, which allow system designers to define the critical parameter set for any technology.
- *Data discipline* — Operational data collection should be constrained to the critical parameter set for the selected technology. If the system performance cannot be explained or controlled on the basis of the critical parameter monitoring, the conceptual site model may require revision. The collection of extraneous data increases costs and may distract system operators from the critical performance issues.
- *Decisiveness* — Armed with a sound understanding of remedy mechanisms and with critical operational monitoring data in-hand, a remedial system operator is prepared to act decisively when the remedy is not performing, as expected.

Remedy failures are often caused by inadequate reagent distribution and, many times, the inadequate distribution goes undiagnosed or untreated, due to inadequate monitoring or failed decision-making. It is much less common that the selected remedial mechanism is not workable, if properly applied. The most common adaptations that are required are adjustment of reagent volumes, concentrations or injection schedule. Only rarely, the selected remedy cannot succeed and an alternative must be applied.

Many of the sites now entering the feasibility study and design queues are large in scale and very complex in site geology and contaminant mix and distribution. Often, these are DNAPL sites in very heterogeneous geological matrices. Adaptive design is essential in these settings and it is often necessary to combine remedial technologies to achieve remedial objectives across the site. The combined remedy concept is gaining popularity as the more challenging sites are undertaken.

Finally, we recommend that service providers exercise utmost candor when developing conceptual site models and remedy strategies, so that stakeholders can frame their expectations with a realistic understanding of the uncertainties and the implications for the cost and duration of clean-up. We encourage stakeholders to expect evolution of the conceptual site model for any site and to view remedy recommendations that are built with no site testing with the greatest measure of skepticism.

16 References

Aller, L., T. W. Bennet, et al. (1989). *Handbook of Suggested Practices for the Design and Installation of Ground Water Monitoring Wells*. Westerville, Ohio, National Ground Water Association.

ASTM (1998). Standard guide for development of ground-water monitoring wells in granular aquifers, American Society for Testing and Materials: 17.

ASTM (2000). Standard Practice for Description and Identification of Soils (Visual-Manual Procedure). West Conshocken, Pennsylvania.

Barkay, T., S. Navon-Venezia, et al. (1999). "Enhancement of solubilization and biodegradation of polyaromatic hydrocarbons by the bioemulsifier alasan." *Applied and Environmental Microbiology* 65(6): 2697-2702.

Bassler, K. E., M. Paczuski, et al. (1999). "Braided Rivers and Superconducting Vortex Avalanches." *Physical Review Letters* 83(19): 3956.

Bear, J. (1972). *Dynamics of Fluids in Porous Media*. New York, Dover Publications, Inc.

Bear, J. (1988). *Dynamics of Fluids in Porous Media*. New York, Dover Publications, Inc.

Blowes, D. W., C. J. Ptacek, et al. (2000). "Treatment of inorganic contaminants using permeable reactive barriers." *Journal of Contaminant Hydrology* 45(1-2): 123-137.

Bohling, G. C., J. J. J. Butler, et al. (2007). "A field assessment of the value of steady shape hydraulic tomography for characterization of aquifer heterogeneities." *Water Resources Research* 43(W05430): 23.

Bohling, G. C., X. Zhan, et al. (2002). "Steady shape analysis of tomographic pumping tests for characterization of aquifer heterogeneities." *Water Resources Research* 38(12): 7.

Burmeister, D. M. (1979). *Suggested method of test for identification of soils*. Symposium on identification and classification of soils.

Burns, S. E. and P. W. Mayne (1998). *Penetrometers for soil permeability and chemical detection*. Atlanta, Georgia, National Science Foundation U.S. Army Research Office.

Butler, J. J. J. (1998). *The Design, Performance, and Analysis of Slug Tests*. New York, Lewis Publishers.

Butler, J. J. J., J. M. Healey, et al. (2001). "Hydraulic tests with direct push equipment." *Groundwater* 40(1): 25-36.

Byrd, R. B., W. E. Stewart, et al. (2002). *Transport Phenomena, Second Edition*. New York, John Wiley & Sons.

Carman, P. C. (1937). "Fluid flow through a granular bed." *Inst. Chem. Eng.* London 15: 150-156.

Carrier, W. D. (2003). "Good-bye Hazen; Hello Kozeny-Carmen." *Journal of Geotechnical and Geoenvironmental Engineering* 129(11): 1054-56.

Carsel, R. F. and R. S. Parrish (1988). "Developing joint probability distribution of soil water characteristics." *Water Resources Research* Vol. 24: p. 755-769.

Charbeneau, R. J. (2000). *Groundwater Hydraulics and Pollutant Transport*. Upper Saddle River, New Jersey, Prentice Hall.

Chen, W., A. T. Kan, et al. (2002). "More realistic cleanup standards with dual-equilibrium desorption." *Ground Water* 40(2): 153-164.

Cirpka, O. A., A. Olsson, et al. (2006). "Determination of transverse dispersion coefficients from reactive plume lengths." *Ground Water* 44(2): 212-221.

Cohen, R. M. and J. W. Mercer (1993). *DNAPL Site Investigation*. Boca Raton, Florida, C.K. Smoley.

Collinson, J. D. (1996). Alluvial sediments. *Sedimentary Environments: Processes, Facies, and Stratigraphy*. H. G. Reading. Malden, Massachusetts, Blackwell Science, Ltd: 37-82.

Corey, A. T. (1994). *Mechanics of Immiscible Fluids in Porous Media*. Highlands Ranch, CO, Water Resources Publications.

Cosler, D. J. (1997). "Ground-water sampling and time-series evaluation techniques to determine vertical concentration distributions." *Ground Water* 35(5): 825-841.

Cosler, D. J. (2004). "Effects of rate-limited mass transfer on water sampling with partially penetrating wells." *Ground Water* 42(2): 203-222.

Crank, J. (1975). *Mathematics of Diffusion*. Oxford, Oxford University Press.

DAEM, D. A. E. M. (2005). SOILPARA. Blacksburg, Virginia, Resources & Systems International.

Darcy, H. (1856). *Les Fontaines Publiques de la Ville de Dijon*. Paris, Dalmont.

Das, B. (1986). *Soil Mechanics Laboratory Manual*. San Jose, California 95103, Engineering Press, Inc.

Dawson, K. J. and J. D. Istok (1991). *Aquifer testing - Design and Analysis of Pumping and Slug Tests*. New York, Lewis Publishers.

de Marsily, G., F. Delay, et al. (2005). "Dealing with spatial heterogeneity." *Hydrogeology Journal* 13: 161-183.

Domenico, P. A. (1987). "An analytical model for multidimensional transport of a decaying contaminant species." *J Hydrology* 91: 49-58.

Driscoll, F. G. (1986). *Groundwater and Wells*. Saint Paul, Minnesota, Johnson Division.

EPA, U. S. (1994). "Alternative Methods for Fluid Delivery and Recovery." EPA/625/R-94/003.

EPA, U. S. (1996). "Soil Screening Guidance: Technical Background Document." (EPA/540/R95   May 1996).

Fetter, C. W. (2001). *Applied Hydrology, Fourth Edition*. Upper Saddle River, New Jersey, Prentice Hall.

FLUTe. (2007). "www.flut.com." Retrieved June 2007, from http://www.flut.com/.

Freeze, R. A. and J. A. Cherry (1979). *Groundwater*. Englewood Cliffs, NJ, Prentice-Hall.

Genuchten, M. T. v. (1980). "A closed-form equation for predicting the hydraulic conductivity of unsaturated soils." *SSSA Journal* Vol. 44: p. 892-898.

Goh, M. C., J. M. Hicks, et al. (1988). "Absolute orientation of water molecules at the neat water surface." *J. Phys. Chem.* 92(18): 5074-5075.

Graton, L. C. and H. J. Fraser (1935). "Systematic packing of spheres - with particular relation to porosity and permeability." *J Geol* 43(8): 785-909.

Harbaugh, A. W., E. R. Banta, et al. (2000). MODFLOW-2000, The U.S. geological survey modular ground-water model — user guide to modularization concepts and the ground-water flow model process. U. S. G. Survey, Department of the Interior: 121.

Hazen, A. (1892). Some physical properties of sands and gravels, with special reference to their use in filtration. M. S. B. o. Health: 539-556.

Hazen, A. (1911). "Discussions on Dams and Foundations." *Transactions of the American Society of Civil Engineers* 72.

Hjulstrom, F. (1935). "Studies of the morphological activity of rivers as illustrated by the River Fyris." *Bulletin of the Geological Institute of Uppsala* 25: 221-527.

Hofmann, J. R. and P. A. Hoffman (1992). "Darcy's law and structural explanation in hydrology." *Proc. 1992 Biennial Meeting of the Philosophy of Science Assoc.* 1(23-35).

Hubbert, M. K. (1937). "Theory of scale models as applied to the study of geological structures." *Geol. Soc. Amer. Bull.* 48: 1459-1520.

Hubbert, M. K. (1940). "The theory of groundwater in motion." *J Geol* 48: 785-944.

Hubbert, M. K. (1956). "Darcy's Law and the field equations of the flow of underground fluids." *Trans. Amer. Inst. Min. Met. Eng.* 207: 222-239.

Hyder, Z., J. J. Butler, Jr., et al. (1994). "Slug tests in partially penetrating wells." *Water Resources Research* 30(11): 2945-2957.

ITRC. (2007). "www.itrcweb.org." from (http://www.itrcweb.org/).

Julian, H. E., J. M. Boggs, et al. (2001). "Numerical simulation of a natural gradient tracer experiment for the natural attenuation study: Flow and physical transport." *Ground Water* 39(4): 534-545.

Kim, S., G. Heinson, et al. (2004). "Electrokinetic groundwater exploration: A new geophysical technique." *Regolith*: 181-185.

Knighton, D. (1987). *Fluvial Forms and Processes*. Baltimore, Maryland, Edward Arnold.

Kruseman, G.P. and N.A. de Ridder, 1990. *Analysis and Evaluation of Pumping Test Data, Second Edition.* International Institute for Land Reclamation and Improvement (ILRI) Publication 47, Wageningen, Netherlands.

Kueper, B. H. and D. B. Mcwhorter (1991). "The behavior of dense, nonaqueous phase liquids in fractured clay and rock." *Ground Water* 29(5): 716-728.

Lamb, H. (1932). *Hydrodynamics*. New York, Dover Publications.

Leeder, M. (1999). *Sedimentology and Sedimentary Basins. From Turbulence to Tectonics*. Oxford, Blackwell.

Lenhard, R. J. and J. C. Parker (1987). "Measurement and prediction of saturation-pressure relationships in three-phase porous media systems." *Journal of Contaminant Hydrology* Vol. 1: p. 407-424.

Li, L., C. H. Benson, et al. (2005). "Impact of mineral fouling on hydraulic behavior of permeable reactive barriers." *Ground Water* 43(4): 582-596.

Li, X. D. and F. W. Schwartz (2004). "DNAPL mass transfer and permeability reduction during in situ chemical oxidation with permanganate." *Geophysical Research Letters* 31(6): L06504.

Li, X. D. and F. W. Schwartz (2004). "DNAPL remediation with in situ chemical oxidation using potassium permanganate. II. Increasing removal efficiency by dissolving Mn oxide precipitates." *Journal of Contaminant Hydrology* 68(3-4): 269-287.

Li, X. D. and F. W. Schwartz (2004). "DNAPL remediation with in situ chemical oxidation using potassium permanganate. Part I. Mineralogy of Mn oxide and its dissolution in organic acids." *Journal of Contaminant Hydrology* 68(1-2): 39-53.

Luthy, R. G., G. R. Aiken, et al. (1997). "Sequestration of hydrophobic organic contaminants by geosorbents." *Environmental Science & Technology* 31(12): 3341-3347.

Mayne, P. W. (2002). "Flow properties from piezocone dissipation tests." Retrieved August 11, 2007, from http://geosystems.ce.gatech.edu/Faculty/Mayne/papers/PiezoDissipation.pdf.

McDonald, M. G. and A. W. Harbaugh (1988). A modular three-dimensional finite-difference ground-water flow model (MODFLOW). U. S. G. Survey, U.S. Government Printing Office: 588.

McWhorter, D. B. (1996). Processes affecting soil and groundwater contamination by DNAPL in low-permeabiity media. *In Situ Remediation of DNAPL Compounds in Low Permeability Media: Fate/ Transport, In Situ Control Technologies, and Risk Reduction*, United States Department of Energy, Oak Ridge National Laboratory. ORNL/TM-13305.

Meinzer, O. E. (1923). *The Occurrence of Groundwater in the United States, With a Discussion of Principles.*

Miall, A. D. (1985). "Architectural-element analysis: a new method of facies analysis applied to fluvial deposits." *Earth Science Review* 22: 261-308.

Murdoch, L. C. (1995). "Forms of hydraulic fractures created during a field test in overconsolidated glacial drift." *Quart. J. Engineering Geol. and Hydrogeol.* 28(1): 23-35.

Muskat, M. (1937). *The Flow of Homogeneous Fluids through Porous Media.* York, Pennsylvania, McGraw-Hill Book Company, Inc.

Myers, K. F., W. M. Davis, et al. (2002). Tri-Service site characterization and analysis system validation of the membrane interface probe. E. Laboratory, US Army Corps of Engineers, Engineer Research and Development Center: 62.

Navier, L. M. H. (1827). "Memoire sur les lois du mouvement des fluides." *Memoires de l'Academie Royale des Sciences* 6.

Newell, C. J., S. D. Acre, et al. (1995). Light non-aqueous fluids, Ralph R. Kerr Environmental Research Laboratory.

Nichols, W. E., N. J. Aimo, et al. (2005). STOMP Subsurface transport over mulitple phase: user's guide. PNNL-15782 (UC-2010). Richland, Washington, Pacific Northwest National Laboratory.

Nielson, D. M. and R. Schalla (1991). *Practical Handbook of Ground-Water Monitoring.* Chelsea, Michigan, Lewis Publishers, Inc.

NRCS (1999). *Soil Taxonomy - A Basic System of Soil Classification for Making and Interpreting Soil Surveys.* Washington, DC 20402, U.S. Government Printing Offce.

Parker, B. L., J. A. Cherry, et al. (2004). "Field study of TCE diffusion profiles below DNAPL to assess aquitard integrity." *Journal of Contaminant Hydrology* 74(1-4): 197-230.

Parker, B. L., R. W. Gillham, et al. (1994). "Diffusive disappearance of immiscible-phase organic liquids in fractured geologic media." *Ground Water* 32(5): 805-820.

Parker, B. L., D. B. McWhorter, et al. (1997). "Diffusive loss of non-aqueous phase organic solvents from idealized fracture networks in geologic media." *Ground Water* 35(6): 1077-1088.

Payne, F. C., S. S. Suthersan, F. C. Lenzo and J. S. Burdick. 2001. Mobilization of Sorbed-Phase Chlorinated Alkenes in Enhanced Reductive Dechlorination. In: Anaerobic Degradation of Chlorinated Solvents. Proc. Internat. In Situ and On-Site Bioremediation Symp. 6(2):53-60.

Payne, F. C. and G. L. Kilmer (1994). "Method of recovering subsurface contaminants." *United States Patent and Trademark Office* U.S. Patent 5,342,147.

Perloff, W. H. and W. Baron (1976). *Soil Mechanics.* New York, John Wiley & Sons.

Potter, M. C. and D. C. Wiggert (1991). *Mechanics of Fluids.* Englewood Cliffs, New Jersey, Prentice Hall.

Pruess, K., C. Oldenburg, et al. (1999). Tough2 user's guide, LBLR LBL-43134. Berkeley, California, Lawrence Berkeley Laboratory.

Puls, R. W., D. W. Blowes, et al. (1999). "Long-term performance monitoring for a permeable reactive barrier at the US Coast Guard Support Center, Elizabeth City, North Carolina." *Journal of Hazardous Materials* 68(1-2): 109-124.

Raudkivi, A. J. (1990). *Loose Boundary Hydraulics.* Elmsford, New York, Pergamon Press, Inc.

Richards, L.A., 1931. "Capillary conductance of liquids through porous mediums." Physics 1:318-333.

Rivett, M. O. and S. Feenstra (2005). "Dissolution of an emplaced source of DNAPL in a natural aquifer setting." *Environ. Sci. Technol.* 39(2): 447-455.

Rivett, M. O., S. Feenstra, et al. (2001). "A controlled field experiment on groundwater contamination by a multicomponent DNAPL: creation of the emplaced-source and overview of dissolved plume development." *Journal of Contaminant Hydrology* 49(1-2): 111-149.

Roberts, A. L., L. A. Totten, et al. (1996). "Reductive elimination of chlorinated ethylenes by zero valent metals." *Environmental Science & Technology* 30(8): 2654-2659.

Robertson, P. K. and R. G. Campanella (1983). "Interpretation of cone penetration tests: sands and clays." *Canadian Geotechnical Journal* 20(4): 719-745.

Sale, T. C. and D. B. McWhorter (2001). "Steady state mass transfer from single-component dense nonaqueous phase liquids in uniform flow fields." *Water Resources Research* 37(2): 393-404.

Sapozhnikov, V. B. and E. Foufoula-Georgiou (1997). "Experimental evidence of dynamic scaling and indications of self-organized criticality in braided rivers." *Water Resources Research* 33(8): 1983-1991.

Schlumberger. (2007). "www.water.slb.com." Retrieved June 2007, from www.water.slb.com.

Schwarzenbach, R. P., P. M. Gschwend, et al. (2003). *Environmental Organic Chemistry, 2nd edition.* Hoboken, New Jersey, John Wiley & Sons.

Shackelford, C. D., T. C. Sale, et al. (2005). *In Situ Remediation of Chlorinated Solvents Using Zero-Valent Iron and Clay Mixtures: A Case History.* Geo-Frontiers 2005, Austin, Texas, ASCE.

Shirazi, M. A. and L. Boersma (1984). "A unifying quantitative analysis of soil texture." *Soil Science Society of America Journal* Vol 48: 142-147.

Siegrist, R. L., M. A. Urynowicz, et al. (2002). "Genesis and effects of particles produced during in situ chemical oxidation using permanganate." *Journal of Environmental Engineering-Asce* 128(11): 1068-1079.

Simons, D. B. and E. V. Richardson (1966). Resistance to flow in alluvial channels, United States Geological Survey Professional Paper 422J: 61.

Solinst. (2007). "www.solinst.com." Retrieved June 2007, from www.solinst.com.

Starr, R. C. and J. A. Cherry (1994). "In situ remediation of contaminated ground water: The funnel-and-gate system." *Ground Water* 32(3): 465-476.

Stokes, G. G. (1845). "On the theories of the internal friction of fluids in motion." *Trans. Cambridge Philosophical Soc.*

Suthersan, S. S. (2002). *Natural and Enhanced Remediation Systems.* Boca Raton, Florida, Lewis Publishers.

Suthersan, S. S. and F. C. Payne (2005). *In Situ Remediation Engineering.* Boca Raton, Florida, CRC Press.

Theodoropoulou, M. A. (2007). "Dispersion of dissolved contaminants in groundwater: from visualization experiments to macroscopic simulation." *Water Air Soil Pollution* 181: 235-245.

Treybal, R. E. (1980). *Mass-Transfer Operations.* New York, McGraw-Hill.

Varljen, M. D., M. J. Barcelona, et al. (2006). "Numerical simulations to assess the monitoring zone achieved during low-flow purging and sampling." *Ground Water Monitoring & Remediation* 26(1): 44-52.

von Engelhardt, W. and W. Tunn, 1954. "Über das Strömen von Flüssigkeiten durch Sandsteine," Contr. to *Minerol. and Petrol.* 4(1-2):12-25.

Yeh, T. C. J. and S. Liu (2000). "Hydraulic tomography: development of a new aquifer test methods." *Water Resources Research* 36(8): 2095-2105.

Zheng, C. and P. P. Wang (1999). MT3DMS: A modular three-dimensional multispecies transport model for simulation of advection, dispersion, and chemical reactions of contaminants in groundwater systems; documentation and user's guide. E. R. a. D. Center, U.S. Army Corps of Engineers: 221.

Zhu, J. and T.-C. J. Yeh (2005). "Characterization of aquifer heterogeneity using transient hydraulic tomography." *Water Resources Research* 41(W07028): 10.

Zhu, J. and T.-C. J. Yeh (2006). "Analysis of hydraulic tomography using temporal moments of drawdown recovery data." *Water Resources Research* 42(W02403): 11.

Zurbuchen, B. R., V. A. Zlotnik, et al. (2002). "Dynamic interpretation of slug tests in highly permeable aquifers." *Water Resources Research* 38(3): 18.

Index

Plates

PLATE 1 Droplets of three fluids on a low-energy surface (l to r): water, 70-percent isopropanol and vegetable oil. The contact angle for the water droplet exceeds 90 degrees, so it is a non-wetting fluid on this surface. The isopropanol and vegetable oil contact angles appear to be less than 90 degrees, although the surface hasn't drawn either fluid fully out of its droplet, indicating they are partially wetting fluids on this surface.

PLATE 2 Well sorted sand, collected at Makaha Beach, Oahu, Hawaii (scale approximate). Particle sizes ranged from 0.5 to 1.5 mm.

PLATE 3 Well sorted sand, collected at Nordhouse Dunes on the eastern shore of Lake Michigan (scale approximate). Grain sizes ranged from 0.2 to 0.5 mm.

PLATE 4 Well sorted sand, collected from the Pacific Ocean intertidal zone at Carmel, California (scale approximate). Grain sizes ranged from 0.2 to 0.5 mm.

PLATE 5 Poorly sorted sand and fine gravel, collected 20 meters below ground surface from an aquifer formation near Valcartier, Quebec, Canada (scale approximate). Particle sizes ranged from approximately 0.01 to 5 mm. The field-determined hydraulic conductivity for this material was 1.5 x 10-1 cm/sec. Grain sizes ranged from 0.2 to 0.5 mm.

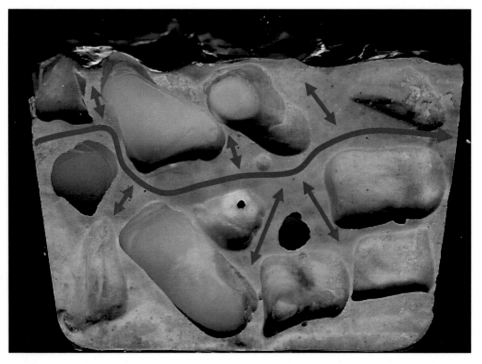

PLATE 6 Cross-section through a physical pore model, showing flow channels and connecting dead-end pore spaces that provide solute mass storage capacity at the pore-scale.

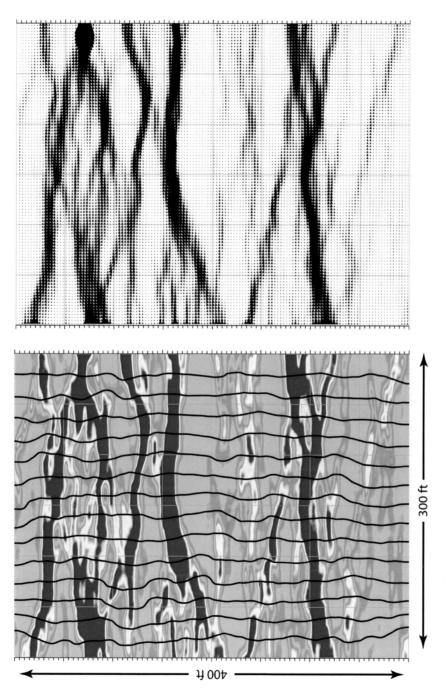

PLATE 7 Heterogeneous porous medium, synthesized to analyze the detectability of high-flow zones using water elevation contours. The relative hydraulic conductivities are shown on the left diagram, ranging over three orders of magnitude (dark blue is 1,000-fold higher conductivity than the light green shaded areas). The grid area is 300 x 400 feet and the water elevation contours, shown in black, are at 0.5-foot intervals. Groundwater flows are highly concentrated in this heterogeneous system, as shown by shading on diagram at the right. In this setting, roughly 70% of the flow occurred in 25% of the cross-sectional area, and more than 90% of the flow occurred in 50% of the cross-section. These flow concentrations are not discernable in the groundwater elevation contours.

400 ft

300 ft

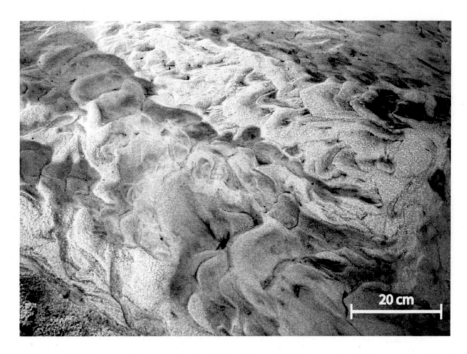

PLATE 8 Fine-scale heterogeneities are being built into this sand-silt stream bed at the mouth of Porter Creek in the Nordhouse Dunes area, on the eastern shore of Lake Michigan (the scale is approximate).

PLATE 9 Glacial till, showing the poorly sorted particle distribution of ice contact deposits.

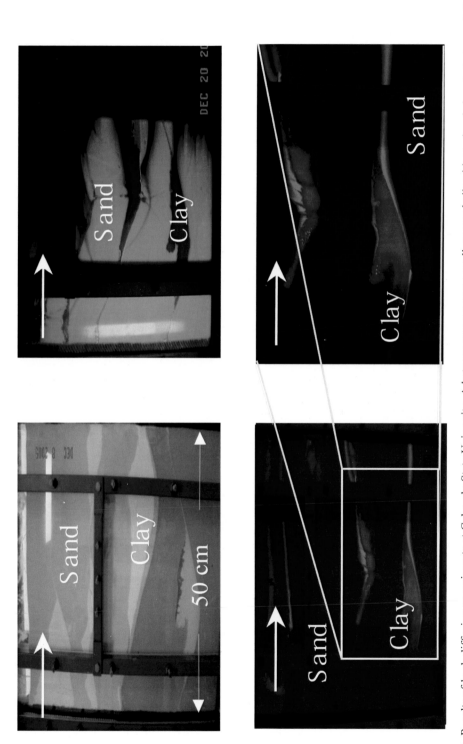

PLATE 10 Results of back-diffusion experiments at Colorado State University. A heterogeneous porous medium was built with sand and clay strata (upper left, in visible light). Groundwater flow was established from left to right and the system was perfused with a fluoresce in solution for 24 days (upper right, in ultraviolet lighting). Clean water was then flushed through the system and fluoresce in that had penetrated the clay through aqueous diffusion formed a persistent source mass that re-contaminated the clean water flush (lower left and expanded view in the lower right). Photos courtesy Ms. Lee Ann Doner and Dr. Thomas Sale, Colorado State University.

PLATE 11 Limestone epikarst in a dimension-stone quarry near Salem, Indiana. Photo courtesy Dr. Ralph Ewers, Eastern Kentucky University (photo used with permission).